SECOND EDITION

Elements of Petroleum Refinery Engineering

OM PRAKASH GUPTA
B.Tech. (Chemical Engineering)
Ex DGM, HRD
Bokaro Steel Plant
Steel Authority of India Limited (SAIL)

KHANNA BOOK PUBLISHING CO. (P) LTD.
Publisher of Engineering and Computer Books
4C/4344, Ansari Road, Darya Ganj, New Delhi-110002
Phone : 011-23244447-48 **Mobile:** +91-9910909320
E-mail : contact@khannabooks.com
Website : www.khannabooks.com

Dear Readers,

To prevent the piracy, this book is secured with HIGH SECURITY HOLOGRAM on the front title cover. In case you don't find the hologram on the front cover title, please write us to at contact@khannabooks.com or whatsapp us at +91-99109 09320 and avail special gift voucher for yourself.

Specimen of Hologram on front Cover title:

Moreover, there is a SPECIAL DISCOUNT COUPON for you with EVERY HOLOGRAM.

How to avail this SPECIAL DISCOUNT:

Step 1: Scratch the hologram

Step 2: Under the scratch area, your "coupon code" is available

Step 3: Logon to www.khannabooks.com

Step 4: Use your "coupon code" in the shopping cart and get your copy at a special discount

Step 5: Enjoy your reading!

Price : ₹ 365.00

ISBN: 978-93-82609-72-8

Elements of Petroleum Refinery Engineering
by O.P. Gupta

First Edition: 2016
Revised Edition: 2019
Reprint Edition: 2022, **2025**

Published by:
Khanna Book Publishing Co. (P) Ltd.
CIN: U22110DL1998PTC095547

Visit us at: www.khannabooks.com
Write us at: contact@khannabooks.com

To view complete list of books, Please scan the QR Code:

Printed by India.

PREFACE TO THE SECOND EDITION

It gives me immense pleasure to introduce 2nd edition of the book titled Elements of Petroleum Refinery Engineering. I have taken the opportunity to present the informations pertaining to petroleum refining in simple, lucid & easily understandable language with special emphasis on crude oil refining technologies and scenario in Indian context. This book is meant for the students, teachers, practicing engineers ,consultants and policy makers on petroleum refining.

Extensive use of internet resources have been made in gathering the relevant informations and the subject matter has been presented in a simple lucid manner for easy understanding of the subject.

This book is targeted to benefit the following:

— Diploma in engineering students.

— Degree in engineering students [B.Tech. (chemical engg.), B.Tech. (petroleum engineering).

— B.Tech. (petrochemical engineering), B.Tech. (Aeronautical Engg.) *etc.*

— AMIE, AMIICHE, Students.

— M.Tech. Students of various disciplines pursuing courses on petroleum refining.

— Faculty members/ Teaching staff of Engineering colleges/IIT's/ NIT's *etc.*

— Practicing petroleum engineers/consultants /refiners in various private sector/public sector undertakings, state/central government departments, NGO's *etc.*

— Students of foreign universities of developing countries pursuing diploma/ degree/postgraduate courses in various engineering disciplines having a paper in petroleum refinery engineering.

New chapter namely 'Objective type Questions & Answers in Petroleum Refinery Engineering' have been added in this edition of the book.

The scope and coverage of the subject matter presented can be gauged from glancing the contents portion of the book to judge its suitability and relevance for a particular target group.

Lot of literatures were consulted while writing this book for making the informations available for various target groups, hence I gratefully acknowledge their assistance & contribution. Latest and updated informations/ data/statistics pertaining to the subject matter has been included in this edition for the benefit of the readers.

Constructive criticisms and suggestion for further improvement of the book are solicited.

Om Prakash Gupta

Contents

Definition of terms used in Petroleum Refinery Engineering

Knowledge of certain preliminary terms is required for understanding the subject of Petroleum Refinery Engineering .Definition of these preliminary terms are given below.

Absorption - The disappearance of one substance into another so that the absorbed substance loses its identifying characteristics, while the absorbing substance retains most of its original physical aspects. Used in refining to selectively remove specific components from process streams.

Acid Treatment - A process in which unfinished petroleum products, such as gasoline, kerosene & lubricating oil stocks, are treated with sulphuric acid to improve color, odour & other properties.

Adsorption - Adhesion of the molecules of gases or liquids to the surface of solid materials.

Aliphatic hydrocarbon Hydrocarbons characterized by open - chain structures: ethane eg, butane, butene, acetyleneetc.

Alkylation - A process using sulphuric or hydrofluoric acid as a catalyst to combine olefins (usually butylene) and isobutane to produce a high octane product known as alkylate.

Aniline Point - The minimum temperature for complete miscibility of equal volumes of aniline and a test sample. This test is an indication of paraffinicity and the ignition quality of diesel.

API Gravity - An arbitrary scale expressing the density of petroleum products.

Aromatics - Hydrocarbon compound that contains atleast one benzene ring.

Asphaltenes - The asphalt compounds soluble in carbon disulphide but insoluble in paraffin naphthas.

ASTM Distillation - A standardized laboratory batch distillation for naphthas and middle distillate at atmospheric pressure.

Atmospheric Tower - A distillation unit operated at atmospheric pressure.

Benzene - An unsaturated, six - carbon ring, basic aromatic compound.

Blending - The process of mixing two or more petroleum products with different properties to produce a finished product with desired characteristics.

Blowdown - The removal of hydrocarbons from a process unit, vessel or line on a scheduled or emergency basis by the use of pressure through special piping & drums provided for this purpose.

Boiling Range - The range of temperature (usually at atmospheric pressure) at which the boiling (or distillation) of a hydrocarbon liquid commences, proceeds and finishes.

Bottoms - Tower bottoms are residues remaining in a distillation unit after the highest boiling point material to be distilled has

been removed. Tank bottoms are the heavy materials that accumulate in the bottom of storage tanks, usually comprised of oil, water & foreign matter.

Catalyst - A material that aids or promotes a chemical reaction between other substances but does not react itself. Catalysts increase reaction speeds and can provide control by increasing desirable reactions & decreasing undesirable reactions.

Catalytic Cracking - The process of breaking up heavier hydrocarbon molecules into lighter hydrocarbon fractions by use of heat & catalysts.

Caustic Wash - A process in which distillate is treated with sodium hydroxide to remove acidic contaminants that contribute to poor odour & stability.

Cetane Number - The percentage of pure cetane in a blend of cetane and alpha methyl naphthalene which matches the ignition quality of a diesel fuel sample.

Coke - A high carbon - content residue remaining from the destructive distillation of petroleum residue.

Coking - A process for thermally converting & upgrading heavy residual into lighter products and by - product petroleum coke. Coking is also the removal of all lighter distillable hydrocarbons that leave a residue of carbon in the bottom of units or as build - up or deposits on equipment & catalysts.

Condensate - The liquid hydrocarbon resulting from cooling vapours.

Conradson Carbon - A test used to determine the amount of carbon residue left after the evaporation & pyrolysis of an oil under specified conditions.

Cracking - The breaking up of heavy molecular weight hydrocarbons into lighter hydrocarbon molecules by the application of heat & pressure with or without the use of catalysts.

Crude Assay - A procedure for determining the general distillation and quality characteristics of crude oil.

Crude Oil - A naturally occurring mixture of hydrocarbons that usually includes small quantities of sulphur, nitrogen & oxygen derivatives of hydrocarbons as well as trace metals.

Cycle Gas Oil - Cracked gas oil returned to a cracking unit.

Deasphalting - Process of removing asphaltic materials from reduced crude using liquid propane to dissolve non asphaltic compounds.

Debutanizer - A fractionating column used to remove butane & lighter components from liquid streams.

Dehydrogenation - A reaction in which hydrogen atoms are eliminated from a molecule. Dehydrogenation is used to convert ethane, propane & butane into olefins (Ethylene, Propylene & Butenes).

Desalting - Removal of mineral salts (most chlorides, e.g. magnesium chloride and sodium chloride) from crude oil.

Desulphurization - A chemical treatment to remove sulphur or sulphur compounds from hydrocarbons.

Dewaxing - The removal of wax from petroleum products (usually lubricating oils & distillate fuels) by solvent absorption, chilling & filtering.

Diethanolamine - A chemical ($C4H11O2N$) used to remove $H2S$ from gas streams.

Distillate - The products of distillation formed by condensing vapours.

Dry Gas - Natural gas with so little natural gas liquids, that it is nearly all methane with some ethane.

Feed stock - Stock from which material is taken to be fed into a processing unit.

Flash Point - Lowest temperature at which a petroleum product will give off sufficient vapour so that the vapour–air mixture above the surface of the liquid will propagate a flame away from the source of ignition.

Flashing - The process in which a heated oil under pressure is suddenly vaporized in a tower by reducing pressure.

Flexicoking - Thermal process in which coke is gasified into fuel gas using steam & air.

Fouling - Accumulation of deposits in condensers, heat exchangers etc.

Fractionating Tower - Process unit that separates various fractions of petroleum by simple distillation, with the column tapped at various levels to separate & remove fractions according to their boiling ranges.

Fuel Gas - Refinery gas used for heating.

Gas Oil - Middle distillate petroleum fraction with a boiling range of about 350–750 ^{0}F, usually includes diesel fuel, kerosene, heating oil & light fuel oil.

Gasoline - A blend of naphthas & other refinery products with sufficiently high octane and other desirable characteristics to be

suitable for use as fuel in internal combustion engines.

Hydrocracking - A process used to convert heavier feedstock into lower boiling, higher value products. The process employs high pressure, high temperature, a catalyst and hydrogen.

Hydrodesulphurization - A catalytic process in which the principal purpose is to remove sulphur from petroleum fractions in the presence of hydrogen.

Hydrogenation - The chemical addition of hydrogen to a material in the presence of a catalyst.

Isomerization - A reaction that catalytically converts straight chain hydrocarbon molecules into branched chain molecules of substantially higher octane number. The reaction rearranges the carbon skeleton of a molecule without adding or removing anything from the original material.

Iso Octane - A hydrocarbon molecule (2,2,4 - trimethylpentane) with excellent antiknock characteristics on which the octane number of 100 is based.

Kerosene - A middle distillate product composed of material of 300–550 ^{0}F boiling range.

Knockout Drum - A vessel wherein suspended liquid is separated from gas or vapour.

Lean Oil - Absorbent oil fed to absorption towers in which gas is to be stripped. After absorbing the heavy ends from the gas, it becomes fat oil. When the heavy ends are subsequently stripped, the solvent again becomes lean oil.

LPG (Liquefied Petroleum Gas) - Liquefied light end gases used for home heating & cooking. The gas contains 85% propane, and the balance is a split between ethane & butane.

MON (Motor Octane Number) - A measure of resistance to self ignition (knocking) of gasoline. It is the percentage by volume of isooctane in a mixture of isooctane & n - heptane that knocks with the same intensity as the fuel being tested.

Napthenes - Hydrocarbons (Cycloalkanes) with the general formula CnH2n, in which carbon atoms are arranged to form a ring.

Naphtha - A general term for low boiling hydrocarbon fractions that are a major component of gasoline. Aliphatic naphtha refers to those naphthas containing less than 0.1% benzene & with carbon numbers from C3 to C16. Aromatic naphthas have carbon numbers from C6 to C16 & contain significant quantities of aromatic hydrocarbons like benzene toluene & xylene.

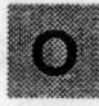

Octane Number - A number indicating the relative antiknock characteristics of gasoline.

Olefins - A family of unsaturated hydrocarbons with one carbon– carbon double bond and the general formula CnH2n.

Paraffins - A family of saturated aliphatic hydrocarbons (alkanes) with the general formula CnH2n+2.

Pour Point - The lowest temperature at which a petroleum oil will flow or pour when it is chilled without disturbance at a controlled rate.

Q

Quench Oil - Oil injected into a product leaving a cracking or reforming heater to lower the temperature & stop the cracking process.

R

Raffinate - The product resulting from a solvent extraction process and consisting mainly of those components that are least soluble in the solvents. The product recovered from an extraction process is relatively free of aromatics, naphthenes & other constituents that adversely affect physical parameters.

Reactor - The vessel in which chemical reactions take place during a chemical conversion type of process.

Reboiler - An auxiliary unit of a fractionating tower designed to supply additional heat to the lower portion of the tower.

Recycle Gas - High hydrogen content gas returned to a unit for reprocessing.

Reduced Crude - A residual product remaining after the removal by distillation of an appreciable quantity of the more volatile components of crude oil.

Reflux - The portion of the distillate returned to the fractionating column to assist in attaining better separation into desired fractions.

Reformate - An upgraded naphtha resulting from catalytic or thermal reforming.

Reforming - The thermal or catalytic conversion of petroleum naphtha into more volatile products of higher octane number. It represents the total effect of numerous simultaneous reactions like cracking, polymerization, dehydrogenation & isomerization.

Regeneration - In a catalytic process, the reactivation of the catalyst sometimes done by burning off the coke deposits under carefully controlled conditions of temperature & oxygen content of the regeneration gas stream.

RON (Research Octane Number) - Same definition as of MON except the engine operates here at 600 rpm.

RVP(Reid Vapour Pressure) - The vapour pressure at 100 ^{0}F of a product determined in a volume of air four times the liquid volume.

Salt Content - Salts in solution that are emulsified with the crude.

Scrubbing - Purification of a gas or liquid by washing it in a tower.

Severity - The degree of intensity of the operating conditions of a process unit.

Smoke Point - A test measuring the burning quality of jet fuels, kerosene & illuminating oils. It is defined as the height of the flame in mm without producing smoke when fuel/kerosene is burned in a standard lamp.

Solvent Extraction - The separation of materials of different chemical types & solubility by selective solvent action.

Sour Gas - Natural gas that contains corrosive, sulphur bearing compounds such as hydrogen sulphide & mercaptans.

Stabilization - A process for separating the gaseous & more volatile liquid hydrocarbons from crude petroleum or gasoline and leaving a stable (less volatile) liquid so that it can be handled or stored with less change in composition.

Straight - Run Gasoline - Gasoline produced by the primary distillation of crude oil. It contains no cracked, polymerized, alkylated, reformed or visbroken stock.

Stripping - The removal (by steam induced vaporization or flash evaporation) of the more volatile components from a cut or fraction.

Sulphuric Acid Treating - A refining process in which unfinished petroleum products such as gasoline, kerosene and lubricating oil stocks are treated with sulphuric acid to improve their colour, odour & other characteristics.

Sulphurization - Combining sulphur compounds with petroleum lubricants.

Sweetening - Processes that either remove obnoxious sulphur compounds (primarily hydrogen sulphide, mercaptans & thiophens) from petroleum fractions or streams, or convert them, as in the case of mercaptans, to odourless disulphides to improve odour, colour & oxidation stability.

Tail Gas - The lightest hydrocarbon gas released from a refining process.

Thermal Cracking - The breaking up of heavy oil molecules into lighter fractions by the use of high temperature without the aid of catalysts.

Turn Around - A planned complete shutdown of an entire process or section of a refinery or of an entire refinery to perform major maintenance, overhaul & repair operations and to inspect, test & replace process materials & equipments.

Vacuum Distillation - The distillation of petroleum under vacuum which reduces the boiling temperature sufficiently to prevent cracking or decomposition of the feedstock.

Visbreaking - A low temperature cracking process used to reduce the viscosity or pour point of straight - run residuum.

Wet Gas - A gas containing a relatively high proportion of hydrocarbons that are recoverable as liquids.

Abberviations

AGO - Atmospheric Gas Oil
ATK - Aviation Turbine Kerosene
BPD - Barrels Per Day
BTX - Benzene, Toluene and Xylene
CCR - Conradson Carbon Residue
CDU - Crude Distillation Unit
DAF - Dissolved Air Floatation
DAO - Deasphalted Oil
DCC - Deep Catalytic Cracking
DEA - Diethanol Amine
DME - Dimethyl Ether
DOC - Dissolved Organic Carbon
EBP - End Boiling Point
EOR - End of Run
ESP - Electrostatic Precipitator
FCC - Fluid Catalytic Cracking
FGD - Flue Gas Desulphurization
Gpm - Gallons Per Minute
GC - Gas Chromatograph
GTL - Gas to Liquid
HAP - Hazardous Air Pollutants
HAZOP - Hazard Operability
HCGO - Heavy Cycle Gas Oil
HCN - Heavy Cycle Naphtha
HDC - Hydrocracking

HDI	-	Hydro-Isomerization
HDN	-	Hydro-Denitrogenation
HDS	-	Hydrodesulphurization
HF	-	Hydrofluoric Acid
HN	-	Heavy Naphtha
HSR	-	Heavy Straight Run
HT	-	Hydrotreating
HTN	-	Hydrotreated Naphtha
HTU	-	Hydrothermal Upgrading
IBP	-	Initial Boiling Point
IGCC	-	Integrated Gasification Combined Cycle
I/O	-	Isobutane/Olefin Ratio
LEL	-	Lower Explosive Limit
LFL	-	Lower Flammability Limit
LHSV	-	Liquid Hourly Space Velocity
LN	-	Light Naphtha
LNG	-	Liquefied Natural Gas
LPG	-	Liquefied Petroleum Gas
LSFO	-	Low sulphur Fuel Oil
LSHS	-	Low Sulfur Heavy Stock Fuel Oil
MDEA	-	Methyl Die Ethanolamine
MEA	-	Monoethanol Amine
MEROX	-	Mercaptans Oxidation
MMSCFD	-	10^6 ft^3/Day
MO	-	Metal Oxide
MON	-	Motor Octane Number
MS	-	Mass Spectroscopy
MSDS	-	Material Safety Data Sheet
MTBE	-	Methyl Ter Butyl Ether
OSHA	-	Occupational Safety and Health Administration
P&ID	-	Piping and Instrumentation Diagram
PIONA	-	Parrafins, Isoparaffins, Olefins and Aromatic
PNA	-	Polynuclear Aromatics
PPE	-	Personal Protective Equipment

PPM	-	Parts Per Million
PTB	-	Pounds Per Thousand Barrels
RME	-	Rapeseed Methyl Ester
RON	-	Research Octane Number
RSH	-	Mercaptans
RVP	-	Reid Vapour Pressure
SCFB	-	Standard Cubic Feet Per Barrel
SCR	-	Selective Catalytic Reduction
SDA	-	Solvent Deasphalting
SG	-	Specific Gravity
SHE	-	Safety, Health and Environment
SNCR	-	Selective Non-Catalytic Reduction
SRU	-	Sulphur Recovery Unit
STP	-	Standard Temperature and Pressure
SV	-	Space Velocity
TBP	-	True Boiling Point
TNT	-	Tri Nitro Toluene
TOC	-	Total Organic Carbon
TSP	-	Total Suspended Particulates
TSS	-	Total Suspended Solids
UEL	-	Upper Explosive Limit
UFL	-	Upper Flammability Limit
VG	-	Viscosity Gravity Constant
VGO	-	Vacuum Gas Oil
VOC	-	Volatile Organic Compound
wt%	-	Weight Percentage

1

Origin, Composition, Classification, Pretreatment & Transportation of Petroleum Crude Oil

Petroleum (or rock oil) is a naturally occurring brown to black oil comprising mainly of hydrocarbons found under the crust of the earth on shore or off shore. It is obtained from the ground either by natural seepage or by drilling wells to various depths. Either petroleum oil flows out itself due to underground gas pressure or these are mechanically pumped out.

Origin of Petroleum

Although several theories were put forward to explain the formation of petroleum, only three widely accepted ones are given below.

Carbide theory. According to this theory, hydrocarbons present in petroleum are formed by the action of water on inorganic carbides. Inorganic carbides, in turn, are formed by the reaction of metal and carbon under high temperature & pressure conditions inside the earth.

$$Ca + 2C \xrightarrow{\text{High temp \& pressure}} \underset{\text{Calcium carbide}}{CaC_2}$$

$$4Al + 3C \xrightarrow{\text{high temp \& pressure}} \underset{\text{Aluminium carbide}}{Al_4C_3}$$

$$CaC_2 + 2H_2O \longrightarrow Ca(OH)_2 + \underset{\text{acetylene}}{C_2H_2}$$

$$Al_4C_3 + 12H_2O \longrightarrow 4Al(OH)_3 + \underset{\text{methane}}{3CH_4}$$

These lower hydrocarbonthen undergo hydrogenation and polymerization to give various types of hydrocarbons (paraffins, aromatic & cycloparaffins).

$$\underset{\text{acetylene}}{C_2H_2} + H_2 \xrightarrow{H_2} \underset{\text{ethylene}}{C_2H_4} \rightarrow \underset{\text{ethane}}{C_2H_6}$$

$$\underset{\text{acetylene}}{3C_2H_2} \xrightarrow{\text{polymerisation}} \underset{\text{benzne}}{C_6H_6}$$

$$3C_2H_4 \xrightarrow{\text{polymerisation}} \underset{\text{cyclohexane}}{C_6H_{12}}$$

$$\underset{\text{ethylene}}{3C_2H_4} \longrightarrow \underset{\text{butene—2}}{CH_3\,CH = CH.CH_3} \xrightarrow{H_2} \underset{\text{butane}}{CH_3.CH_2.CH_2CH_3}$$

However, this theory fails to explain the following facts:

— Presence of nitrogen and sulphur compounds.
— Presence of chlorophyll
— Presence of optically active compounds.

Engler theory. According to Engler, petroleum is of animal origin. Engler (1900) suggested that petroleum is formed by the decay and decomposition of marine animals under high pressure and temperature. The sulphur dioxide gas given out by the volcanoes beside the sea-side kills of fish & other sea animals which go on piling beside the volcano. After hundreds of years, these animals start decomposing under the influence of high pressure and temperature to form petroleum.

Engler's theory is supported **by the following facts:**

Experimental destructive decomposition of fish oil & other animal fats under high pressure &temperature gives a product similar to natural petroleum.

— Presence of brine or sea water together with petroleum.
— Presence of nitrogen and sulphur compounds.
— Presence of optically active compounds.
— Presence of fossils (the remains of animals or plants which died long ago) in the petroleum area.

However, this theory fails to account for the presence of chlorophyll (green-colouring matter of plants) in petroleum. Moreover, it does not explain the presence of coal deposits in the vicinity of the oil fields.

Modern theory. According to modern views, petroleum is believed to be formed by the decay & decomposition of marine animals as well as that of vegetable organism of the pre-historic forests, *i.e.* it is of animal as well as of plant origin. It is thought that due to some upheavals or earthquakes, these prehistoric forests, and sea animals got buried under the crust of earth. Due to action of prolonged action of high temperature &pressure in the interior of earth for ages, the biological matter decomposed into petroleum.

This modern theory explains the presence of brine and coal in the vicinity of petroleum. Presence of brine is explained on the basis of animal origin while the presence of coal is explained by plant origin. It also explains the presence of N & S compounds, chlorophyll and optically active compounds.

Difference Between Origins of Petroleum Oil and Coal:

- Petroleum oil was formed mainly from sea plants & animals decaying under strongly reducing conditions while the coal was formed mainly from land plants decaying under mildly reducing conditions.
- Coal seams remained static where formed/deposited while oil can migrate under the effects of pressure&temperature so that the location of the existing deposits may not be the location of the initial accumulation of oil forming debris. The source rock is sedimentary and mainly or entirely of marine origin.

Composition of Petroleum. A typical composition (by weight %) of petroleum is given below:

Carben = 84-87%	Hydrogen = 11-15%	Sulphur = 0.1-3%
Nitrogen = 0.1-1.5%	Oxygen = 0.3-1.8%	

The principal components of petroleum are hydrocarbons (paraffins, naphthenes, aromatics, olefins etc.), small amounts of sulphur, nitrogen & oxygen compounds as impurities and some inorganic compounds & metals (vanadium & platinum in traces).

Sulphur compounds found in petroleum are hydrogen sulphide, thiophenes, mercaptans. Oxygen occurs in combined form in alcohols, phenols, resins and organic acids present in petroleum. Nitrogen compound include pyridines, quinolines, pyroles etc. Inorganic compounds present in petroleum are salt, clay and sand etc.

Typical properties of crude oil

Specific gravity = 0.80 - 0.95, viscosity at 37.8°C = 2.3 – 23 cst, carbon/hydrogen ratio = 6-8, pour point = 18-30 °C, wax content = 6 - 12%, flash point = 15- 27°C, water content = 0.1 - 1.5%, Salt content = 1.5 - 8%, Ash content = 0.004 - 0.006%.

Nature and Characteristics of Crude Oils

It is a complex mixture containing many different hydrocarbons compounds that vary in appearance and composition from one oil field to another. Crude oils range in consistency from thick to tar like liquids and in color from clear to black.

- An "Average" crude oil contains about 84% carbon, 14% hydrogen, 1 -3% sulfur, and less than 1% each of nitrogen, oxygen, metals, & salts. Crude oils are generally classified as paraffinic, naphthenic, or aromatic, based on the predominant proportion of similar hydrocarbon molecules. Mixed-base crudes have varying amounts of each type of hydrocarbon. Refinery crude base stocks usually consist of mixtures of two or more different crude oils.
- Relatively simple oil assays are used to classify crude oils as paraffinic, naphthenic, aromatic, or mixed. One assay method (United Sates Bureau of Mines) is based on distillation, and another method (UOP "K" factor) is based on gravity and boiling points. More comprehensive crude assays determine the value of the crude (i.e., its yield & quality of useful products) and processing parameters. Crude oils are usually grouped according to yield structure.
- Crude oils are also defined in terms of API (American Petroleum Institute) gravity. The higher the API gravity, the lighter the crude. For example, light crude oils have high API gravities and low specific gravities.'Crude oils with low carbon, high hydrogen, and high API gravity are usually rich in paraffins and tend to yield greater proportions of gasoline &light petroleum products; those with high carbon, low hydrogen, and low API gravities are usually rich in aromatics.
- Crude oils that contain appreciable quantities of hydrogen sulfide or other reactive sulfur compounds are called "sour, " Those with less sulfur are called "sweet."
- Crude oil is a mixture of hydrocarbon molecules, that contains from one to 60 carbon atoms. The properties of hydrocarbons depend on the number &arrangement of the carbon and hydrogen atoms in the molecules.

- Hydrocarbons containing up to four carbon atoms are usually gases, those with 5 to 19 carbon atoms are usually liquids, and those with 20 or more are solids. The crude oil refining process uses chemicals, catalysts, heat& pressure to separate and combine the basic types of hydrocarbon molecules naturally found in crude oil into groups of similar molecules. The refining process also rearranges their structures and bonding patterns into different hydrocarbon molecules and compounds.
- Sulfur may be present in crude oil as hydrogen sulfide (H2S), as compounds (e.g. mercaptans, sulfides, disulfides, thiophenes, etc.) or as elemental sulfur. Each crude oil has different amounts & types of sulfur compounds, but the proportion, stability, & complexity of the compounds are greater in heavier crude-oil fractions. Hydrogen sulfide is a primary contributor to corrosion in refinery processing units. Other corrosive substances are elemental sulfur & mercaptans. Moreover, the corrosive sulfur compounds have an obnoxious odor.

Pyrophoric iron sulfide are formed from the corrosive action of sulfur compounds on the iron & steel used in refinery process equipment, piping, and tanks. The combustion of petroleum products containing sulfur compounds produces undesirables such as sulfuric acid and sulfur dioxide. Catalytic hydrotreating processes such as hydrodesulfurization remove sulfur compounds from refinery product streams. Sweetening processes either remove the obnoxious sulfur compounds or convert them to odorless disulfides, as in the case of mercaptans.

- Oxygen compounds such as phenols, ketones, and carboxylic acids occur in crude oils in varying amounts.
- Nitrogen is found in lighter fractions of crude oil as basic compounds, and in heavier fractions of crude oil as non basic compounds that may also include traces of copper, vanadium, and/or nickel. Nitrogen oxides can form in process furnaces. The decomposition of nitrogen compounds in catalytic cracking & hydrocracking processes forms ammonia and cyanides that can cause corrosion.
- Metals, including nickel, iron, and vanadium are often found in crude oils in small quantities and are removed during the refining process. Burning heavy fuel oils in refinery furnaces & boilers can leave deposits of vanadium oxide &nickel oxide in furnace boxes, ducts, and tubes. It is also desirable to remove trace amounts of arsenic, vanadium, & nickel prior to processing as they can poison certain catalysts.

- Crude oils often contain inorganic salts such as sodium chloride, magnesium chloride, and calcium chloride in suspension or dissolved in entrained water (brine). These salts must be removed or neutralized before processing to prevent catalyst poisoning, equipment corrosion, and fouling. Salt corrosion is caused by the hydrolysis of some metal chlorides to hydrogen chloride (HCl) and the subsequent formation of hydrochloric acid when crude is heated. Hydrogen chloride may also combine with ammonia to form ammonium chloride (NH4Cl), which causes fouling and corrosion.
- Some crude oils contain naphthenic (organic) acids, which may become corrosive at temperatures above 230°C when the acid value of the crude is above a certain level.

Major Oil Refinery Products and their Characteristics

Gasoline. The most important refinery product is motor gasoline, a blend of hydrocarbons with boiling ranges from ambient temperatures to about 205°C. The important qualities for gasoline are octane number, volatility, and vapor pressure. Additives are often used to enhance performance and provide protection against oxidation & rust formation.

Kerosene. Kerosene is a refined middle-distillate petroleum product that finds considerable use as a jet fuel and in cooking and space heating. When used as a jet fuel, some of the critical qualities are freeze point, flash point, and smoke point. Commercial jet fuel has a boiling range of about 190-275°C and military jet fuel 55-288°C. Kerosene with less-critical Specifications, is used for lighting, heating, solvents, and blending int diesel fuel.

Liquified Petroleum Gas (LPG). LPG, which consists principally of propane & butane, is produced for use as fuel and is an intermediate material in the manufacture of petrochemicals. The important specifications for proper performance include vapor pressure and control of contaminants.

Distillate Fuels. Diesel fuels and domestic heating oils have boiling ranges of about 205-370°C. The desirable qualities required for distillate fuels include controlled flash & pour points, clean burning, no deposit formation in storage tanks, and a proper diesel fuel cetane rating for good starting & combustion.

Residual Fuels. Many marine vessels, power plants, commercial building and industrial facilities use residual fuels or combinations of residual& distillate fuels for heating and processing. The two most critical specifications of residual fuels are viscosity and low sulfur content for environmental control.

Coke and Asphalt. Coke is almost pure carbon with a variety of uses from electrodes to charcoal briquets. Asphalt, used for roads & roofing materials, must be inert to most chemicals and weather conditions.

Solvents. A variety or products, whose boiling points and hydrocarbon composition are closely controlled, are produced for use as solvents. These include benzene, toluene, and xylene.

Petrochemicals. Many products derived from crude oil refining, such as ethylene, propylene, butylene, and isobutylene, are primarily intended for use as petrochemical feedstock in the production of plastics, synthetic fibres, synthetic rubbers, and other products.

Lubricants. Special refining processes produce lubricating oil base stocks. Additives such as demulsifiers, antioxidants, and viscosity improvers are blended into the base stocks to provide the characteristics required for motor oils, industrial greases, lubricants, & cutting oils. The most critical quality for lubricating-oil base stock is a high viscosity index, which provides for greater consistency under varying temperatures.

Detection of Petroleum Deposit

Petroleum deposits are detected by visual, geophysical, geological and drilling methods. Visual method includes oil seepages at the surface of the earth.

Geophysical method includes the measurement of density, elasticity, magnetic & electrical properties of the rock in the crust of the earth from which the inference are drawn whether the petroleum deposit occurs there or not.

Geological method includes the measurement of age and nature of rocks inside the earth's crust which gives an idea of oil deposit.

Drilling in the crust of earth is the final test when the petroleum deposits have been proved by either of the earlier methods.

Production of Petroleum

Petroleum almost always occurs along with gas called natural gas. After drilling in the crust of earth both natural gas & oil flow up through pipe under pressure initially. When the oil pressure decreases, then the residual oil is either sucked by a pump or pressure is created by injecting compressed gas or high pressure water through a pipe bored by the side of oil delivery pipe so that it can flow out.

When the oil well contains both oil & gas it is called ***wet well*** and if it contains only gas then it is called a ***dry well***.

A sketch of typical oil pool formation inside the crust of earth is given in Figure below. Hard cap rock & hard rock at the bottom, both are impermeable to oil & gas. Geological studies indicates that petroleum was not formed in the pools where it is found today. Deposits of crude petroleum were made near seashores, but the action of the surrounding water gradually shifted the location of oil pool. After a period of many centuries, the oil was forced through layers of porous rock strata until it became trapped under a dome capped by hard rock. This makes it impossible for the gas or oil to escape and the water keeps the pool of water under pressure.

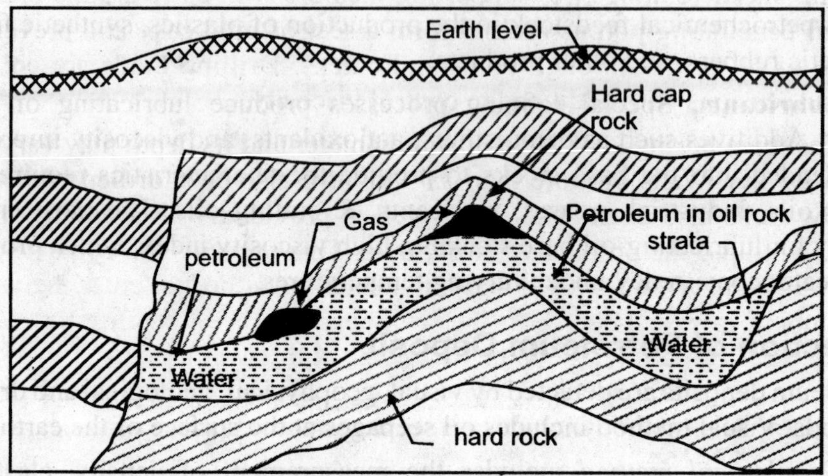

Fig. Typical oil-pool formations

In the centre, pool water has forced the petroleum and gas up into a dome, or *anticline,* in the porous oil rock layer. Hard cap rock prevents the escape of the oil. A *fault,* at the left, or a *stratigraphic trap,* as shown at the right, similarly traps oil pools between water under high pressure and hard rock strata.

For digging an oil well, the rotary drilling system, utilizes the principle of torsional grinding or cutting. The drilling rig consists of four main parts, viz. (a) **derrick**—a tall steel structure equipped with hoisting gear for raising and lowering the drilling equipment, (*b*) ***drill column or stem*** - this consists of a main section of steel pipes (*drill pipe*) with at its lower end some extra heavy pipe *drill collars*), to which the bit is attached; (c) ***rotating mechanism*** at the upper end of the drill cloumn, *i.e.* at the surface, is a hollow bar of square or hexagonal cross-section (the *kelly* or *grief stem*), and this slides through a square or hexagonal hole in the machine (*rotary table*) which provides the rotary movement; (*d*) ***mud circulation system*** — at the top of the kelly is the so-called *swivel* through which a thin mud (drilling fluid) passes from a flexible hose; the bottom half of the swivel revolves with the kelly while the top does not, and to it the flexible hose is attached; the top part of the swivel is attached also to the hoisting mechanism.

The hoisting system is controlled from a *working platform* built within the derrick above the ground level. The mud is pumped down the drill pipe through the holes of the bit and returned to the pump section *via* the annular space between the drill pipe and the walls of the bore hole.

An efficient mud circulation is required to drill an oil well rapidly. The drilling fluid serves four principal functions, *viz.* (*a*) cooling the bit, (*b*) flushing out the cuttings, (c) supporting the bore wall by hydraulic pressure, and (*d*) penetrating porous rock to form a seal of mud cake and prevent the escape of oil or gas through pores and cracks. Drilling muds are colloidal clay-suspensions with additions to control viscosity.

The main boring is of 20 to 30 cm diameter and its depth may vary from 1.5 to 4.5 km. Wells of more than 7.5 km have also been drilled while quite shallow wells also exist. The entire well is provided with a casing of steel pipes right upto the productive work layer. This prevents the collapse of the bore walls and ensures that other extraneous fluids do not enter the well. A steel tubing is suspended from the well-head up to oil-bearing rock and is provided with control valves and outlet pipe at the wall head.

Oil is normally under pressure in the reservoir. The source of the pressure may be a body of compressed gas above the oil or a mass of water underlying the oil. These pressures cause the oil to flow towards the *foot* of the well and then drive it up the tubing to the surface (*natural flow*). When oil pressure is low or it falls on production, the lifting is done by sending a high pressure gas through the annulus between the casing and the tubing or by providing a suitable pumping device at the lower end of the tubing. Another method of secondary recovery of oil is by *water flooding*; water is injected into the reservoir rock through separate wells on the periphery of the oil field.

The total production from each well is piped to the *gathering station.* The gas is separated by directing the fluid against baffles in cylindrical pressure vessels. The water and solids are separated either by settling out in tanks or by centrifuging. The gas is called *wet natural gas.* It is processed for the recovery of natural gasoline and the crude is sent to the refinery.

Petroleum Refining Operations in Oil Refineries

Petroleum refining begins with the distillation, or fractionation of crude oils into separate hydrocarbon groups. The resultant products are directly related to the characteristics of the crude processed. Most distillation products are further converted into more usable products by changing the size and structure of the hydrocarbon molecules through cracking, reforming, and other conversion processes. These converted products are then subjected to various treatment and separation processes such as extraction, hydrotreating, and sweetening to remove undesirable constituents and improve product quality. Integrated refineries incorporate fractionation, conversion, treatment, and blending operations and may also include petrochemical processing.

Petroleum refining processes and operations can be separated into five basic areas:

— **Fractionation** (distillation) is the separation of crude oil in atmospheric and vacuum distillation towers into groups of hydrocarbon compounds of differing boiling-point ranges called "fractions" or "cuts."

— **Conversion** processes change the size and/or structure of hydrocarbon molecules. These processes include:

- Decomposition (dividing) by thermal and catalytic cracking;
- Unification (combining) through alkylation and polymerization; and
- Alteration (rearranging) with isomerization and catalytic reforming.
- Treatment processes are intended to prepare hydrocarbon streams for additional processing and to prepare finished products. Treatment may include the removal or separation of aromatics and naphthenes as well as impurities and undesirable contaminants. Treatment may involve chemical or physical separation such as dissolving, absorption, or precipitation using a variety and combination of processes including desalting, drying, hydrodesulfurzing, solvent refining, sweetening, solvent extraction, and solvent dewaxing.

— **Formulating and Blending** is the process of mixing and combining hydrocarbon fractions, additives, and other components to produce finished products with specific performance properties.

— **Other Refining Operations** include: light-ends recovery; sour-water stripping; solid waste & wastewater treatment; process water treatment & cooling; storage & handling; product movement; hydrogen production; acid & tail-gas treatment; and sulfur recovery.

Pre-Treatment of Oil at Oil Field Before Refining

Oil & gas when they come out of oil field are separated. The natural gas is compressed to liquify it which is used for heating of domestic &industrial ovens. Petroleum oil is made free of water, sediments & salts present in it. It is then made free of some dissolved gases into it by the process called '*stabilisation*'. It is then sent to oil refineries for separation into various petroleum products by distillation mainly and auxiliary operations.

Crude oil as it comes out of well may contain upto 25% water, salts ($MgCl_2$, $CaCl_2$, NaCl etc.) upto 2000-5000 mg/litre & sediments upto 1-1.5%. For refining crude oil, the salt content in it should be < 50 mg/litre and water

< 0.3%. Excessive water in crude requires extra heat for its distillation, increases its cost of transportation and forms emulsion which absorb materials like resin (hence emulsion breakers are to be used). Salt in crude oil causes scaling, corrosion & reduces heat transfer co-efficient during its processing. Sediments present in crude causes erosion and scaling.

In mechanical method of separation of impurities from crude oil, it is subjected to centrifuging, filtration & settling after heating it to 120-160°C at 6-8 atm. pressure.

In physico-chemical method, emulsion breakers are added. But they are costly and cause corrosion & sludge formation.

Dehydration of Crude Oil. In *electrical dehydration* method of crude oil shown below in Figure (widely practised in Russia and America), 10% water is added to crude and it is heated before passing it through an electrical dehydrator which has two concentric metal plates as electrodes in which 30000 volts is supplied thereby separating water from crude.

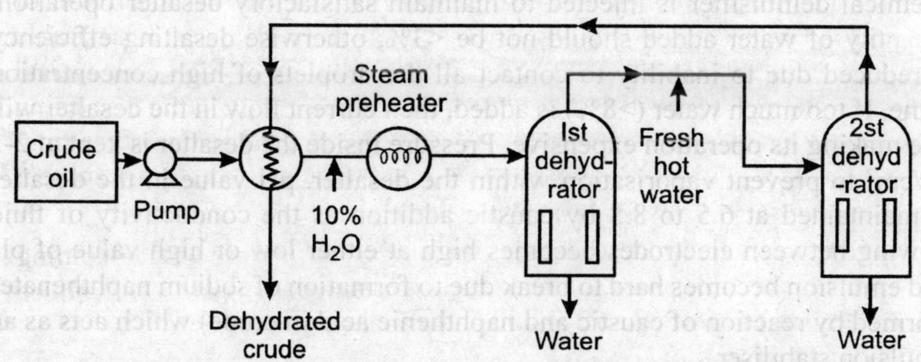

Fig. A scheme diagram for electrical dehydration of crude oil.

Details of dehydrators are given below:

First dehydrator → T = 90-95°C, P = 6.8 atm.

Second dehydrator →T = 80-90°C, P = 5.5-6 atm.

In Russian practice, however, spherical dehydrators are used whose separation efficiency is more.

Electrical Desalting of Crude Oil. All crude oil contain salt water together with sediment. Salt content in crude oil is reported as *pounds of salt* (*measured as sodium chloride*) *per thousand barrel of oil* (*PTB*) and it varies from 10 to 200 PTB. Besides being present as brine during its production in the oil well, crude oil gets contaminated with salt water when shipped in tankers as ballast water mixes into the crude oil during shipment and may be emulsified with crude oil while pumping it to shore installation.

Desalting of crude oil is done electrically in which, first an emulsion of crude oil & water is formed and then the emulsion is broken by means of an electrical field. Crude oil is heated to 120- 130°C to reduce its viscosity (<40SUS) to produce good emulsion. Water (3-8% of crude oil) is then added in the crude oil and the mixture is then passed through a mixing valve across which a pressure drop of about 1.5 kg/cm^2 is maintained for proper emulsification of crude oil & water. The emulsified crude oil is fed to the electrical desalter, where it is subjected to high DC voltage (20000- 24000 Volts) electrical field which results in coalescence of water droplets and its separation from the emulsion together with the salt from crude oil. Desalted crude overflows while water along with salt & crude sludge is removed from the bottom of the desalter. Salt content of crude oil can be brought down to a level of 3 PTB or to the extent of 95% removal. Proper pressure drop (usually 1.5 kg/cm^2) through the mixing valve should be maintained to create a good emulsion. However, maximum pressure drop (usually 4 kg/cm^2) should not be exceeded otherwise an unbreakable emulsion may be formed. Sometimes, chemical demulsifier is injected to maintain satisfactory desalter operation. Quantity of water added should not be <3%, otherwise desalting efficiency is reduced due to inability to contact all the droplets of high concentration brine. If too much water (>8%) is added, then current flow in the desalter will rise making its operation expensive. Pressure inside the desalter is kept at 2-3 kg/cm^2 to prevent vaporisation within the desalter. pH value in the desalter is maintained at 6.5 to 8.5 by caustic addition as the conductivity of fluid flowing between electrodes becomes high at either low or high value of pH and emulsion becomes hard to break due to formation of sodium naphthenates (formed by reaction of caustic and naphthenic acid in crude) which acts as an emulsion stabiliser.

A flow sheet of electrical desalting process of crude oil is given below in Figure.

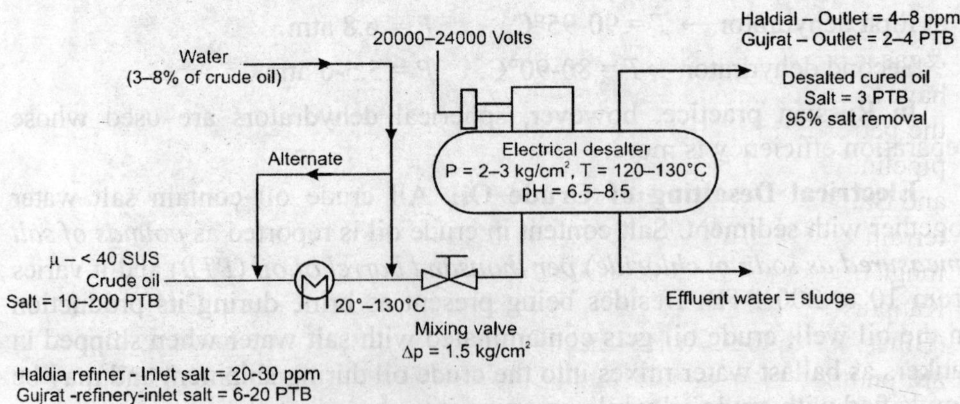

Fig.: Electrical Desalting Process of crude oil.

Stabilisation of Crude Oil

Removal of dissolved gases from crude oil by heating it, is called its stabilisation. Gas accompanied with crude oil must be removed to avoid breathing loss (loss of gasoline, if gas is not removed from oil during pre refining). Breathing loss occurs due to changes of humidity of ambient air and day & night temperature change also during emptying & filling of crude tanker. Breathing loss is about 0.4-0.75 kg/m^3 of tank volume per month.

A scheme for stabilisation of crude oil is shown below in the Figure.

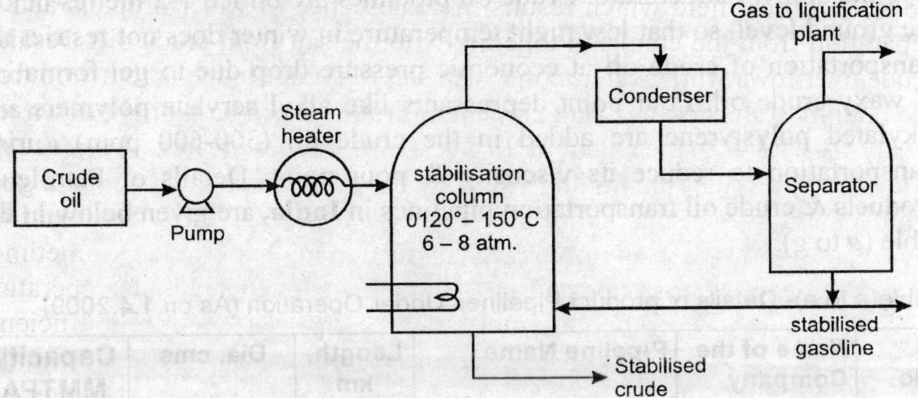

Fig.: Stabilisation of crude oil.

Transportation of Crude Oil

Most of the crude oil is transported from oil fields to refineries by pipelines. However, it is also transported by rail road (tank cars), tankers, tank truck and barges (small tankers). Underwater super tankers (in marine ships) are used for inter-continental transport of crude oil.

Trans-Arabian underground pipeline for crude transportation has a diameter of 75 cms equipped with 6000 hp turbine pumping 340000 bbl/day at a pressure of about 35-60 kg/cm^2. The cheapest mode of transporting oil &gas is by taking them through pipelines. Several large networks of pipelines have been built in India. The Gauhati product pipeline, is designed to transport the petroleum products from Gauhati Refinery. Rudrasagar-Gauhati-Barauni pipeline system transports crude oil from Rudrasagar oilfields to Gauhati and Barauni refineries. It brings oil from Assam through 1200 km of difficult terrain crossing many rivers. The Barauni refinery pipeline system is very long. It is in two sections, Barauni-Kanpur and Barauni-Haldia. Barauni to Kanpur is a products line,, i.e. finished products are pumped from Barauni (Bihar) to Kanpur. At Haldia port, crude oil brought by tankers from abroad are pumped to Barauni. Gujarat pipeline system, handles gas, crude oil and refined products. Cambay- Dhuvaran, Ankleshwar-Baroda and Ankleshwar-Uttaranpipelines are for naturalgastransportation; Ankleshwar-Koyali line is for crude oil and Koyali-Ahmedabadpipeline is a product line.

Transportation Pipeline for Crude Oil & its Products in India

Underground pipeline for the transportation of crude oil and its products are laid in various parts of India. Middle east crude oil having pour point below 0°C and wax content upto 7% pose no problem in its transportation through pipeline at low temperature. Crude oil pipelines are buried 1-2 metres below the ground level, so that low night temperature in winter does not restrict the transportation of crude oil at economic pressure drop due to gel formation in waxy crude oil. Pour point depressants like alkyl acrylate polymers and alkylated polystyrene are added in the crude oil (300-600 ppm) during transportation to reduce its viscosity & pour point. Details of Petroleum products & crude oil transportation pipelines in **India,** are given below in the table (*a* to *g*).

Table(a to e)- Details of product Pipelines Under Operation (As on 1.4.2009)

S. No.	Name of the Company	Pipeline Name	Length, km	Dia, cms	Capacity MMTPA
(a)	Indian Oil Corporation (IOC)				
1.		Guwahati-Siliguri Pipeline	435	20	1.40
2.		Koyali-Ahmedabad Pipeline	116	20	1.10
3.		Koyali-Navagam Pipeline	78	35	1.80
4.		Koyali-Dahaj Pipeline	103	35	2.60
5.		Koyali-Sanganer Pipeline	1056	45/25	4.10
6.		Barauni-Kanpur Pipeline	745	50/30	3.50
7.		Haldia-Mourigram Rajbandh Pipeline	277	30	1.35
8.		Haldia-Barauni Pipeline	525	30	1.25
9.		Mathura-Jalandhar Pipeline	763	40/35/30/25	3.70
10.		Mathura-Tundla Pipeline	56	40	1.20

S. No.	Name of the Company	Pipeline Name	Length, km	Dia, cms	Capacity MMTPA
11.		Panipat Rewari Pipeline	155	30	1.50
12.		Panipat Bhatinda Pipeline	219	35	1.50
13.		Digboi-Tinsukia Pipeline	75	20/15	1.00
14.		Chennai-Trichy-Madurai Pipeline	683	35/30	1.80
15.		Koyali-Ratlam Pipeline	265	40	2.00
16.		ATF pipeline to AFS Chennai	95	20	2.18
17.		ATF pipeline to BIAL	36	20	1.56
		Total	5682		33.54
(b) HPCL (Hindustan Petroleum)					
1.	Corporation Limited)	Mumbai-Pune-Solapur Pipeline	508	35/30	3.67
2.		Visakh-Vijaywada-Secundrabad Pipeline	572	45/40/35	5.38
3.		Mundra-Delhi Pipeline	1054	45/40	5.00
		Total	2134		14.05
(c) BPCL (Bharat Petroleum					
1.	Corporation Limited)	Mumbai-Manmad-Mangalya- Bijwasan Pipeline	1384	45/40/35/20	6.00
		Total	1384		6.00
(d) Petronet					
1.		Mangalor-Hassan-Bangalore Pipeline	362	60/50	5.60
2.		Cochin-Coimatore-Karur Pipeline	292	35/30	2.00
		Total	654		7.60
(e) Oil India Ltd. (OIL)					
1.		Numaligarh-Siligure Pipeline	660	40	1.72
		Total	660		1.72
		Total 24 Nos. of Product Pipelines are in operation	10514		62.91

Table (b): Details of Petroleum Product Pipelines Under Implementation (As on 1-4-2009)

S. No.	Name of company	Pipeline Name	Length, km	Dia, cms	Capacity MMTPA
(a)	Indian oil Corpution (IOC)				1.45
1.		Chennai-Bangalore Pipeline	290	35/30	1.45
2.		Mathura-Bharatpur Spur Pipeline	21	20	-
3.		Branch Pipeline to Hazira	94	30	-
4.		Bijwasan-Panipat Naphatha Pipeline	118	25/190	0.80
5.		Hook up of Tikrikalan with MJPL	8	35	-
		Total	531		2.25
(b)					
1.		Bina-Kota pipeline	265	45	2.80
		Total	265		2.80
		Total 6 Nos. of product pipeline are under implementation	796		5.05

Table (c): Details of crude oil pipeline Under operation (As on 1-4-2009)

S. No	Name of Compeny	Pipeline Name	Length, km	Dia, cms	Capacity, MMTPA
(a)	Indian oil Corporation (IOC)				
1.		Salaya-Mathura Pipeline	1870	105/70/60	21.00
2.		Haldia-baraunia crude Pipeline	943	45	7.50
3.		Mundra-Panipat Crude oil Pipeline	1174	70/40	6.00
		Paradip-Haldia Pipelines	330	120/75	11.00
		Total	4317		45.50

S. No	Name of Company	Pipeline Name	Length, km	Dia, cms	Capacity, MMTPA
(*b*)	Oil India Limited				
1.		Duliajan-Barauni Crude oil trunk Pipeline	1417	40/35	3.75
2.		Feeder lines	61	20/25/30/40	3.50
		Total	1478		7.25
		Total 5 Nos. of crude oil Pipelines are in operation	5795		52.75

Table (d): Details of Crude Oil Pipelines Under Implementation (As on 1.4.2009)

S. No.	Name of the company	Pipeline Name	Length, (km)	Dia, cms	Capacity (MMTPA)
(*a*)	Indian Oil Corporation (IOC)				
1.		Augmentation of Mundra- Panipat Pipeline	20	55	3.00
		Total	20		3.00
(*b*)	BPCL				
1.		Vadinar-Bina Pipeline	935	60	6.00
		Total	935		6.00
		Total 2 Nos. of crude pipelines are under implementation	955		9.00

Table (e): Details of LPG Pipelines in Operation (As on 1.4.2009)

S No.	Name of the company	Pipeline Name	Length, (km)	Dia, cms	Capacity (MMTPA)
(*a*)	GAIL				
1.		GAIL LPG Pipeline Network	1922	-	3.80
		Total	1922	-	3.80

S No.	Name of the company	Pipeline Name	Length, (km)	Dia, cms	Capacity (MMTPA)
(*b*)	Indian Oil				
1.		Panipat-Jalandhar Pipeline	275	25	0.70
		Total	275		0.70
		Total 2 LPG pipeline are in operation	2197		4.50

MMTPA-Million Metric Tonnes per Annun

Table (f): Details of NG (Natural Gas) Pipelines in Operation (As on 1.4. 2009)

S. No.	Name of the company	Pipeline Name	Length, (km)		Capacity (MMSCMD)
(*a*)	GAIL (Gas Authority of India Limited)	GAIL NG Pipeline Network	6700		148.00
		Total	6700		148.00
		Total NG Pipeline is Under Implementation	6700		148.00

MMSCMD-Million Metric Standard Cubic Meter Per Day

S. No.	Name of the company	Pipeline Name	Length, (km)	Dia, cms	Capacity (MMSCMD)
(*a*)	Indian Oil Corporation (IOC)	Dadri-Panipat R-LNG pipeline	132	75	10
		Total	132	-	10

LNG-Liquefied Natural Gas

Table: Growth of Petroleum Product, LPG & Crude Oil Pipelines in India during last five years

Oil Company	Name of the Pipeline	Capacity* (MMTPA 1.04.2010	Length (KM) 1.04.2010 2005-06	Throughput ('000' Tonne)				
				2006-07	2007-08	2008-09	2009-10$_p$	
	Product Pipelines							
IOC	Barauni-Patna-Kanpur (incl Gawaria Lucknow branchline)	5.30	745	4180	4050	3870	3580	4000
IOC	Guwahati-Siliguri	1.40	435	1260	1230	1310	1510	1800
IOC	Haldia-Barauni	1.30	525	870	960	940	820	#
IOC	Haldia-Mourigam-Rajbandh	1.40	277	1760	1900	1580	1290	2500
IOC	Koyali-Ahmedabad	1.10	116	520	540	610	540	500
IOC	Koyali-Ratlam	2.00	265					100
IOC	Panipat-Ambala-Jalandhar (incl K'kshetra-Roorkee-Najibabad branchline)	3.5{	434					2600
IOC	Panipat-Delhi (Sonipat-Meerut branchline)		182					700
IOC	Mathura-Delhi	3.70	147					2500
IOC	Panipat-Bhatinda	1.50	219		1290	1440	1510	1500
IOC	Mathura-Tundla	1.20	56	250	250	240	280	300
IOC	Koyali-Viramgam-Sidhpur-Sanganer	4.10	1056	2060	3010	3180	3240	3000

IOC	Panipat-Rewari	1.50	155	910	1240	1380	1420	1500
IOC	Chennai-Trichi-Madurai	1.80	683	130	820	1370	1690	1900
IOC	Koyali-Dehej	2.60	103	-	180	900	800	1000
IOC	Chennai ATF	0.20	95			0	30	200
IOC	Chennai-Bangalore	1.50	290					0
HPCL	Mumbai-Pune	3.70	508	2780	3240	3360	3370	3400
HPCL	Vizag-Vij ayawada-Secunderabad	5.40	572	2880	3500	3850	3880	3800
HPCL	Mundra-Delhi	3.80	1054			620	3390	4800
BPCL	Mumbai-Manmad-Mangliya-Piyala-Bijwasan	11.80	1389	3792	3790	8010	10490	11200
OIL	Numaligarh-Siliguri	1.70	654.30					800
PCCK	Cochin-Coimbatore-Karur (PCCK)	3.30	292.5			1360	1560	1500
PMHB	Mangalore-Hasan-Bangalore	2.10	363	1012	1440	2141	2450	2500
	Subtotal	**65.90**	**10615.8**	**22404**	**27440**	**36161**	**41850**	**52100**
	LPG Pipelines							
IOC	Panipat-Jalandhar	0.70	274				600	400
GAIL	Jamnagar-Loni	2.50	1355	1910	2030	2232	2090	2400
GAIL	Vizag-Vijayawada-Secunderabad	0.70	623	320	460	522	650	700
	Subtotal	**3.90**	**2252**	**2230**	**2490**	**2754**	**3340**	**3500**

	Crude Pipeline							
OIL	Duliajan-Digboi-Bongaigaon-Barauni	8.40	1193	5990	6000	6070	6220	6200
IOC	Salaya-Mathura-Panipat (Incldg Loop lines)	21.00	1870	20300	21910	21680	22650	22400
IOC	Haldia-Barauni/Paradip barauni	11.00	1302	7040	7060	7240	8680	11600
IOC	Mundra-panipat	8.40	1194	-	3460	6940	6910	7400
ONGC	Mumbai High-Uran-Trunk Pipeline	15.63	204					8600
ONGC	Heera-Uran-Trunk Pipeline	11.50	81					5530
ONGC	Kalol-Nawagam-Koyali	8.54	129.31					5170
ONGC	MHN-NGM Trunk Line	2.26	77					2180
ONGC	CTF, Ankleshwar to Koyali Oil Pipeline (AKCL)	2.00	94.8					2090
ONGC	Lakwa-Moran Oil Line	1.50	18					430
ONGC	Geleki-Jorhat Oil Line	1.50	48.47					650
ONGC	NRM to CPCL	0.74	6					240
ONGC	KSP-WGGS to TPK Refinery	0.08	13.5					30
ONGC	GMAA EPT to S. Yanam Unloading Terminal	0.09	4					120
	Subtotal	92.64	6235.08	33330	38430	41930	44460	72640
	Grand Total	162.44	19103	57964	68360	80845	89650	128240

Notes: p: provisional #: Included in Haldia-Mourigram-Rajbandh

Hazira, Bijapur, Jagdishpur (HBJ) natural gas pipeline passing through Gujarat, Madhya Pradesh, Rajasthan and Uttar Pradesh for transportation of natural gas from Gujarat (Hazira) oil field, feeds natural gas to some fertiliser plants and some thermal power plants enroute.

Classification of Petroleum

Depending on the *nature of hydrocarbons* present in it, crude petroleum oil is classified as:

- paraffinic crude oil (residue left on distillation is paraffin wax)
- naphthenic crude oil (residue left on distillation contains asphalt plus little of paraffin wax)
- asphaltic (aromatic) crude oil (residue on distillation contains asphaltic materials)
- mixed crude oil (containing all paraffinic, naphthenic, and aromatic constituents)

Indian crudes are mostly mixed crude oil. Naharkatia (Assam) crude contains 40% paraffins, 25% naphthenes and 35% aromatics.

Paraffins having general formula C_nH_{2n+2} are saturated hydrocarbons (where $n = 1$ to 35) *e.g.* methane (CH_4), propane (C_3H_3), pentane (C_5H_{12}) etc.

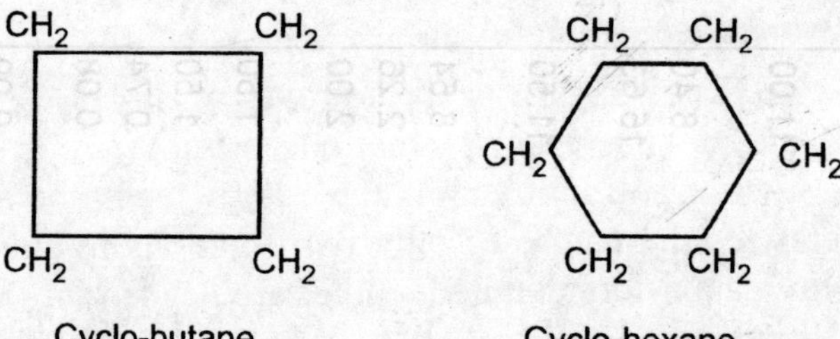

Cyclo-butane Cyclo-hexane

When $n = 1$ to 5, then constituents are gases *e.g.* C_4H_{10} (butane); when $n =$ 5 to 15, then constituents are liquids, *e.g.* $C_{10}H_{22}$ (decane) & when $n = 16$ and above then constituents are solids/semisolids *e.g.* $C_{17}H_{36}$ (wax).

Naphthenes having general formula C_nH_{2n} are saturated ring compounds (where $n = 1$ to 9) *e.g.* C_4H_8 (cyclo-butane), C_6H_{12} (Cyclohexane) etc.

Another hydrocarbon group having the same general formula C_nH_{2n} but straight chain (instead of ring structure as in case of naphthenes) is called *olefin e.g.* ethylene (C_2H_4), propylene (C_3H_6) butylene (C_4H_8) etc.

Aromatics having general formula C_6H_{2n-6} ($n \geq 6$) contain 6 carbon atoms in the form of a hexagon shaped ring and are unsaturated ring compounds *e.g.* benzene (C_6H_6), toluene $C_6H_5CH_3$), naphthalene ($C_{10}H_8$), anthracene ($C_{14}H_{10}$).

Benzene Toluene Napthalene Authracene

CH_3

Lower hydrocarbons in these groups are generally gases, intermediate ones are liquids and higher ones are solids/semi-solids.

The above *group crude classification* is based on the principle that the nature of the residue determines the nature of the whole crude which is of limited validity. Moreover the grouping of the bulk of the crudes under mixed base also gives an incomplete picture. These considerations have led to the development of *Lane and Garton classification* which is the most widely used in which two arbitrary fractions are recovered from the crudes on the basis of boiling range; (a) 250-275°C at 1 atmosphere pressure and (b) 275-300°C at a pressure of 40 mm of mercury. The former is termed Key Fraction 1 and the latter Key Fraction 2. Each fraction is classified as paraffin, intermediate or naphthene according to specific gravity, and the crude is classified on the nature of the key fractions as given in table below.Following this system, a crude oil is called *wax free* or *wax bearing* according as the cloud point of the key fraction 2 is below or above (- 15°C) respectively. Lane and Garton have found that 85% of the crudes examined fell into the three simple classes *viz.* paraffin, intermediate or naphthene. Paraffin-naphthene or intermediate-naphthene crudes were not found to exist.

Lane and Garton classification is an improvement upon the four group classification in as much as it recognised that one type of hydrocarbon might predominate in one fraction and another might predominate in another fraction. However it characterizes the nature of only two fractions by the method of using specific gravity data. More detailed studies are needed for the complete characterization of crudes.

Table: Lane and Garton Classification of Petroleum

Class	Key Fraction 1 Sp.gr. (15.5° /15. 5°C)	Key Fraction 2 Sp.gr. (15.5° /15.5°C)
Paraffin	≤0.8251	≤0.8762
Paraffin-intermediate	-do-	0.934 - 0.8762
Paraffin-nephthene	-do-	≥ 0.934
Intermediate	0.8251 - 0.8602	0.934 - 0.8762
Intermediate-paraffin	-do-	≤ 0.8762
Intermediate-naphthene	-do-	> 0.934
Naphthene	≥ 0.8602	≥0.934

Class	Key Fraction 1 Sp.gr. (15.5° /15. 5°C)	Key Fraction 2 Sp.gr. (15.5° /15.5°C)
Naphthene-paraffin	-do-	<0.8762
Naphthene-intermediate	-do-	0.934 - 0.8762

Another crude oil classification based on its specific gravity, API gravity and viscosity is as follows:

Type of crude oil	Specific gravity	°API gravity	Viscosity, centipoise (m.Pa. S)
Conventional or light crude oil	< 0.934	> 20	—
Heavy crude oil	0.934-1.000	10 to < 20	10000 (maximum)
Extra heavy crude oil	> 1.0	< 10	10000 (maximum)
Tar sand bitumen or natural asphalt	> 1.0	< 10	> 10000

Crude oil classification using specific gravity, API gravity and viscosity is shown below diagramatically.

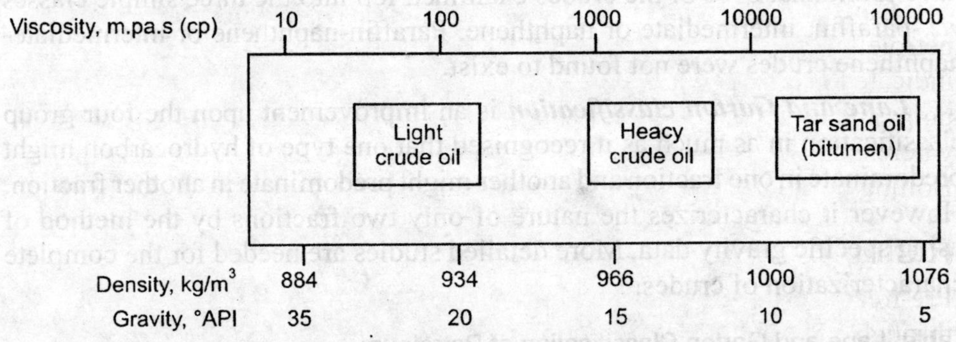

General classification of crude oil by viscosity, density and API gravity

Classification of crude oil based on its correlation index (CI).- Developed by U.S. Bureau of Mines, CI = 0 to 15 for paraffnic crude, CI = 15 to 50 for naphthenic or mixed crude and CI= > 50 for asphaltic (aromatic) crude. CI is defined as $CI = \frac{48460}{T_B} + 473.7\,G - 456.8$,

where, T_B = average boiling point, °R and G = specific gravity at 15.5°C.

Characteristics of Paraffinic and asphaltic crude oil. These two types of crudes yield products with different specification, properties &yields. The properties of naphthenic crudes are intermediate between that of the paraffinic& asphaltic crude. Table below gives the salient features of these two types of crudes.

Table - Characteristics of Paraffinic and Asphaltic Crude Oil

Properties and yields	Paraffinic crude	Asphaltic crude
Density	low	high
Yield of gasoline (petrol)	high	low
Octane No.(quality) of petrol	low	high
Sulphur content	low	high
Hydrogen/carbon ratio	high	low
Smoke point (quality) of Kerosene	high (better)	low
Cetane No.(quality) of diesel	high	low
Freezing point of fuel oil	high	low (better)
Yield of lubricants	high	low
Viscosity index of lubricating oil	high (better)	low
Yield of wax	high	low
Yield of bitumen/asphalt	low	high
Odour	sweet	pungent
Colour	light	dark

Constituents of Petroleum. Following commercial constituents of petroleum are recovered by its refining which has got different boiling ranges, indicated in *parenthesis.* The products which are lighter *i.e* having low boiling point and molecular weight are separated from top of the distillation column whereas the high boiling point, heavier products are separated at the bottom. Boiling point ranges of various products in increasing order is given below.

Refining of Petroleum

Pre-treated crude oil is heated in pipes in a furnace and its vapors are passed through a tall cylinderical fractionating column which has a number of plates inside it. As the petroleum vapor moves upward, the higher boiling point fractions condense &fall back and only lower boiling point vapors move to higher plates where further condensation takes place and some more liquid falls back. As the freshly condensed liquid comes to the lower plates it gets heated & vaporised and again condenses on higher plates. This process of condensation & vaporisation takes place many times causing separation of the constituents of petroleum according to their boiling points. Thus, the higher boiling point fractions condense towards the lower part of the column and the lower boiling point fractions towards the upper part.

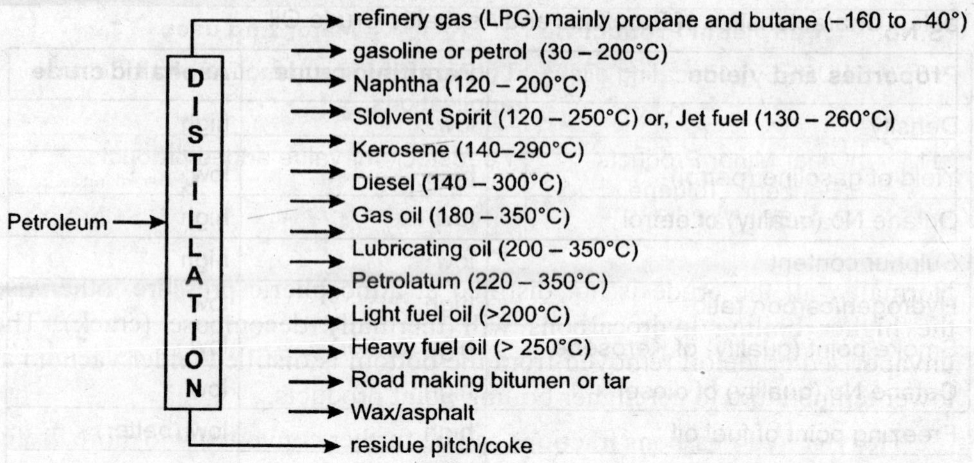

Major End uses of Petroleum Products are given below:

S.No.	Petroleum Products	Major End uses
1.	LPG (Liquefied Petroleum Gas)	Domestic fuel. Also for Industrial application where technically essential. Now permitted as auto fuel, used for cutting/welding work .industries.
2.	NAPTHA/NGL	Feedstock/fuel for Fertilizer Units, feedstock for petro-chemical sector and fuel for Power Plants, substitute fuel in steel plants for coke oven gas gas.
3.	Petrol (Gasoline)	Fuel for passenger cars, taxies, two & three wheelers
4.	ATF (Aviation Turbine Fuel)	Fuel for aircrafts.
5.	SKO (Superior kerosene oil)	Fuel for cooking & lighting.
6.	HSD (High speed diesel)	Fuel for transport sector (Railways Road), Agriculture (tractors pump sets, threshers, etc.) and Captive Power Generation.
7.	LDO (Light diesel oil)	Fuel for agricultural pump sets, small industrial units, start up fuel for power generation
8.	FO/LSHS (Furnace oil/ Low Sulfur Heavy stock)	Secondary fuel for Thermal Power Plants, Fuel/ feedstock for fertilizer plants, industrial units.
9.	BITUMEN	Surfacing of roads.

S.No.	Petroleum Products	Major End uses
10.	LUBES (Lubricating oils)	Lubrication for automotive and industrial applications.
11.	Other Minor Products (Benzene, Toluene, Paraffin Wax etc.	Feedstock for value added products.

Normally, all the crude is not distilled at atmospheric pressure, otherwise the higher boiling hydrocarbons will thermally decompose (crack). The unvaporised crude oil removed from the bottom is distilled under vacuum at lower temperature to get higher boiling point products.

Heavier hydrocarbons fractions can be further converted into more useful lighter products by processes such as cracking, reforming, visbreaking etc.

General Scheme of Petroleum Distillation in a refinery is shown below.

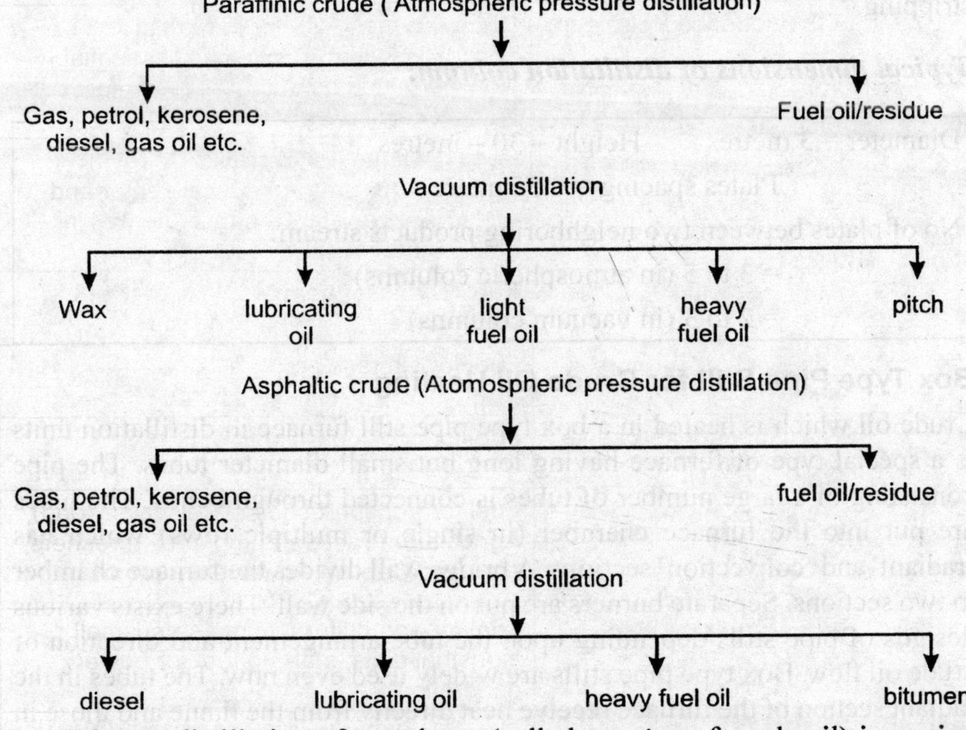

Primary distillation of petroleum (called *topping* of crude oil) is carried out first under *atmospheric pressure* and the residue (called reduced crude) is then distilled under vacuum. Maximum temperature upto which crude can be heated without cracking in primary distillation is 400°C. Steam is used for vaporisation of high boiling constituents. Heavier fractions boiling upto 600°C is recovered in vacuum distillation units.

Primary distillation recovers fractions upto gas oils while vacuum distillation recovers vacuum gas oil, raw lubricating oil, distillates and asphalt or bitumen. The actual distillation practice varies from plant to plant depending upon type of crude oil and products demand. Different crudes yield fractions/ products of different quality &quantity and their (product's) boiling range overlap each other. Primary distillation involves both high pressure (3-4 kgf/ cm^2) & atmospheric pressure distillation. High pressure distillation produces gases & light distillates boiling upto 90°C & its bottom product is heated to 400°C and then sent to atmospheric pressure distillation until where naphtha, petrol, kerosene & gas oil are recovered. The bottom reduced crude (from atmospheric distillation column) is reheated to 400°C and sent to the vacuum (30-80 mm Hg) distillation column where heavy gas oil raw lubricating oil, heavy fuel oil, bitumen, asphalt & wax are produced. Paraffin wax is made from vacuum gas oil while micro crystalline wax is produced/recovered from heavy lubricating oil distillates or residue from paraffinic crude oil. Lighter component impurities present inside stream products are removed by steam stripping.

Typical dimensions of distillation column:

Diameter = 3 metres   Height – 30 – metres
Plates spacing = 0.55 metre
No of plates between two neighboring products stream:
= 3 to 5 (in atmospheric columns)
= 2 to 3 (in vacuum columns)

Box Type Pipe Still for Crude Oil Heating

Crude oil which is heated in a box type pipe still furnace in distillation units is a special type of furnace having long but small diameter tubes. The pipe consisting of a large number of tubes is connected through bends. The tubes are put into the furnace chamber (in single or multiple rows) which has 'radiant' and 'convection' sections. A bridge wall divides the furnace chamber in two sections. Separate burners are put on the side wall. There exists various designs of pipe stills depending upon the tube arrangement and direction of crude oil flow. Box type pipe stills are widely used even now. The tubes in the radiant section of the furnace receive heat directly from the flame and those in the convection section recover heat from the hot gases travelling to the stack. The two sections are separated from each other by a row of tube. The crude oil is pumped into the pipe inlet at the convection section. Its final heating to 400°C takes place when the preheated crude oil passes through the radiant section tubes. High flow rate of crude oil in the pipe ensures high heat transfer rate by convection & radiation as well as prevents local overheating which might cause cracking & coke deposition .

Typical design features of a pipe still heater is given below:

Heat Input capacity = 5 to 20 x 10^6 kcal/hr
Heat release rate = 20000 to 45000 kcal/hr/m^3 of furnace volume
Heat flow = 15000 to 40000 kcal/hr/m^2 of circumferential tube area
Proportion of heat transfer by radiation = 55 to 60%

A box type pipe still is shown below in Figure.

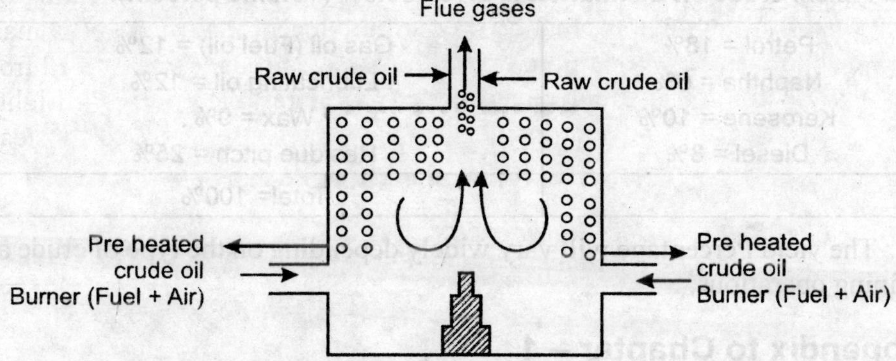

Uses of Petroleum Products

Gasoline (***Petrol***) - Fuel for spark ignition internal combustion engine (motor/aviation), dry cleaning of clothes.

Naphtha - Used as solvent, paint thinner, blending of motor fuel. Its most important use is in the production of H_2 (used in making nitrogeneous fertiliser) by its steam reforming. It is used as a fuel gas (naphtha vapor) in steel plants.

Jet Fuel - Fuel for jet plane, aviation turbine engines.

Kerosene - Domestic *illuminant* and tractor fuel.

Diesel - Fuel for diesel engines. (compression ignition engine)

Gas oil - It is gasified for fuel gas production. It is a fuel for industrial furnace. It is also used as blend for heavy fuel oil.

Lubricating oils - Used as a lubricant in machines and engines.

Petrolatum - It is also a lubricant used as a base material for grease manufacture.

Light fuel oil - Used as a fuel in industrial furnaces.

Heavy fuel oil - Used as furnace fuel after blending with light fuel oil or naphtha to reduce its viscosity and improve fluidity & atomisation characteristics. It is also cracked to produce gasoline, diesel, gas, light fuel oil etc.

Bitumen or Tar - Used for road making as binder and for moisture proof coating.

Wax - It is used for making candles, waxed papers & cartons, match stick coating, rust prevention etc.

Residue Pitch - For making road and as industrial fuel & electrodes.

Typical Products Yields in Crude oil Distillation. The yield of products of a typical crude oil distillation is given below (volume percent):

Petrol = 18% Naphtha = 6% Kerosene = 10% Diesel = 8%	Gas oil (Fuel oil) = 12% Lubricating oil = 12% Wax = 9% Residue pitch = 25%
	Total= 100%

The yield Percentage will vary widely depending on the type of crude and refining operations.

Appendix to Chapter – 1

1.1: Composition of crude oil

2.2: Characterization of crude oil

Appendix 1.1: Composition of Crude Oil

Petroleum (Crude oil) consists of mainly carbon (83-87%) and hydrogen (12-14%) having complex hydrocarbon mixture like paraffins, naphthenes, aromatic hydrocarbons, gaseous hydrocarbons (from CH_4 to C_4H_{10}). Table Below gives more details about composition of petroleum. Besides, crude oil also contains small amount of non hydrocarbons (sulphur compounds, nitrogen compounds, oxygen compounds) and minerals. Heavier crudes contains higher sulphur percentage. Depending on predominance of hydrocarbons, petroleum is classified as paraffin base, intermediate base or naphthenic base.

Hydrocarbons components

S.N.	Hydrogen Family	Distinguishing characteristics	Major hydrocarbons	Remarks
1	Paraffins (Alkanes)	Straight carbon chain	Methane, ethane, propane, butane, pentane, hexane	General formula C_nH_{2n+2} Boiling point increases as the number of carbon atom increases. With number of carbon 25-40, paraffin becomes waxy.
2	Isoparaffins (Iso alkanes)	Branched carbon chain	Isobutane, Isopentane, Neopentane, Isooctane	The number of possible isomers increases as in geometric progression as the number of carbon atoms increases.
3	Olefins (Alkenes)	One pair of carbon atoms	Ethylene, Propylene	General formula C_nH_{2n+2} Olefins are not present in crude oil, but are formed during process. Undesirable in the finished product because of their high reactivity. Low molecular weight olefins have good antiknock properties.
4	Naphthenes	5 or 6 carbon atoms in ring	Cyclopentane, Methyl cyclopentane, Dimethyl cyclopentane, cyclohexane, 1, 2dimethyl cyclohexane.	General formula C_nH_{2n+2}-$2R_n$ Rn is number of naphthenic ring The average crude oil contains about 50% by weight naphthenes. Naphthenes are modestly good components of gasoline.
5	Aromatics	6 carbon atom in ring with three around linkage.	Benzene, Toluene, Xylene, Ethyl Benzene, Cumene, Naphthaline	Aromatics are not desirable in kerosene and lubricating oil. Benzene is carcinogenic and hence undesirable part of gasoline.

Non Hydrocarbons components

	Non-hydrocarbons	Compounds	Remarks
6	Sulphur compounds	Hydrogen sulphide, Mercaptans	Undesirable due to foul odour 0.5% to 7%
7	Nitrogen compounds	Quinotine, Pyradine, pyrrole, indole, carbazole	The presence of nitrogen compounds in gasoline and kerosene degrades the colour of product on exposure to sunlight. They may cause gum formation normally less than 0.2.
8	Oxygen compounds	Naphthenic acids, phenols	Content traces to 2%. These acids cause corrosion problem at various stages of processing and pollution problem.

The majority of crude oil is alkanes, cycloalkanes (naphthenes), aromatics, polycyclic aromatics, S-containing compounds, etc.
Gasoline: branched alkanes
Diesel: linear alkanes

Alkanes

normal — $CH_3–CH_2–R$

branched — $CH_3–CH_2–CH(CH_3)–R$

cycloalkanes (Naphthenes)

Alkylcyclopentanes

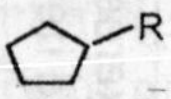

Alkycycclohexanes

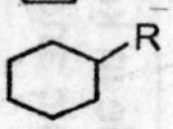

Bicycloalkanes

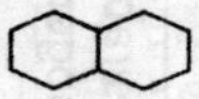

Aromatics

Alkylbenzenes

Aromatic-cycloakanes

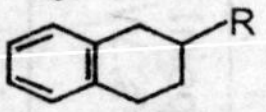

Fluorenes

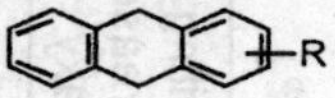

Binuclear aromatics

Examples of alkanes, cycloalkanes and aromatics present in crude oil.

Heavier crude contains more polycyclic aromatics which lead to carboneceous deposits called "coke"

Phenanthene

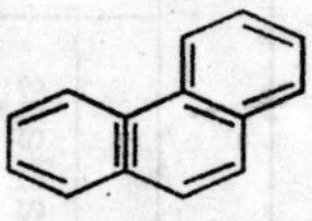

1, 2–Benzanthracene

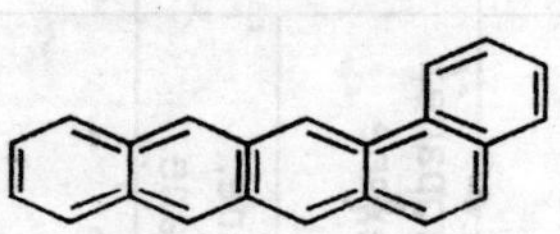

Pyrene

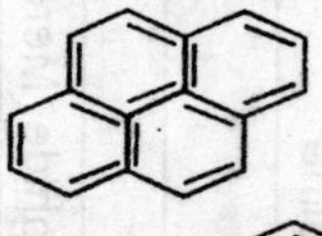

3,4–Benzopyrene

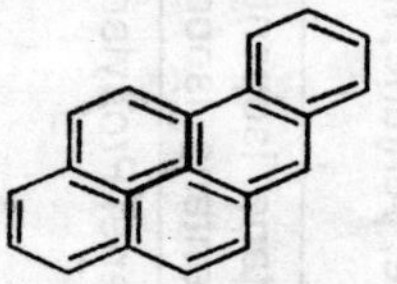

Chrysene

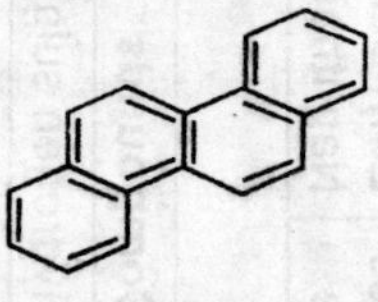

Examples of polycyclic, polynuclear aromatics in crude oil

Note

- Crude oils are complex mixtures containing many different hydrocarbon compounds that vary in appearance and composition from one oil field to another. Crude oils range in consistency from water to tar like solids, and in color from clear to black. An "average" crude oil contains about 84% carbon, 14% hydrogen, 1 - 3% sulfur, and less than 1 % each of nitrogen, oxygen, metals, & salts. Crude oils are generally classified as paraffinic, naphthenic, or aromatic, based on the predominant proportion of similar hydrocarbon molecules. Mixed-base crudes have varying amounts of each type of hydrocarbon. Refinery crude base stocks usually consist of mixtures of two or more different crude oils.

- Relatively simple crude oil assays are used to classify crude oils as paraffinic, naphthenic, aromatic, or mixed. One assay method (United State Bureau of Mines) is based on distillation, and another method (UOP "K" factor) is based on gravity & boiling points. More comprehensive crude assays determine the value of the crude (i.e., its yield & quality of useful products) and processing parameters. Crude oils are usually grouped according to yield structure.
- Crude oils are also defined in terms of API (American Petroleum Instituted) gravity. The higher the API gravity, the lighter the crude for example, light crude oils have high API gravity & low specific gravities. crude oils with low carbon, high hydrogen, and high API gravities are usually rich in paraffins and tend to yield greater proportions of gasoline & light petroleum products; those with high carbon, low API gravities are rich in aromatics.
- Crude oils that contain appreciable quantities of hydrogen sulfide or other reactive sulfur compounds are called "sour" those with less sulfur are called "sweet." Some exceptions to this rule are West Texas crudes, which are always considered "sour" regardless of their H_2S content, and Arabian high-sulfur crudes, which are not considered "sour" because their sulfur compounds are not highly reactive.

Crude oil is a mixture of hydrocarbon molecules, which are organic compounds of carbon and hydrogen atoms that may include from one to 60 carbon atoms. The properties of hydrocarbons depend on the number and arrangement of the carbon and hydrogen atoms in the molecules.

Hydrocarbons containing up to four carbon atoms are usually gases, those with 5 to 19 carbon atoms are usually liquids, and those with 20 or more are solids. The refining process uses chemicals, catalysts, heat, & pressure to separate and combine the basic types of hydrocarbon molecules found in crude oil into groups of similar molecules. The refining process also rearranges their structures & bonding patterns into different hydrocarbon molecules and compounds. Therefore it is the type of hydrocarbon (paraffinic, naphthenic, or aromatic) rather than its specific chemical compounds that is significant in the refining process.

Three Principal Groups or Series of Hydrocarbon Compounds that Occur Naturally in Crude Oil are:

i **Paraffins.** The paraffinic series of hydrocarbon compounds found in crude oil have the general formula C_nH_{2n+2} and can be either straight chains (normal) or branched chains (isomers) of carbon atoms. The lighter, straight-chain paraffin molecules are found in gases and paraffin waxes. Examples of straight-chain molecules are methane, ethane, propane and butane, (gases containing from one to four carbon atoms) and pentane & hexane (liquids with five & six carbon atoms) .The branched-chain (isomer) paraffins are usually found in heavier fractions of crude oil and have higher octane numbers than normal paraffins. These compounds are saturated hydrocarbons, with all carbon bonds satisfied, that is, the hydrocarbon chain carries the full complement of hydrogen atoms.

ii **Aromatics.** Are unsaturated ring-type (cyclic) compounds which react readily because they have carbon atoms that are deficient in Hydrogen. All aromatics have at least one benzene ring (a single-ring compound, characterized by three double bonds alternating with three single bonds between six carbon atoms) as part of their molecular structure. Naphthalenes are fused double ring aromatic compounds. The most complex aromatics, Polynuclears (three or more fused aromatic rings), are found in heavier fractions of crude oil

iii **Naphthenes.** They are saturated hydrocarbon groupings with the general, formula C_nH_{2n}, arranged in the form of closed rings (cyclic) and found in all fractions of crude oil except the very lightest. Single-ring naphthenes (monocycloparaffins) with five & six carbon atoms predominate, while two-ring naphthenes (dicycloparaffins) are found in the heavier ends of naphtha.

Other Hydrocarbons:

i. **Alkenes** are mono-olefins with the general formula c_nH_{2n} and contain only one carbon-carbon double bond in the chain. The simplest alkene is ethylene, with two carbon atoms joined by a double bond and four hydrogen atoms. Olefins are usually formed by thermal & catalytic cracking and rarely occur naturally in unprocessed crude oil.

ii. **Dienes and Alkynes.** Dienes, also known as diolefins, have two carbon-carbon double bonds. The alkynes, another class of unsaturated hydrocarbons, have a carbon-carbon triple bond within the molecule. Both these series of hydrocarbons have the general formula c_nH_{2n-2} Diolefins such as 1, 2-butadiene & 1, 3-butadiene, and alkynes such as acetylene, occur in C_5 and lighter fractions from cracking. The olefins, diolefins, and alkynes are said to be unsaturated because they contain less than the amount of hydrogen necessary to saturate all the valences of the carbon atoms. These compounds are more reactive than paraffins or naphthenes and readily combine with other elements such as hydrogen, chlorine, & bromine.

Nonhydrocarbons:

i **Sulfur Compounds.** Sulfur may be present in crude oil as hydrogen sulfide (H_2S), as compounds (e.g. mercaptans, sulfides, disulfides, thiophenes, etc.) or as elemental sulfur. Each crude oil has different amounts & types of sulfur compounds, but as a rule the proportion, stability, & complexity of the compounds are greater in heavier crude oil fractions. Hydrogen sulfide is a primary contributor to corrosion in refinery processing units. Other corrosive substances are elemental sulfur and mercaptans. Moreover, the corrosive sulfur compounds have an obnoxious odor.

Pyrophoric iron sulfide results from the corrosive action of sulfur compounds on the iron & steel used in refinery process equipment, piping, and tanks. The combustion of petroleum products containing sulfur compounds produces undesirables such as sulfuric acid and sulfur dioxide. Catalytichydrotreating processes such-as hydrodesulfurization remove sulfur compounds from refinery product streams. Sweetening processes either remove the obnoxious sulfur compounds or convert them to odorless disulfides, as in the case of mercaptans.

ii **Oxygen Compounds.** Oxygen compounds such as phenols, ketones and carboxylic acids occur in crude oils in varying amounts.

iii **Nitrogen Compounds.** Nitrogen is found in lighter fractions of crude oil as basic compounds, and more often in heavier fractions of crude oil as non basic compounds, that may also include trace metals such as copper, vanadium, and/or nickel. Nitrogen oxides can form in process furnaces. The decomposition of nitrogen compounds in catalytic cracking &hydrocracking processes forms ammonia and cyanides that can cause corrosion.

iv **Trace Metals.** Metals, including nickel, iron & vanadium are often found in crude oils in small quantities and are removed during the refining process. Burning heavy fuel oils in refinery furnaces & boilers can leave deposits of vanadium oxide &nickel oxide in furnace boxes, ducts, & tubes. It is also desirable to remove trace amounts of arsenic, vanadium, & nickel prior to processing as they can poison certain catalysts.

v **Salts.** Crude oils often contain inorganic salts such as sodium chloride, magnesium chloride, and calcium chloride in suspension or dissolved in entrained water brine. These salts must be removed to neutralized before processing to prevent catalyst poisoning, equipment corrosion, & fouling. Salt corrosion is caused by the hydrolysis of some metal chlorides to hydrogen chloride (HCl) and subsequent formation of hydrochloric acid when crude is heated. Hydrogen chloride may also combine with ammonia to form ammonium chloride (NH4Cl), which causes fouling & corrosion.

vi **Carbon Dioxide.** Carbon dioxide may result from the decomposition of bicarbonates present in or added to crude, or from steam used in the distillation process

vii **Naphthenic Acids.** Some crude oils contain naphthenic (organic) acids which may become corrosive at temperatures above 230°C when the acid value of the crude is above a certain level.

Oil and natural gas were formed hundred years ago from the prehistoric plant and animals, it is believed that hydrocarbon formed by the thermal maturation of organic matter buried deep in earth, over the millions of years under extreme pressure &high temperature; these organic mattergot converted to hydrocarbons consisting of oil & gas. Hydrocarbons are present in the variety of forms: koregen, asphalt, crude oil, natural gas, condensates, and coal in solid form.

Oil and production includes exploration, drilling, extraction stabilization. The underground traps of oil and gas are called reservoir. Various types of traps are structural traps, stratigraphic traps and combination traps. Most reservoir contain water also along with oil and gas. Crude oil reserves are classified as proven, probable and possible reserves. Earlier finding of oil & gas was a matter of luck and hit & miss process. Tools used for oil & gas exploration are based & dependent on gravity change, magnetic field change, time, change and electrical resistance. With advent of three dimensional seismic technology (which is based on the sound waves). which identify the subsurface formation by reflection of sound, there has been much improvement in identification of oil & gas traps and reservoirs. Seismic technology significantly improves the method of estimating the oil & gas deposits. Next step after exploration is the drilling of exploratory well. Drilling may be vertical drilling or horizontal drilling. Drilling may be performed on-shore or off-shore. Horizontal drilling and hydrofacturing has resulted in economical and more productive drilling of shale gas which was not economical with conventional vertical drilling.

Appendix 1.2: characterization of crude oil

Assessing the refining value of a crude oil requires a full description of the crude oil and its components, involving scores of properties. However, two properties are especially useful for quickly classifying and comparing crude oils: *API gravity* (a measure of *density*) and *sulfur* content.

API Gravity (Density)

The *density* of a crude oil indicates how light or heavy it is, as a whole. Lighter crudes contain higher proportions of small molecules, which the refinery can process into gasoline, jet fuel, and diesel. Heavier crudes contain higher proportions of large molecules, which the refinery can either (1) use in heavy industrial fuels, asphalt, & other heavy products or (2) process into smaller molecules that can go into the transportation fuels products.

In the refining industry, the density of an oil is usually expressed in terms of *API gravity,* a parameter whose units are degrees (° *API*) - e.g., 35° API. API gravity varies inversely with density (i.e., the lighter the material, the higher its API gravity). By definition, water has API gravity of 10°.

Figure below indicates the quality of A typical *light* crude (35°API) and a typical *heavy crude* (*25*° API)in terms of their natural yields of light gases, gasoline components, distillate (mainly jet fuel & diesel) components, and heavy oils. It also shows the average demand profile for these product categories in the developed countries.

Typical Natural Yields of Light & Heavy Crude Oils

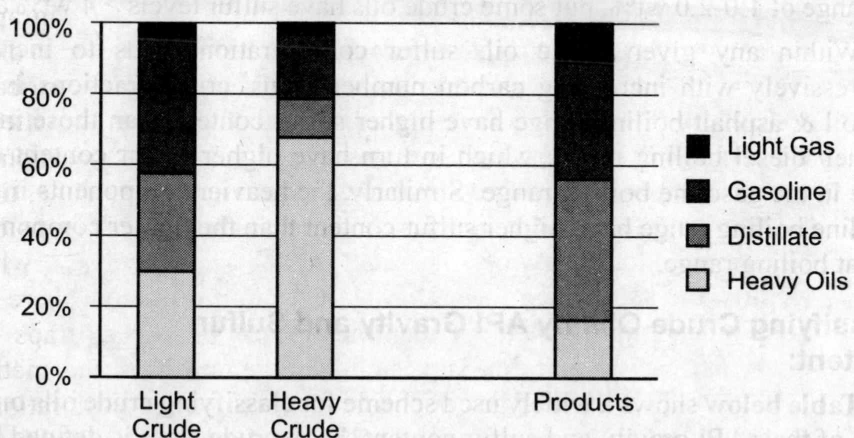

The natural yields of the heavy oils from both the light & the heavy crudes exceed the demand for heavy refined products, and the natural yield of heavy oil from the heavy crude is more than twice that of the light crude. These general characteristics of crude oils imply that (1) refineries must be capable of converting at least some, of the heavy oil into light products, and (2) the heavier the crude, the more of this conversion capacity is required to produce any given product slate.

Sulfur Content

Of all the hetero-elements in crude oil, sulfur has the most important effects on refining.

- Sufficiently high sulfur levels in refinery streams can (1) deactivate ("poison") the catalysts that promote desired chemical reactions in certain refining processes, (2) cause corrosion in refinery equipment, and (3) lead to air emissions of sulfur compounds, which are undesirable& are subject to regulatory environmental controls.
- Sulfur in vehicle fuels leads to undesirable vehicle emissions of sulfur compounds and interferes with vehicle emission control systems that are directed at regulated emissions such as volatile organic compounds, nitrogen oxides, & particulates.

Refineries must have the capability to remove sulfur from crude oil and refinery streams to the extent needed to mitigate these unwanted effects. The higher the sulfur content of the crude, the greater the required degree of sulfur control and the higher the associated cost.

The sulfur content of crude oil and refinery streams is usually expressed in weight percent (wt%) or parts per million by weight (ppmw). In the refining industry, crude oil is called *sweet* (low sulfur), if its sulfur level is less than a threshold value (e.g., 0.5 wt% (5, 000 ppmw)) and *sour* (high sulfur), if its sulfur level is above a higher threshold. Most sour crudes have sulfur levels in the range of 1.0-2.0 wt%, but some crude oils have sulfur levels > 4 wt% also.

Within any given crude oil, sulfur concentration tends to increase progressively with increasing carbon number. Thus, crude fractions in the fuel oil & asphalt boiling range have higher sulfur content than those in the jet fuel/ diesel boiling range, which in turn have higher sulfur content than those in the gasoline boiling range. Similarly, the heavier components in, the gasoline boiling range have higher sulfur content than the lighter components in that boiling range.

Classifying Crude Oils by API Gravity and Sulfur Content:

Table below shows a widely used scheme for classifying crude oils on the basis of their API gravity and sulfur content. Each crude class is defined by a range of API gravity and a range of sulfur content; the names of the categories indicate these ranges in qualitative terms.

Another Table below lists some important crude oils in the world oil trade and indicates the API gravity/sulfur classification for each of these crudes.

Typical Approximate Characteristics & Properties and Gasoline Potential of Various Crudes is Tabulated Below:

S. N	Crude Oil Class	Property Range	
		Gravity (°API)	Sulfur (wt.%)
1.	Light Sweet	35–60	0–0.5
2.	Light Sour	35–60	>0.5
3.	Medium Medium Sour	26–35	0–1.1
4.	Medium Sour	26–35	>1.1
5.	Heavy Sweet	10–26	0–1.1
6.	Heavy Sour	10–26	>1.1

S. N	Crude Oil	Country of Origin	Crude Oil Class	Properties	
				Gravity (°API)	Sulfur (wt.%)
1.	Bernt	U.K	Light Sweet	40.0	0.5
2.	West text Intermediate	U.S.A		39.8	0.3
3.	Arabian Extra Lt. Export	Saudi Arabia	Light Sour	38.1	1.1
4.	Daqing	China	Medium Medium Sour	33.0	0.1
5.	Forcados Export	Nigeria		29.5	0.2
6.	Arabian light Export	Saudi Arabia	Medium Sour	34.0	1.9
7.	Kuwait Export Blend	Kuwait		30.9	2.5
8.	Marlim Export	Brazil	Heavy Sweet	20.1	0.7
9.	Cano Limon	Colombia		25.2	0.9
10.	Oriente Export	Ecuador	Heavy Sour	25.0	1.4
11.	Maya heavy Export	Mexico		21.3	3.4

Typical Approximate Characteristics & Properties And Gasoline Potential of Various Crudes is Tabulated Below:

S. N	Crude Source	Paraffins (% vol)	Aromatics (% vol)	Naphthenes (% vol)	Sulfur (% wt)	API gravity (approx.)	Naphtha/ gasoline	
							yield (% vol)	Octane no (typical)
1.	Nigerian -light	37	9	54	0.2	36	28	60
2.	Saudi -light	63	19	18	2	34	22	40
3.	Saudi -Heavy	60	15	25	2.1	28	23	35
4.	Venezuela -Heavy	35	12	53	2.3	30	2	60
5.	Venezuela -Heavy	52	14	34	1.5	24	18	50
6.	USA -Midcont. Sweet	–	–	–	0.4	40	–	–
7.	USA -W Texas Sour	46	22	32	1.9	32	33	55
8.	North Sea- Brent	50	16	34	0.4	37	31	50

The physical characterization of the crude oil in terms of viscosity, density, boiling point curves is equally important. These properties are also indicative of the quality of the product as well as the feed. Therefore, in petroleum processing, obtaining any intermediate or a product stream with a defined characterization of several properties indicates whether it is diesel or petrol or any other product. The product characterization is illustrated ☺ now with an example. Aviation gasoline is characterized using ASTM distillation. The specified temperatures for vol% distilled at 1 atm are 70°C maximum for 10 % volume, 105°C maximum for 50% volume and 135°C maximum for 90% volume. This means that any product obtained in the refinery process and meets these ASTM distillation characteristics represents aviation gasoline product. However, other important properties such as viscosity, density, aniline point, sulphur, density are as well measured to fit within a specified range and to conclude that the produced stream is aviation gasoline.

Important characterization properties- Numerous important feed and product characterization properties in refinery engineering include:

- API gravity
- Watson Characterization factor
- Viscosity
- Sulfur content
- True boiling point (TBP) curve
- Pour point
- Flash and fire point
- ASTM distillation curve
- Octane number

API gravity. API gravity of petroleum fractions is a measure of density of the stream. Usually measured at 60 °F, the API gravity is expressed as °API = 141.5/specific gravity – 131.5, where specific gravity is measured at 60 °F.

According to the above expression, 10°API gravity indicates a specific gravity of 1 (equivalent to water specific gravity). Higher values of API gravity indicate lower specific gravity and therefore lighter crude oils or refinery product and vice-versa. As far as crude oil is concerned, lighter API gravity value is desired as more amount of gas fraction, naphtha &gas oil can be produced from the lighter crude oil than with the heavier crude oil. Therefore, crude oil with high values of API gravity are expensive due to their quality.

Watson characterization factor. The Watson characterization factor is usually expressed as $K = (TB)^{1/3}$/specific gravity, where TB is the average boiling point in degrees R(Rankine) taken from five temperatures corresponding to 10, 30, 50, 70 and 90 volume % vaporized. Typically Watson characterization factor varies between 10.5 and 13 for various crude stream. A highly paraffinic crude typically possesses a K factor of 13. On the other hand, a highly naphthenic crude possesses a K factor of 10.5. Therefore, Watson characterization factor can be used to judge the quality of the crude oil in terms of the dominance of the paraffinic or naphthenic components.

Viscosity. Viscosity is a measure of the flow properties of the refinery stream. Viscosity is measured in terms of centistokes (termed as cst) or saybolt seconds or redwood seconds. Usually, the viscosity measurements are carried out at 100 °F and 210 °F. Viscosity is a very important property for the heavy products obtained from the crude oil. The viscosity acts as an important characterization property in the blending units associated to heavy petroleum products. Typically, viscosity of these products is specified to be within a specified range and this is achieved by adjusting the viscosities of the streams entering the blending unit.

Sulfur content. Usually, crude oil has both organic & inorganic sulphur in which the inorganic sulphur dominates the composition. Typically, crude oils with high sulphur content are termed as sour crude whereas crude oils with low sulphur content are termed as sweet crude. Typically, crude oil sulphur content consists of 0.5 - 5wt% of sulphur. Crude with sulphur content lower than 0.5 wt % are termed as sweet crudes. It is estimated that about 80 % of world crude oil reserves are sour. The sulphur content in the crude oil is responsible for numerous hydrotreating operations in the refinery process. Strict & tighter legislations enforce the production of various consumer petroleum products with low quantities of sulphur (in the range of ppm). Presently, India is heading towards the generation of diesel with Euro III standards that indicates that the maximum sulphur content is about 500 ppm in the products, This indicates that large quantities of inorganic sulphur needs to be removed from the fuel. Inorganic sulphur from various intermediate product streams is removed using hydrogen as hydrogen sulphide.

A typical refinery consists of good number of hydrotreaters to achieve the desired separation. The hydrotreaters in good number are required due to the fact that the processing conditions for various refinery intermediate process streams are significantly different and these streams cannot be blended together as well due to their diverse properties which were achieved .

TBP/ASTM distillation curves. The most important characterization properties of the crude oil/intermediate product streams are the TBP/ ASTM distillation curves. Both these distillation curves are measured at 1 atm pressure. In both these cases, the boiling points of various volume. fractions are measured. However, the basic difference between TBP curve and ASTM distillation curve is that while TBP curve is measured using batch distillation apparatus consisting of no less than 100 trays & very high reflux ratio, the ASTM distillation is measured in a single stage apparatus without any reflux. Hence the ASTM does not indicate a good separation of various components and indicates the operation of the laboratory setup far away from the equilibrium.

Pour point -When a petroleum product is cooled, first a cloudy appearance of the product occurs at a certain temperature. This temperature is termed as the cloud point. Upon further cooling, the product will cease to flow at a temperature. This temperature is termed as the pour point. Both pour & cloud points are important properties of the product streams as far as heavier products are concerned. For heavier products, they are specified in a desired range and this is achieved by blending appropriate amounts of lighter intermediate products.

Flash and fire point. Flash & fire point are important properties that are relevant to the safety & transmission of refinery products. Flash point is the temperature above which the product flashes forming a mixture capable of inducing ignition with air. Fire point is the temperature well above the flash point where the product could catch fire. These two important properties are always taken care in the day to day operation of refinery.

Octane number. Though irrelevant to the crude oil stream, the octane number is an important property for many intermediate streams that undergo blending later on to produce automotive gasoline, diesel etc. Typically gasoline tends to knock the engines. The knocking tendency of the gasoline is defined in terms of the maximum compression ratio of the engine at which the knock occurs. Therefore, high quality gasoline will tend to knock at higher compression ratios and vice versa. Iso-octane is considered as the barometer for octane number comparison. While iso-octane is given an octane number of 100, & n-heptane is given an octane number of 0. Therefore, the octane number of a fuel is equivalent to a mixture of a iso-octane & n-heptane that provides the same compression ratio in a fuel engine. Thus an octane number of 80 indicates that the fuel is equivalent to the performance characteristics in a fuel engine fed with 80 vol % of isooctane and 20 volume% of n-heptane. Octane numbers are very relevant in the reforming, isomerisation and alkylation processes of the refining. These processes enable the successful reactive transformations to yield long side chain paraffins & aromatics that possess higher octane numbers than the feed constituents which do not consist of higher quantities of constituents possessing straight chain paraffins and non-aromatics (naphthenes).

2

Indian Crude Oil & Oil Refineries

The major *oil fields in India* are in Assam (Naharkatiya) and Gujarat (Cambay and Ankleswar). Besides, Bombay high is producing good amount of crude oil.The *Naharkatiya crudes* have a relative density of 0.857, sulphur 0.17% & high wax content (7.8%) and contain very little water & salt. Characterisation of the crudes shows that they are of mixed nature.

The oil fields of India yield mixed-base crudes. The crudes from the *Assam fields* has high aromatic content (20-25%) of gasoline, a common feature for Burma and other South-East Asian crudes. The yield of straight run gasoline is about 17-18% of the crude. The Assam crudes are also highly paraffinic containing high melting waxes but are deficient in good quality kerosene.

The *Cambay crude* has been found to be associated with substantial amounts of water, has a relative density of 0.863 ,sulphur content of 0.16% and gives somewhat low yields of lighter distillates. The dehydrated oil has high wax content (10.6%). The oil is predominantly paraffinic in nature.

The *Ankleswar crude* is a light oil containing high amounts of lighter & middle boiling fractions and only negligible amount of water. In these respects, it differs very much from the Cambay oil though both are paraffinic base crudes. The Ankleswar crude has relative density of 0.803, sulphur 0.14% and wax 7%.

The crudes from the *Gujarat oil fields* give gasoline with a much smaller content of aromatic hydrocarbons (3-5%). These are very highly paraffinic and wax-bearing. Straight run gasoline content from Cambay crude is low (9-10%). The yield of total distillates from this crude is much lower than from Naharkatiya crude. Gujarat crudes are nearer to the Middle East crudes than to the Assam crudes. The sulphur content of Indian crudes is low (0.15%).

Exploration for oil in all prospective sedimentary basins of India is continuing. The areas which are still under survey are : Kutch regions of Gujarat, Brahmaputra valley, place of Punjab, Uttar Pradesh & Bihar and the adjoining Himalayan foot hills, Rajastan, Cauveri, Mahanadi & Godavari deltas, the Andaman & Nicobar Islands and off'Cambay, Kutch & Madras coast.

Of the total requirement of crude in India, only one-third is produced within the country and the rest is imported from the Middle East countries (Iran, Arabia, Iraq, Kuwait, etc.) & from Indonesia in the East. The Government of India has a share in rostam crude from its joint venture in the Persian Gulf. USSR also exports crude oil to India.

The crudes from the eastern oil fields in Assam resemble those from the Far Eastern countries (like Indonesia, Borneo, Burma) in aromatic contents. The Ankleswar crudes, on the other hand, are similar to those of Darius & Aghajari crudes of Iran and other light Arabian crudes. The Middle East crudes vary in relative density from 0.825 to 0.953 and the sulphur contents are as high as 4.7%. The Kuwait crude has a relative density of 0.871 and a sulphur content of 2.5%.

The total estimated reserve of crude in India is about 200 million tons spread over an area of 1.04 x 10^6 km^2. Exploration of oil in India is done by Oil and Natural Gas Corporation Ltd. (ONGC) and Oil India Limited (OIL). Most of the refineries in Public sector are under Hindustan Petroleum, Indian Oil Corporation and Bharat Petroleum. Refineries are located at Gauhati (Assam), Barauni (Bihar), Baroda (Gujarat), Cochin, Haldia, Mathura, Bongaigaon (Assam), Madras, Visakhapatnam etc. Few more refineries are under construction including at Mangalore and Karnal (Haryana).

Some of the properties of three Indian crudes are tabulated below.

Properties of Indian Crudes

Characteristics	Cambay (Gujarat)	Naharkatiya (Assam)	Bombay High off shore
Specific gravity (15.5°/15.5°C)	0.8638	0.8574	0.8324
Water content % volume	2.60	0.06	-
Salt content, % weight	0.014	nil	-
Sulphur content, % weight	0.13	0.017	0.05
Kinematic viscosity at 37.8°C, CSt	35.6	4.91	5.10
Calorific Value, kcal/kg	10,800	-	-
Pour Point °C,	35	30-32	27

Characteristics	Cambay (Gujarat)	Naharkatiya (Assam)	Bombay High off shore
Distillation characteristics:			38
I.B.P., °C	62.5	41.0	22
Upto 150°C, % volume	9.2	23.0	33
150—300°C, % volume	26.5	32.5	-
300–465°C, % volume	35.25	32.2	-
Total upto 465°C, % volume	70.95	87.7	-
Fraction, upto 150°C:			
Sp. gr. (15.5°/15.5°C)	0.7920	0.8516	-
fraction, 150-300°C :			
sp. gr., (15.5°15.5°C)	0.7195	0.7471	-
fraction: 300-450°C :			
Sp. gr, (15.5°/15.5°C)	0.8698	0.9273	-
Sulphur, % weight	0.05	0.24	-
Wax content, % weight	32.2	24.4	—
Short residue (>465°C) :			
Sp. gr. (15.5°/15.5°C)	0.9554	1.0050	-
Penetration number	50	16	–
Softening Point (R & B), °C	55	67	-
Wax (300–465°C) on the basis of crude, % weight	10.5	7.8	12

Table below gives details of crude oil and natural gas fields in India as on 1.4.2010.

Company / State	Basin	No. of Oil Fields	No. of Gas Fields	No. of Oil and Gas Fields	Total* No. of Fields
A. Oil & Natural		28	131	218	377
Gas Corporation Ltd.					
(a) Onshore					
Gujarat	Cambay	7	8	86	101
Assam	Upper Assam	5	1	30	36
	A&AA	0	4	2	6

Company / State	Basin	No. of Oil Fields	No. of Gas Fields	No. of Oil and Gas Fields	Total* No. of Fields
Rajasthan	Jodhpur	0	7	0	7
Andhra Pradesh	K.G. Basin	1	38	11	50
Tamil Nadu	Cauvery	7	10	11	28
Tripura	A&AA	0	11	0	11
Nagaland	A&AA	1	0	2	3
(b) Offshore					
East Coast	Cauvery Offshore	0	0	3	3
	KG Offshore (Shallow)	1	7	7	15
	KG Offshore (Deep)	0	11	2	13
	Mahandi	0	3	0	3
	Andaman	0	1	0	1
West Coast	Cambay	0	0	2	2
	Mumbai Offshore	6	30	62	98
	Kutch	0	0	0	0
B. Oil India Ltd.		1	3	15	19
Assam	Upper Assam	0	0	14	14
Arunachal Pradesh	Upper Assam	0	0	1	1
Rajasthan	Jaisalmer	0	3	0	3
	Bikaner Nagaur	1	0	0	1
C. Private/JVCs		27	11	9	47
(a) Onshore					–
Gujarat	Cambay	18	4	5	
Arunachal Pradesh	AAFB	1	0	0	1
Assam	Assam	1	0	0	1
Rajasthan		5	2	0	7
West Bengal	Ranigunj (South)	0	1	0	1
(b) Offshore					
East Coast	Cauvery	1	1	0	2
	KG Basin	1	1	1	3
West Coast	Mumbai	0	1	1	2
	Cambay	0	1	2	3
Grand Total		56	145	242	443

Notes: A&AA: Assam & Assam Araken.
JVCs-Joint Venture Companies

Petroleum Refining in India

The chronological development of petroleum oil industry in India has been given below:

— First oil well of the world drilled in Pennsylvania in 1859 by Drake.

— First Indian oil well was struck at Makum in Assam in 1867.

— A very small refinery was set up at Margherita in Assam in 1893 to process the Makum crude.

— Systematic exploration and production in India started in 1899 after oil was struck at Digboi and Assam Oil Co. was formed under the Aegis of Burmah Oil Co.

— Oil refinery in Digboi was commissioned in 1901 with initial refining capacity of 0.25 million ton crude per year.

— From 1901 to 1947 no further development in exploration or refining took place.

— At the end of 1946-47, total oil consumption in undivided India stood at 2.6 million metric tons with refining capacity of only 0.5 million metric tons .Nation's oil trade was totally import oriented and dominated by the foreign oil companies till then.

— Formation of ONGC in 1957 was done.

— ONGC struck oil in 1957.

— Second agency, Oil India Ltd. was set up in the joint sector in 1959 with participation by Burmah Oil Co. and The Govt, of India to explore the oil fields of Naharkatia in Assam.

— In 1961, AOC in private sector and ONGC & oil India in public sector produced 0.513 million metric tons of crude.

— In 1970 the crude production reached 6.8 million mt.

— In 1980s oil was struck in Bombay high.

— Between 1980 and 1984 crude oil production shot up from 9.4 million metric tons per annum to 27.9 million metric tons per annum.

— Self sufficiency in crude production increased from 29.7% in 1980 to 70% in 1984.

— In mid-fiftees three coastal refineries were set up by the foreign controlled oil distributing companies - two on the west coast and one on the east. These refineries processed imported crudes.

— Government controlled oil refineries were set up between 1962 and 1965 at Noonmati, Barauni and Koyali - all based on indigenous crude (of Assam and Gujarat).

— Later 3 more coastal refineries (two on south coast and one in the east) were set up based on imported crude.

— In 1958, Indian Refineries Ltd. was set up.

— In 1959, Indian Oil Company Ltd. was set up for marketing of petroleum products.

— Both merged to form Indian Oil Corporation Ltd. in 1964.

— In 1970, Indo Burmah Petroleum Co. Ltd. (which was established in 1909 in Burmah) was acquired by Indian Oil Corporation Ltd.

— In 1974, ESSO was nationalised to form Hindustan Petroleum Corporation Ltd.

— In 1976, Burmah Shell was nationalised and Bharat Petroleum Corporation Ltd. was formed in 1977.

— In 1978, Caltex India was nationalised and merged with Hindustan Petroleum Corporation Ltd.

— Oil India Ltd. was nationalised in 1981.

— Assam Oil Co. was nationalised in 1981 and merged with IOC.

— Latest oil refineries have been set up at Karnal, Mangalore, Numligarh, Panipat and Jamnagar.

— Three more new oil refineries of 9 MT capacity each are to be set up at Bhatinda, Paradeep and Nagapattinum by Indian Oil Corporation.

India's requirements of crude oil is more than its refining capacity and production hence, the balance is to be imported. India has gone for collaboration with Iran in the exploration and production of crude oil thereby leading to the production of oil in Rustam oil fields and discovery of a new oil field at Rakhash in the off shore areas of Persian Gulf. Crude oils are imported from middle east countries, Russia, Indonesia etc. Existing oil refineries can meet the present requirements of all products except kerosene and lubricating oils. Lubricating oils are made in the refineries at Digboi, Barauni, Madras, Haldia etc.

The details of petroleum oil refineries in India has been given below in the Table.

Crude Oil Refineries in India

To meet the growing demand of petroleum products, the refining capacity in the country has gradually increased over the years by setting up of new refineries in the country as well as by expanding the refining capacity of the existing refineries. As of April, 2009 there are a total of 20 refineries in the country comprising 17 (seventeen) in the Public Sector and 3 (three) in the

Private Sector. The country is not only self sufficient in refining capacity for its domestic consumption but also exports petroleum products substantially. The total refining capacity in the country as on 1.10-2009 stands at 179.956 MMTPA(Million Metric Tons Per Year). The company-wise location and oil refining capacity of the oil refineries as on 1.10.2009 is given in Table below.

Table: Company-wise location and oil refining capacity of Indian oil refineries (as on 1.10.2009)

S.No.	Name of the Company	Location of the Refinery	Capacity, MMTPA*
1.	Indian Oil Corporation Limited (IOC)	Guwahati, Assam	1.00
2.	IOC	Barauni, Bihar	6.00
3.	IOC	Koyali, Vadodara, Gujarat	13.70
4.	IOC	Haldia, West Bengal	6.00
5.	IOC	Mathura, Uttar Pradesh	8.00
6.	IOC	Digboi, Assam	0.65
7.	IOC	Panipat, Haryana	12.00
8.	IOC	Banaigaon, Assam	2.35
9.	Hindustan Petroleum Corporation Limited (HPCL)	Mumbai, Maharashtra	5.50
10.	HPCL, Visakh	Visakhapatnam, Andhra	7.50
11.	Bharat Petroleum Corporation Limited (BPCL)	Mumbai, Maharashtra	12.00
12.	BPCL, Kochi	Kochi, Kerala	9.50
13.	Chennai Petroleum Corporation Limited (CPCL)	Manali, Tamil Nadu	9.50
14.	CPCL, Nagapattnam	Nagapattnam, Tamil Nadu	1.00
15.	Numaligarh Refinery Ltd. (NRL)	Numaligarh, Assam,	3.00
16.	Mangalore Refinery & Petrochemicals Ltd. (MRPL)	Mangalore, Karnataka	9.69
17.	Tatipaka Refinery (ONGC)	Tatipaka, Andhra Pradesh	0.066
18.	Reliance Industries Ltd. (RIL); Private Sector	Jamnagar, Gujarat	33.0
19.	Reliance Petroleum Limited (SEZ); Private Sector	Jamnagar, Gujarat	29.00
20.	Essar Oil Limited (EOL); Private Sector	Jamnagar, Gujarat	10.50
	Total		179.956

*Milion Metric Tonnes per Annum

Expansion of Existing Oil Refineries

Capacity expansions planned during XIth Five Year Plan has been indicated in Table below.

Table: Capacity expansion plans of oil refineries in India during 11th five year plan

S.No.	Name of the company	Location of the Refinery	Increase in Capacity **(MMTPA)*
1.	Indian Oil Corporation Limited (IOCL)	Panipat, Haryana	3.0
2.	IOCL	Haldia West Bengal	1.5
3.	Hindustan Petroleum Corporation Limited (HPCL)	Mumbai, Maharashtra	2.40
4.	Hindustan Petroleum Corporation Limited (HPCL)	Visakhapatnam, Andhra Pradesh	2.50
5.	Chennai Petroleum Corporation Limited (CPCL)	Manali, Tamil Nadu	1.60
6.	BPCL, Kochi Refineries Ltd.	Kochi, Kerala	2.0
7.	Mangalore Refinery & Petrochemicals Ltd. (MRPL)	Mangalore, Karanataka	5.31
8.	Oil & Natural Gas Corporation Limited (ONGC)	Tatipaka, Andhra Pradesh	0.066
9.	Essar Oil Limited	Vadinar, Gujarat	5.50
	Total		23.876

*Million Metric Tonnes Per Annum (MMTPA)

New Oil Refineries

New grassroots refineries coming up during XIth Five Year Plan is indicated in Table below.

Table: New grassroot oil refineries coming up during 11th five year plan

S. No.	Name of the Company	Location of the Refinery	Capacity (MMTPA)
1.	Indian Oil Corporation Limited (IOCL)	Paradip, Orisa	15
2.	Bharat Oman Refineries Limited (BORL)	Bina, Madhya Pradesh	6
3.	Hindustan Mittal Energy Limited (HMEL)	Bathinda, Punjab	9
4.	Essar Oil Limited, (EOL)	Jamnagar, Gujarat	18
5.	Nagarjuna Oil Corporation Limited (NOCL)	Cuddalore, Tamil Nadu	6
		Total	54

Table -Growth of Installed oil refinning Capacity and oil Refinery Crude Throughput in India during last five years

Refinery / Location	Installed Capacity			Refinery Crude Throughput			
	as	on	2005-06	2006-07	2007-08	2008-09	2009-10*
	1.4.2009	1.4.2010					
(a) Public Sector	105468	111886	96946	108172	112541	112223	112117
IOC, Guwahati, Assam	1000	1000	864	839	920	1076	1078
IOC, Barauni, Bihar	6000	6000	5553	5469	5634	5940	6184
IOC, Koyali, Gujarat	13700	13700	11543	12953	13714	13852	13206
IOC, Haldia, West Bengal	6000	7500	5502	5836	5715	6042	5686
IOC, Mathura, Uttar Pradesh	8000	8000	7938	8883	8033	8601	8107
IOC, Digboi, Assam	650	650	615	586	564	623	600
IOC, Panipat, Haryana	12000	12000	6507	9435	12821	13070	13615
Total IOC	47350	48850	38522	44001	47401	49204	48476
BPCL, Mumbai, Maharashtra	12000	12000	10298	12030	12746	12262	12516
HPCL, Mumbai, Maharashtra	5500	5500	6249	7419	7409	6652	6965
HPCL, Visakh, Andhra Pradesh	7500	8300	7980	9377	9409	9155	8796
Total HPCL	13000	13800	14229	16796	16818	15807	15761
KRL, Kochi, Kerala	7500	9500	6939	7742	8134	7739	7875
CPCL, Manali, Tamil Nadu	9500	9500	9680	9784	9802	9718	9580
CPCL, Narimanam, Tamil Nadu	1000	1000	682	618	464	418	517
Total CPCL	10500	10500	10362	10402	10266	10136	10097

Refinery / Location	Installed Capacity		Refinery Crude Throughput				
	as	on	2005-06	2006-07	2007-08	2008-09	2009-10*
BRPL, Bongaigaon, Assam	2350	2350	2356	2067	2020	2163	2220
NRL, Numaligarh, Assam	3000	3000	2133	2504	2568	2251	2619
ONGC, Tatipaka, Andhra Pradesh	78	66	93	94	63	84	55
MRPL, Mangalore, Karnataka	9690	11820	12014	12536	12525	12577	12498
(b) Private Sector	72500	72500	33163	38379	43562	48549	47916
RPL, Jamnagar, Gujarat	33000	33000	33163	36616	36931	35663	34415
RPL (SEZ), Jamnagar, Gujarat$	29000	29000	—	—	—	—	—
ESSAR Oil Ltd. Vadinar**	10500	10500	–	1763	6631	12913	13501
Total (a+b)	177968	184386	130109	146551	156103	160772	160033

Note: **: Provisional **: Essar Oil Ltd. Commissioned on 24. 11. 2006 & Started Production from Decemeber, 2006.

$: RPL (SEZ), Jamnagar, Gujarat Commissional on 25. 12. 2008 & Started Production from January, 2009.

— CPCL and BRPL are subsidiaries of IOC.

— NRL is subsidiary of BPCL.

— MRPL is subsidiary of ONGC.

Total crude oil refinning capacity in India currently in 2009 is about 180 million tons per year. Table (a to *e*) below gives the crude oil & petroleum products production, import reserve, & consumption pattern of India as well as international petroleum statistics.

Table (a)- Growth of Indian Petroleum Industry during last 5 years

Item	Unit	2005-06	2006-07	2007-08	2009-09	2009-10*
1. Reserves : (Balance Recoverable)						
(i) Crude Oil	Mn. Tonne	786	756	725	769	775
(ii) Natural Gas	Bn.Cub. Mtr.	1101	1075	1055	1050	1074
2. Consumption						
(i) Crude Oil	Mn. Tonne	130.11	146.55	156.1	160.77	160.3
(ii) terms of refinery crude throughput)						
(ii) Petoleum Products (excl. RBF)	"	113.21	120.75	128.95	133.40	138.20
3. Production						
(i) Crude Oil	"	32.19	33.988	34.117	33.51	33.69
(ii) Petoleum Products	"	119.75	135.26	144.93	150.52	149.65
(iii) LPG from natural gas	"	2.19	2.09	2.06	2.16	2.24
4. Imports and Exports						
(i) Grosh Imports:						
(a) Qty: Crude Oil	Mn. Tonne	99.41	111.5	121.67	132.78	159.26
LNG	"	5.06	6.81	8.32	8.06	8.83
Pet. Products	"	13.44	17.66	22.46	20.33	14.66
Total (a)	"	117.91	135.97	152.45	161.17	182.75
(b) Value: Crude Oil	Rs. Billion	1717.02	2190.29	2726.99	3482.88	3753.78
LNG	"	33.66	56.5	71.97	95.48	93.44
Pet. Products	"	279.72	411.6	610	642.69	337.53
Total (b)	"	2030.40	2658.39	3408.96	4221.05	4184.75
Pet. Imports as per DGCI&S	"	1946.4	2585.72	3206.54	4199.68	4115.79
(ii) Exports:						
(a) Qty:						
Pet. Products	Mn. Tonne	23.46	33.62	40.78	38.57	50.97
(b) Value:						
Pet. Products	Rs. Billion	499.74	810.94	1107.89	1210.86	1440.37
(iii) Net Imports:						

Item	Unit	2005-06	2006-07	2007-08	2009-09	2009-10*
(a) Qty: Crude Oil	Mn. Tonne	99.41	111.50	121.67	132.78	153.25
LNG	"	5.06	6.81	8.32	8.06	8.83
Pet. Products	"	-10.02	-15.96	-18.32	-18.24	-36.31
Total (a)	"	94.45	102.35	111.67	122.60	131.78
(b) Value: Crude Oil	Rs. Billion	1717.02	2190.29	2726.99	3482.88	3753.78
LNG	"	33.66	56.5	71.97	95.48	93.44
Pet. Products	"	-212.10	-399.34	-497.89	-568.17	-1102.84
Total (b)	"	1530.66	1847.45	2301.07	3010.19	2744.38
(iv) Unit Value of Crude						
Oil imports (gross)	Rs./MT	17272	19644	22413	26230	24494
5. India's Total Exports	Rs. Billion	4564.18	5717.79	6558.64	8407.55	8451.25
6. Pet. Imports as percentage of India's total exports	%	44.5	46.5	52.0	50.2	49.5
(i) Gross Imports						
(ii) Net Imports	%	33.5	32.3	35.1	35.8	32.5
7. Contribution of Oil Sector to Centre/ State Resources						
(i) Royalty from crude oil	Rs. Billion	50.67	58.57	65.44	71.55	NA
(ii) Royalty from gas	"	8.637	10.75	14.87	16.24	NA
(iii) Oil Development Cess	"	51.96	71.77	71.56	68.86	65.59
(iv) Excise & Custom duties	"	631.43	718.93	783.73	705.57	717.67
(v) Sales Tax	"	459.34	539.49	564.45	633.49	649.99
(vi) Dividend	"	718.93	115.27	94.74	98.61	67.63
8. Natural Gas:						
(i) Gross Production	Bn.Cub. Mtr.	32.202	31.747	32.417	32.849	47.51
(ii) Utilisation	"	31.325	30.791	31.478	31.770	46.49

Notes: *: Provisional ! As on 1st April of initial year. ,N.A. Not Available.

Table (b) India's Imports/ Exports of Crude Oil and Petroleum Products

Items	2005-06		2006-07		2007-08		2008-09		2009-10*	
	Qty.	Value	Qty.	Value	Qty.	Value	Qty.	Value	Qty.	Value
Gross Import										
A. Crude Oil	99409	171702	111502	219029	121672	272699	132775	348288	159259	375378
B. LNG	5060	3366	6810	5650	8320	7197	8060	9548	8828	9344
C. Petroleum Products										
I. Light Distillates	6181	14855	8005	21369	9485	29221	8001	27617	4886	15525
1. LPG	2883	7042	2278	5766	2832	8700	2360	7916	2718	8329
2. MS	486	1286	421	1326	328	1137	397	1530	385	1264
3. Naphtha	2331	5400	5306	14277	5983	18470	5023	17410	1734	4942
4. Propane/ MTBE	481	1127	0	0	342	914	221	761	49	989
II. Middle Distillates	1846	4532	2392	6938	5443	17556	4215	16318	3519	9325
1. ATF	2	12	2	17	3	20	4	34	3	27
2. SKO	1044	2617	1423	4250	2489	8324	1423	6480	985	2909
3. HSD	800	1903	967	2671	2951	9212	2788	9804	2531	6390
4. Others	0	0								
III. Heavy Ends	5413	8585	7263	12853	7533	14223	8111	20334	6258	8903

Items	2005-06		2006-07		2007-08		2008-09		2009-10*	
	Qty.	Value	Qty.	Value	Qty.	Value	Qty.	Value	Qty.	Value
1. FO/ LSHS	1434	2099	2983	5174	1187	2211	1637	3465	762	1626
2. Lubes/ Others HEI	3979	6486	4280	7679	6346	12012	6474	16869	5496	7278
Total (C)	13440	27972	17660	41160	22461	61000	20327	64269	14662	33753
Grand Total (A+B+C)	117909	203040	135972	265839	152453	340896	161162	422105	182749	418475
Exports										
Petroleum Products										
I. Light Distillates	7769	17384	12344	32581	13937	42220	13141	43094	19804	61508
1. LPG	53	164	112	342	99	401	109	455	131	491
2. MS	2417	5843	3615	10191	4258	13614	5427	17767	9762	31170
3. Naphtha	5066	10787	8411	21431	9297	27365	7601	24854	9911	29848
4. TAME	0	0	0	0	0	0	0	0	0	0
5. Reformate	233	590	206	617	283	841	4	18	0	0
II. Middle Distillates	11453	26102	15171	39993	18931	54919	18469	63428	23094	64405
1. SKO	121	371	150	541	137	492	3701	13498	46	154
2. HSD/LDO	8504	18654	11369	29198	14308	40871	77	362	18460	50920
3. ATF	2828	7077	3652	10254	4486	13556	14691	49568	4588	13331

Items	2005-06		2006-07		2007-08		2008-09		2009-10*	
	Qty.	Value	Qty.	Value	Qty.	Value	Qty.	Value	Qty.	Value
III. Heavy Ends	4239	6488	6109	8520	7911	13650	6962	14564	8076	18123
1. FO/LSHS	1815	2268	3759	4988	4718	6811	6118	12132	5173	10501
2. VGO/ Lubes	993	1577	859	1826	1498	3236	219	829	111.2	321
3. Coke/ Bitumen/ White Oil/ Paraffin	339	302	66	77	409	614	177	391	324.3	619
4. Others	1092	2341	1425	1629	1286	2989	448	1212	2467	6683
Total	23461	49974	33624	81094	40779	110789	38572	121086	50947	144037
Net Imports										
A. Crude Oil	99409	171702	111502	219029	121672	272699	132775	348288	159259	375378
B. LNG	5060	3366	6810	5650	8320	7197	8060	9548	8828	9344
C. Petroleum Products	–10021	–22002	–15964	–39934	–18318	–49790	–18245	–56817	–36312	–110284
Grand Total	94448	1530660	102348	184745	111674	230106	122590	301019	131775	274439

Notes:

TAME: Tertiary Amyl Methyl Ether. *: Provisional.

Heavy Ends others: Includes Bitumen, Lube Oil Base Stock, Low Sulphur Waxy Residue. Carbon Black

Feed Stock, Rubber processing oil & Paraffin Wax.

SKO-Superior Kerosene oil, FO-Furnace Oil, LSHS-Low sulfur heavy stock furnace oil

HSD-High speed diesel, LDO-Light diesel oil, ATF-Aviation turbine fuel

AREA	2005	2006	2007	2008	2009	2010
Crude Oil (Million Metric Tonnes) Onshore	376	387	357	403	405	614
Offshore	410	369	368	366	369	587
Total	786	756	725	769	775	1201
Natural Gas (Billion Cubic Metres) Onshore	340	330	270	264	287	829
Offshore	761	745	785	786	787	608
Total	1101	1075	1055	1050	1074	1437

Note: The oil and natural gas reserves (proved and indicated) data relate to 1st April of each year.

MTBE-Methyl tertiary butyl ether

Table -(d) Consumption of Petroleum Products during five years in India

(000' Tonne)

Products	2005-06	2006-07	2007-08	2008-09	2009-10*
1.Light Distillates of which	33660	37076	38556	39908	39086
LPG	10456	10849	12165	12193	13121
Mogas	8647	9286	10332	11257	12818
Naphtha	12194	13886	13294	13875	10239
NGL	0	0	0	0	0
Others	2363	3055	2766	2583	2908
2. Middle Distillates of which	54423	57595	62823	66364	71198
SKO	9541	9505	9365	9303	9304
ATF	3296	3983	4543	4455	4627
HSDO	40191	42896	47669	51668	56320
LDO	883	720	667	552	457
Others	512	491	579	386	490
3. HeavyEnds of which	25129	26078	27567	27128	27911

Products	2005-06	2006-07	2007-08	2008-09	2009-10*
Furnace Oil @	8921	9257	9469	9267	9105
LSHS/HHS	3904	3361	3248	3170	2484
Lubes/Greases	2081	1900	2290	2149	2657
Bitumen	3508	3832	4506	4705	4919
Petroleum Coke	4928	5440	5950	5870	6750
Paraffin Wax	268	303	241	199	211
Other Waxes	105	65	65	3	78
Others	1414	1920	1799	1765	1707
Total	113212	120749	128946	133400	138196
Refinery Fuel	9141	10920	11751	11912	11607
GrandTotal	122353	131669	140697	145312	149803

Notes: Consumption includes sales by all companies and direct private imports.

@ : LSHS sales by pvt. parties included in FO. Break-up to these two products is not available *: Provisional.

Table -(e) Primary Commercial Energy Consumption in India during last five years

Sources of commercial energy	Unit	2004-05	2005-06	2006-07	2007-08	2008-09	2009-10*
1. Petroleum Products	MMT	120.17	122.35	131.67	140.70	145.31	149.80
2. Natural Gas (net)	BCM	30.78	31.33	30.79	31.48	31.77	46.49
3. Coal	MMT	382.61	407.04	430.83	457.08	492.76	325.87$
4. Lignite	MMT	30.34	30.06	31.29	33.98	32/42	16.32$
5. Electricity (incl. Non-utilities)	Bn.KWH	665.80	697.40	752.50	813.10	842.80	NA

Notes: *: Provisional $: April-November, 2009 BCM: Billion Cubic Metres (10^{12} m^3) , NA: not available

Table (f)- International Petroleum Statistics during last 5 years

Item	Unit	2005	2006	2007	2008	2009*
1. World Reserves: Oil % Gas at year end						
(i) Crude Oil	Bn. Tonne	165.0	164.5	168.6	170.8	181.7
of which OPEC	"	123.5	123.6	127.6	129.8	140
(ii) Natural Gas	Tr.Cu.ft.	6361	6405	6263	6534	6621
of which OPEC	"	3118	2961	3259	3295	3335
2. World Prodn. of Oil & Gas						
(i) Crude Oil	Mn. Tonne	3898.6	3910	3901.4	3934.7	3820.5
of which OPEC	"	1666.7	1673.7	1654.4	1703.8	1574.7
Cii) Natural Gas Net	Bn.Cu.Mtr.	2779.5	2880.2	2954.7	3060.8	2987
of which OPEC	"	501	522.2	546.7	572.3	580.7
3. World Oil Refinery						
(i) Refinery capacity at year end	Mn.	4266	4331	4378	4414	4512
(ii) Crude throughput	Tonne/ year Mn. Tonne	3693	3718	3762	3744	3670
4. World Oil Consumption						
(i) World Total	Mn. Tonne	3877.8	3916.2	3969.5	3959.9	3882.1
(ii) Main consuming areas	"	2257	2262	2711	2698	2612.3
(iii) Consumption in main areas as % of world	%	58.2	57.8	68.3	68.1	67.3
5. World Imports/ Exports						
(i) Crude Oil	Mn. Tonne	1885	1933	1984	1970	1892.5
(ii) Pol. Products	"	576	658	717	728	714
Total	"	2461	2591	2701	2698	2606.5
6. Export of Crude Oil/ Pol. from OPEC (i) Crude Oil (ii) Pol. Products Total	" " "	1161 224.6 1385.6	1173 229.1 1402.1	1197 220.1 1417.1	1190 219.5 1409.5	1102 204.5 1306.5
7. World Primary Energy Consumption						
(i) Oil	Mn. Tonne Oil Eqv.	3861	3911	3939	3960	3882.1
(ii) Natural Gas		2512	2558	2652	2717	2653.1

Item	Unit	2005	2006	2007	2008	2009*
(iii) Coal	»	2904	3039.1	3184	3286	3278.3
(iv) Hydro-Electric		658.5	684.3	696	731	740.3
(v) Nuclear Energy	»	626.8	634.9	622	620	610.5
Total	it	10562.3	10827.3	11094.2	11315.2	11164.3

Bn Tonne-Billion tonne, Tr cu.ft-Trillion cubic feet, Bn.Cu. Mtr.-Billion cubic metre

Note: Includes USA, Germany, France, Italy, U.K., Japan, China, Russia, India and South Korea.

OPEC-Oil Producing and Exporting Countries

Appendix to Chapter 2

2.1: Global oil consumption and refining capacity.

2.2: World oil capacity countrywise.

2.3: Important crude oil source for india

2.4: Worldwide crude quality in recent years

2.5: Characteristics of various indian crude oil

2.6: World's crude oil largest refiners (based on total refining capacity)

2.7: World's largest oil refineries (based on single lane)

Appendix 2.1: Global Oil Consumption and Refining Capacity(2009) million barrel/day(mbpd)

Region	Oil consumption	Refining capacity	Remarks
Asia Pacific	25.99	26.81	Just matching
North america	22.83	21.13	Deficit
Europe and Eurasia	19.37	24.92	Surplus
South & central America	5.65	6.69	Surplus
Middle east	7.15	7.86	Surplus
Africa	3.08	3.26	Surplus
Total	84.04	90.66	Surplus

Appendix 2.2: World oil refining capacity country wise (2009)

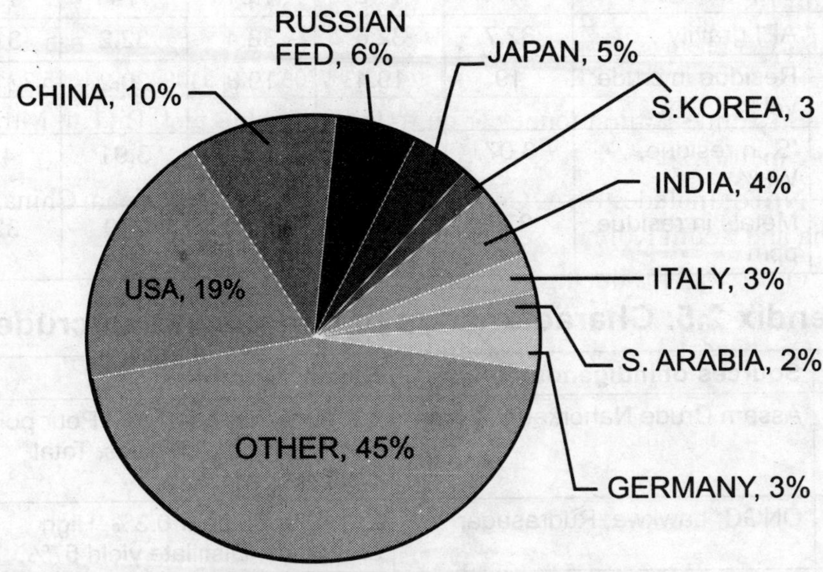

Total: 90.7 mbpd (4533 MMTPA)

Appendix 2.3: Imported crude oil sources for india

1	Middle East	Kuwait, Dubai, S. Arabia(Arab Mix, Arab medium), Iraq, Abu Dhabi, UAE(upper Zakum, Murban, UM Shaif) Iran, Kuwait: Ratawai, Egypt (Suez Mix, Zeit mix),
2	Mediterranean	Libya - Es Sider
3	West Africa	Nigeria-Bonny Light, Eseravos, Forcados, Penington, Quaiboe) Angola, -Cabinda, Palanca, Girassol Eq. Guinea -Ceiba, Zaffiro Congo -Nikossa, Kitina
4	Far East	Malaysia -Labuan, Miri Light Australia -Barrow Island, Cooper Basin,Chalis Brunei -Seria Light

Appendix 2.4: Worldwide Crude quality in recent years

S. N	Properties	1985	1990	1995	1999	2010
1	Sulphur, Wt %	1.14	1.12	1.31	1741	1.51
2	API gravity	32.7	32.6	32.4	32.2	31.8
3	Residue in crude, vol %	19	19.4	19.8	20.2	21.3
4	'S' in residue, Wt %	3.07	3.26	3.61	3.91	4.0
5	Metals in residue, ppm	275	286	297	309	320

Appendix 2.5: Characteristics of various Indian crude oil

S. N	Sources of indigenous crude	Salient features
1	Assam Crude Nahorkatia/ Moran	31 °API, Sulphur 0.3%, Pour point +30 °C, High aromatics, Total distillate yield 65%.
2	ONGC, Lawkwa, Rudrasagar	27 ^{0} API, Sulphur 0.3%, High aromatics. Distillate yield 57%.
3	Ankleshwar Crude	48 °API, Sulphur 0.1%, Pour point +18 °C, Distillate yield 80-82% (Light distillates 24%, Middle distillate 47%), Wax content 9.9%, total sulphur 0.02%.
4	North Gujarat Crude	28 °API, Sulphur content 0.1%, Pour point +27 °C, Distillate yield low 33-35%, High organic acidity.
5	Bombay High Crude	38 -°API; Sulphur 0.2%, Pour point +30 °C, Distillate yield 65-70% (Light distillate 24%, Middle distillates 46%), High aromatics.
6	Narimanam Crude	46 °API, Sulphur 0.1%, Pour point 3 °C, Distillate yield 80%.
7	KG Basin Ravva Crude	36 °API, Sulphur 0.1%, Pour point +30 °C, Distillate yield 61%.

Note-: Global Sources of Energy/Fuel/ is tabulated below :

S. N	Parameters	Coal	Crude Oil	Natural Gas
1	Reservoirs	930 billion tonnes	1900 billion barrels	180 trillion m^3
2	Expected life based on current consumption, years	120	70	60
3	Major reservoirs located at	USA, Russia, China, India, Australlia	OPEC, Russia, USA	OPEC, Brazil, USA

Crude Oil based Industries are as follows :

- Exploration of Crude Oil (E&P Industries)
- Refining of Crude Oil (Petroleum Refineries)
- Chemical production of hydrocarbons (Petrochemical Industries)

Reasons for India as a favourite location for New oil Refineries are given below:

- **Location:** Sea Coast and land locations are appropriate for the transportation of crude oils as well as export of finished products.
- **Manpower**: Availability of skilled and economic manpower.
- **Market:** Market is quite big from population point of view, expansion of chemical industries & significant increase in usage of vehicles.
- High rate of return/short pay-back-period.
- Environmental regulations.

Major shift in oil Refinery Technologies include the following points

- Heavy crude oils are to be treated (high impurities & average molecular weights).
- Light fractions yield decreases during primary fractionations that leads to further processing of heavy factions on large scale.
- Loads on catalytic cracking, hydrocracking, coking etc increased.
- Because of stringent environmental regulations; gasoline, diesel & jet fuel fractions are required to undergo various purification processes to meet high quality standards.
- More emphasis on hydrocracking and propylene product from FCC (fluid catalytic cracking)
- All above factors leads the existing refineries to go under revamping, and new refineries are being installed with aim of processing worst quality crude oils.

- Current trend is to locate refinery at/near oil exploration sites to minimize transportation.
- Integrated Refinery with petrochemical complexes reduce the transportation and also increases the overall profit margin.

Indian oil Refineries scenario:

- Current installed capacity: 194 MMTA (Million metric ton per anum)
- Capacity addition by 2013: 40 MMTA & by 2016: 100 MMTA
- Total refining capacity by **2017** would be **340 MMTA** (Including refineries installed outside India as a joint-ventures between Indian company & other foreign country)

In india presently about 80% crude oil is imported. **Capacity expansion undertaken** : 3.18 at Mangalore Refinery & 7.5 MMTA at Essar Oil, Jamnagar:: Total 10.5 MMTA

New oil refineries coming up in india is tabulated below.

S.NO.	Name of the Company	Location of the Refinery	Capacity (MMTPA)	Expected date of Commissioning
1.	Indian Oil CorporationLimited (IOCL)	Paradip, Orissa	15	2016
2.	Hindustan Mittal Energy Limited (HMEL)	Bathinda, Punjab	9	2013
3.	Nagarjuna Oil Corporation Limited (NOCL)	Cuddalore, TimilNadu	6	2014
		Total	30	

Current Location & Refining Capacity of Indian Oil Refineries are tabulated below.

S.No.	Name of the company	Location	Capacity. : MMTPA*
1.	Indian Oil Corporation Limited (IOCL)	Guwahati, Assam	1.00
2.	Indian Oil Corporation Limited (IOCL)	Barauni, Bihar	6.00
3.	Indian Oil Corporation Limited (IOCL)	Koyali, Vadodara, Gujarat	13.70
4.	Indian Oil Corporation Limited (IOCL)	Haldia, West Bengal	7.50

S.No.	Name of the company	Location	Capacity. : MMTPA*
5.	Indian Oil Corporation Limited (IOCL)	Mathura,Uttar Pradesh	8.00
6.	Indian Oil Corporation Limited (IOCL)	Digboi, Assam	0.65
7.	Indian Oil Corporation Limited (IOCL)	Panipat, Haryana	15.00
8.	Indian Oil Corporation Limited (IOCL)	Bongaigaon, Assam	2.35
9.	Hindustan Petroleum Corporation Limited(IOCL)	Mumbai, Maharashtra	6.50
10.	Hindustan Petroleum Corporation Limited (HPCL)HPCL,visakh	Visakhapatnam, Andhra Pradesh	8.30
11.	Bharat Petroleum Corporation Limited (BPCL)	Mumbai, Maharashtra	12.00
12.	Bharat Petroleum Corporation Limited (BPCL)	Kochi, Kerala	9.50
13.	Chennai Petroleum Corporation Limited	Manali, Tamil Nadu	10.50
14.	Chennai Petroleum Corporation Limited (CPCL)	Nagapattnam, Tamil Nadu	1.00
15.	Numaligarh Refinery Ltd	Numalighrh, Assam	3.00
16.	Mangalore Refinery & Petrochemical Ltd. [MRPL)	Mangalore, Karnataka	11.82
17.	Tatipaka Refinery (OMGC)	Tatipaka, Andhra Pradesh	0.066
18.	Bharat Petroleum Corporation Limited & Company, Joint venture, bina	Bina, Madhya Pradesh	6.00
19.	Reliance Industries Ltd. (RIL); Private Sector	Jamnagar, Gujarat	33.00
20.	Reliance Petroleum Limited (SEZ); Private Sector	Jamnagar, Gujarat	27.00
21.	Essar Oil Limited (EOL); Private Oil Sector	Jamnagar, Gujarat	10.50
	TOTAL		193.386

MMTPA – Million metric ton per anum , SEZ-Special Economic Zone

Appendix 2.6: World's Largest Crude Oil Refiners (Based on Total Refining Capacity)

Rank	Company	Crude Capacity, Thousand Barrels per Calendar Day
1.	Exxon Mobil Corporation	5,632
2.	Royal Dutch/Shell (Netherlands)	4,599
3.	Sinopec	3,811
4.	BP (United Kingdom)	3,328
5.	ConocoPhillips	2,696
6.	Petroleos de Venezuela.S.A. (Venezuela)	2,678
7.	Total	2,655
8.	Valero Energy Corporation	2,596
9.	China National Petroleum Company (China)	2,440
10.	Saudi Arabian Oil Company (Saudi Arabia)	2,433

Appendix 2.7: World's largest oil refineries (based on single lane)

Rank	Company	Location	Capacity, barrels per day
1.	Paraguana Refining Center	Cardon/Judibana, Falcon, Venezuela	940,000
2.	SK Corp.	Ulsan, South Korea	817,000
3.	GS Caltex Corp.	Yeosu, South Korea	750,000
4.	Reliance Petroleum Ltd.	Jamnagar, India	660,000
5.	ExxonMobil Refining &Supply Co.	Jurong/Pulau Ayer Chawan, Singapore	605,000
6.	Reliance Industries Ltd.	Jamnagar, India	580,000
7.	S-Oil Corp.	Onsan, South Korea	565,000
8.	ExxonMobil Refining & Supply Co.	Baytown, Texas, USA	560,500
9.	Saudi Arabian Oil Co. (Saudi Aramco)	Ras Tanura, Saudi Arabia	550,000

Rank	Company	Location	Capacity, barrels per day
10.	Formosa Petrochemical Co.	Mailiao, Taiwan	540,000
11.	ExxonMobil Refining & Supply Co.	Baton Rouge, Louisiana, USA	503,500
12.	Hovensa LLC	St. Croix, Virgin Islands, USA	500,000
13.	Kuwait National Petroleum Co.	Mina Al-Ahmadi, Kuwait	466,000
14.	Shell Eastern Petroleum (Pte) Ltd.	Pulau, Bukom, Singapore	462,000
15.	BP PLC	Texas City, Texas, USA	451,250

NOTE- World's largest refinery at single location: RIL (Reliance Indian Limited) Jamnagar, INDIA. , Gujrat,

3

Processing of Crude Oil-Refinery Operation & Products

The process of distillation is the most common method adopted for separating the constituents of crude oil into compounds or groups of compounds having industrial uses. Crude oil consists of a complex mixture of hydrocarbons widely differing in boiling points. Distillation is done to separate the crude oil into the basic fractions like motor gasoline kerosene, gas oil and fuel oil. The order in which these components emerge as crude oil is distilled is given below in Table.

Table: Components of Crude Oil Distillation

Components	Approximate Boiling Range, °C	Uses
Gases	–162 to –42	Domestic cooking, Heating furnaces in refinery plants.
Gasoline/light naphtha	–1 to 150	Motor Fuel (straight run gasoline)
Heavy naphtha	150–205	Catalytic reformer feed, jet fuel, solvent
Kerosene	205–250	Illuminant and fuel.
Light Gas Oil	250–315	High speed diesel oil, fuel oil, absorbent
Heavy Gas Oil	315–430	Catalytic cracking feed.

Components	Approximate Boiling Range, °C	Uses
Vacuum gas oil	430–540	Lubricating oil, catalytic cracking feed
Fuel Oil/residuum	> 595	Fuel for furnaces, asphalt, bitumen, coke making feed

The proportions of these fractions vary considerably with the source of crude oil but each fraction is identified by its boiling range which overlap considerably and hence nomenclature of fraction also changes accordingly. The crude oil is pumped from the tankers to the crude oil tanks, where it is allowed to settle and the water is drained off. Water associated with crude is highly objectionable as it gives rise to sudden pressure changes in the plant, deposits salts on equipment where it vaporizes off, dissolves harmful acids and causes deterioration of equipment by way of corrosion.

The crude oil is pumped from the tanks to the primary column of the distillation unit by centrifugal pumps. The pressure at the discharge of these pumps is maintained high enough to force the crude oil through preheat exchangers and a preheat furnace before it enters the column.

All oil streams from the columns like kerosene, gas oils & residue are at high temperatures and their heat is transferred to the incoming crude oil in heat exchangers. This effects a great fuel economy in the process as this heat would otherwise have to be obtained by burning fuel oil or gas.

Bulk of the heat to the crude oil is given by oil or refinery gas burnt in tube still heaters. Each heater has a convection section in which the oil is gradually heated by the fuel gas and a radiant section where the tubes are exposed to radiant heat and the oil is brought upto the required temperature. The oil flows through straight tubes joined at their ends by expanding them into box headers. The air for combustion is provided by a blower. A tall stack sucks the hot gases and discharges them into the atmosphere. The heater outlet temperature is maintained at the required level by varying the quantity of gas or oil burnt as fuel. The air to the furnace is regulated by dampers. The furnaces are operated at a temperature which would not lead to deterioration of brick work or excessive coke formation in the tubes. Emergency valves are provided to drop out the contents of the furnace tubes quickly, if the flow of oils fails for any reason. Stagnant oil in the tubes will be converted into coke, if the heating medium is not cut out promptly.

The crude oil, partly vaporized, enters the primary tower which is a conventional distillation column with a number of plates. Gasoline & gas rise to the top, are condensed in the overhead condenser and flow into an accumulator. The heavier components such as gas oil and residue fall to the bottom of the column. In order to provide sufficient liquid flow down the column, to effect fractionation, a part of the condensed liquid from the overhead accumulator is pumped back to the column, as reflux. Within certain limits, greater the reflux, better is the quality of the overhead product. Sea water (salty water) is used to condense the overhead gasoline vapours which is in abundance, where the refineries are situated on the sea shore. To minimise corrosion by sea water, aluminium brass tubes are used. In places where water is fresh, direct contact condensers are used. This type is, however unsuitable with salty water, because the latter does not separate very well from oils. The primary tower is operated under pressure to retain and dissolve substances like butane which improve the antiknock value of gasoline.

The straight run gasoline from the primary tower contains certain very low boiling compounds which render it unsuitable for storage under atmospheric conditions. The stabilisation of gasoline is effected by preheating and fractionating it under pressure to get a product with the desired vapour pressure. A portion of the kerosene is drawn off as a side stream from the primary tower. The maximum temperature upto which crude oil can be heated without cracking is 400°C, which sets the limit of the maximum amount of distillate that can be collected by atmospheric distillation.

The reduced crude at the bottom of the primary tower consists of kerosene, gas oils and residue. It is pumped through a furnace into the secondary tower. The components of reduced crude are all high boiling materials and would require very high temperature to vaporise them. High temperature will destroy the quality of oil by cracking it and will prove very corrosive to the material of the tower itself. The boiling points are therefore lowered by applying a vacuum to the tower by the use of steam jet ejectors. Operation at a pressure of 300 mm. Hg absolute lowers boiling points well enough to vaporize all the oils with a heater outlet temperature of approximately 320°C.

Kerosene and light & heavy gas oils are drawn off as side streams. The last two liquids still contain some low boiling material which lower the flash point and render them unsafe for storage. They are led into columns called strippers where steam or oil heat is applied to drive away the light material. Steam reduces the partial pressure of oil vapour and causes the low boiling components to be driven off. The residue from the secondary tower is pumped away to storage or to the feed preparation unit where it is split into asphalt for bitumen manufacture and waxy distillate for catalytic cracker feed.

A typical distillation operation carried out in an Indian oil refinery is shown below in the Figure.

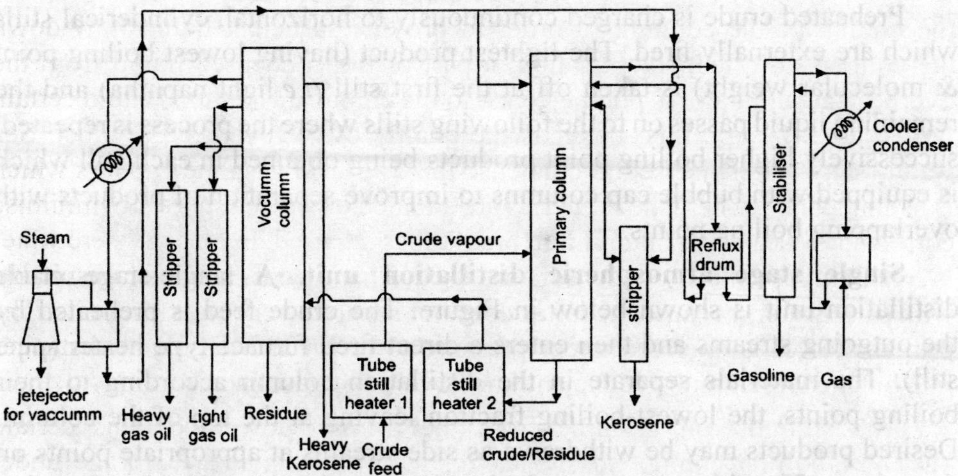

Fig. Flow-sheet of a Typical Indian Oil Refinery for Crude Distillation.

Various crude oil distillation systems

There are four ways in which the crude oil is distilled *i.e.*

- *Shell still distillation* (Obsolete)
- *Single stage distillation unit* (Atmospheric pressure)
- *Two-stage distillation unit* (Atmospheric and high pressure units)
- *Three-stage distillation unit* (High pressure, atmospheric Pressure & vacuum units)

Shell still distillation of crude oil. It is the oldest and obsolete method of crude distillation in which distillation is accomplished in a battery of shell stills as shown in the Figure below.

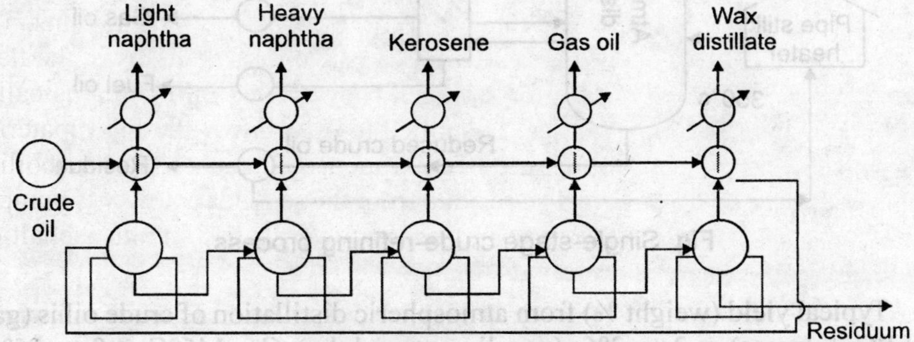

Fig. Battery of shell stills.

Preheated crude is charged continuously to horizontal, cylinderical stills which are externally fired. The lightest product (having lowest boiling point & molecular weight) is taken off at the first still (*i.e.* light naphtha) and the remaining liquid passes on to the following stills where the process is repeated, successively higher boiling point products being obtained in each still which is equipped with bubble cap columns to improve separation of products with overlapping boiling points.

Single stage atmospheric distillation unit. A single-stage crude distillation unit is shown below in Figure. The crude feed is preheated by the outgoing streams and then enters a direct fired furnace type heater (pipe still). The materials separate in the distillation column according to their boiling points, the lowest-boiling fraction leaving at the top of the column. Desired products may be withdrawn as side streams at appropriate points on the column. The side streams are further fractionated in small columns called strippers. Here, steam is used to free the cut from its more volatile components so that the initial boiling point of the products can be adjusted to its desired value. It is also called straight run distillation of crude oil.

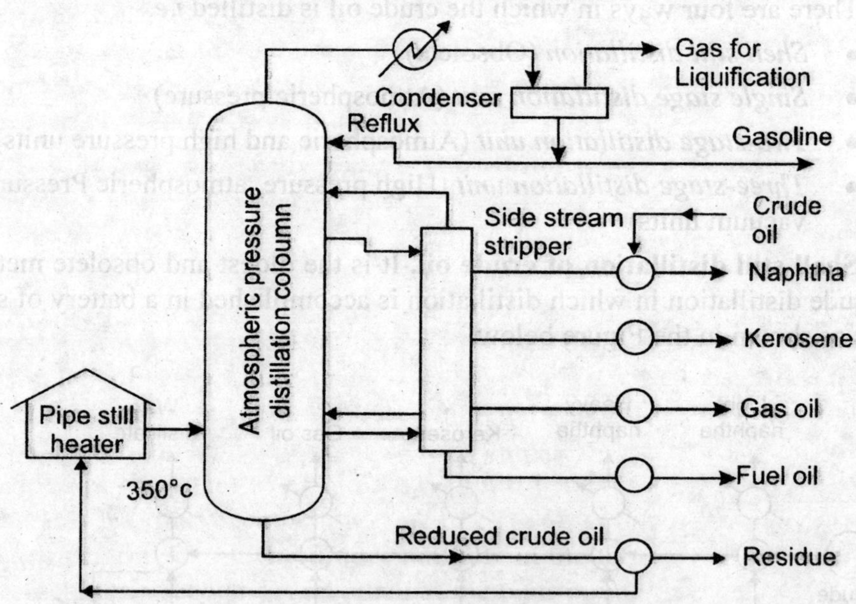

Fig. Single-stage crude-refining process.

Typical yield (weight %) from atmospheric distillation of crude oil is (gas + LPG + losses) = 2 to 3%, (gasoline + naphtha) C_5 –145°C = 8 to 15%, Kerosene (145 –250°C) = 15 to 20%, High speed diesel oil (270 –370°C) = 20 to 35%, reduced crude (>370°C) = 30 to 45%.

However capacity of single stage distillation unit for processing crude oil is limited which calls for addition of pressure units. Moreover, high boiling point products can not be separated as operating it at higher temperature may result in cracking of products. Therefore vacuum units are added to get the high boiling point products vaporised and separated at lower temperature. This is also called crude *topping* or atmospheric distillation of crude oil as the pressure in the distillation column is slightly above atmospheric.

Two stage crude distillation unit. This scheme of distillation includes a *primary tower* which operates at about *3–4 atmospheric pressure* and a *secondary tower* operating at *atmospheric pressure* together with stabiliser. This scheme is used when crude is to be separated into six to ten narrow cuts *(i.e.* products with very close and overlapping boiling points). Two/ three side streams*(e.g.* gasoline, naphtha) are withdrawn from primary tower while the overhead gas product becomes feed to stabiliser. Primary tower bottoms becomes the feed for secondary tower from which side streams like kerosene, diesel, gas oil etc. are withdrawn at appropriate points after passing through strippers to remove the light components.

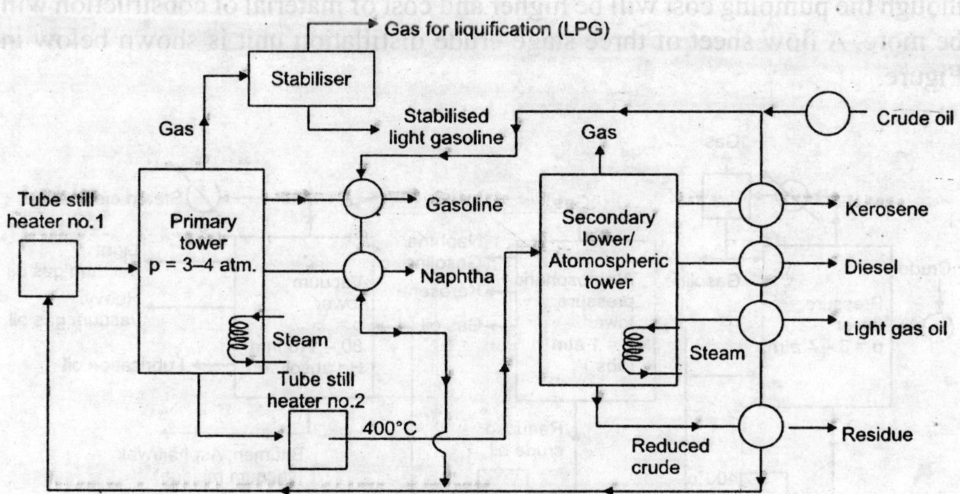

Fig. Two-stage crude distillation unit.

Three stage crude distillation unit. This system comprises of all the three *i.e.* pressure, atmosphere & vacuum towers. Because some products of crude oil are having very high boiling temperature or heat sensitive so that they can not be distilled at atmospheric pressure. Since boiling point is reduced as the pressure is lowered *(i.e.* vacuum is created), such materials are distilled under vacuum. Vacuum distillation is used for the production of lubricating oils, asphalts and feed to catalytic cracking unit (*i.e.* heavy gas oil). Hence, a vacuum tower is added to a two stage distillation unit for the production of above products thus making it a three stage unit. Light vacuum gas oil (LVGO) is a blending component for light diesel oil (LDO) and a feedstock for catalytic cracker/hydrocracker. Heavy vacuum gas oil (HVGO) is a feedstock for visbreaker and catalytic cracker/hydrocracker. Vacuum residue is used for bitumen production as well as is a feed component for visbreaker.

The residue from the atmospheric tower is fed at about 425°C to the vacuum tower, where pressure is maintained at 80-110 mm Hg absolute. The overhead gas oil fraction is removed and the side streams are withdrawn through strippers. The bottom product is tar or asphalt.

Vacuum distillation units are similar to atmospheric crude distillation unit except that their diameter is larger due to larger crude vapor volumes at lower pressure (vacuum). 10 metre diameter vacuum columns are in operation, steam jet ejector or barometric condensers are used for maintaining desired vacuum in the vacuum tower. Pressurised units can handle larger throughput of crude though the pumping cost will be higher and cost of material of construction will be more. A flow sheet of three stage crude distillation unit is shown below in Figure.

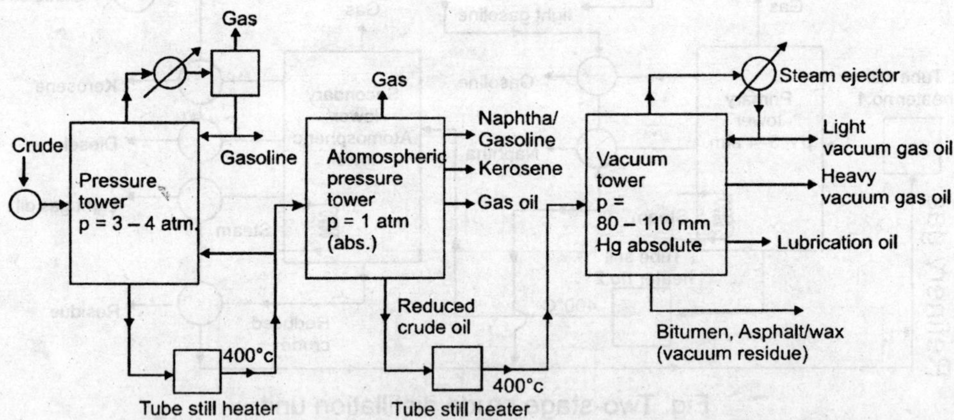

Fig. Three stage processing of crude-oil General scheme.

A detailed typical layout scheme for a 2 million tons per year crude oil processing unit is shown below in the Figure which includes all the three stages (pressure, atmospheric & vacuum) of distillation. Tower details have been given below in the table and the products are designated by their boiling point ranges.

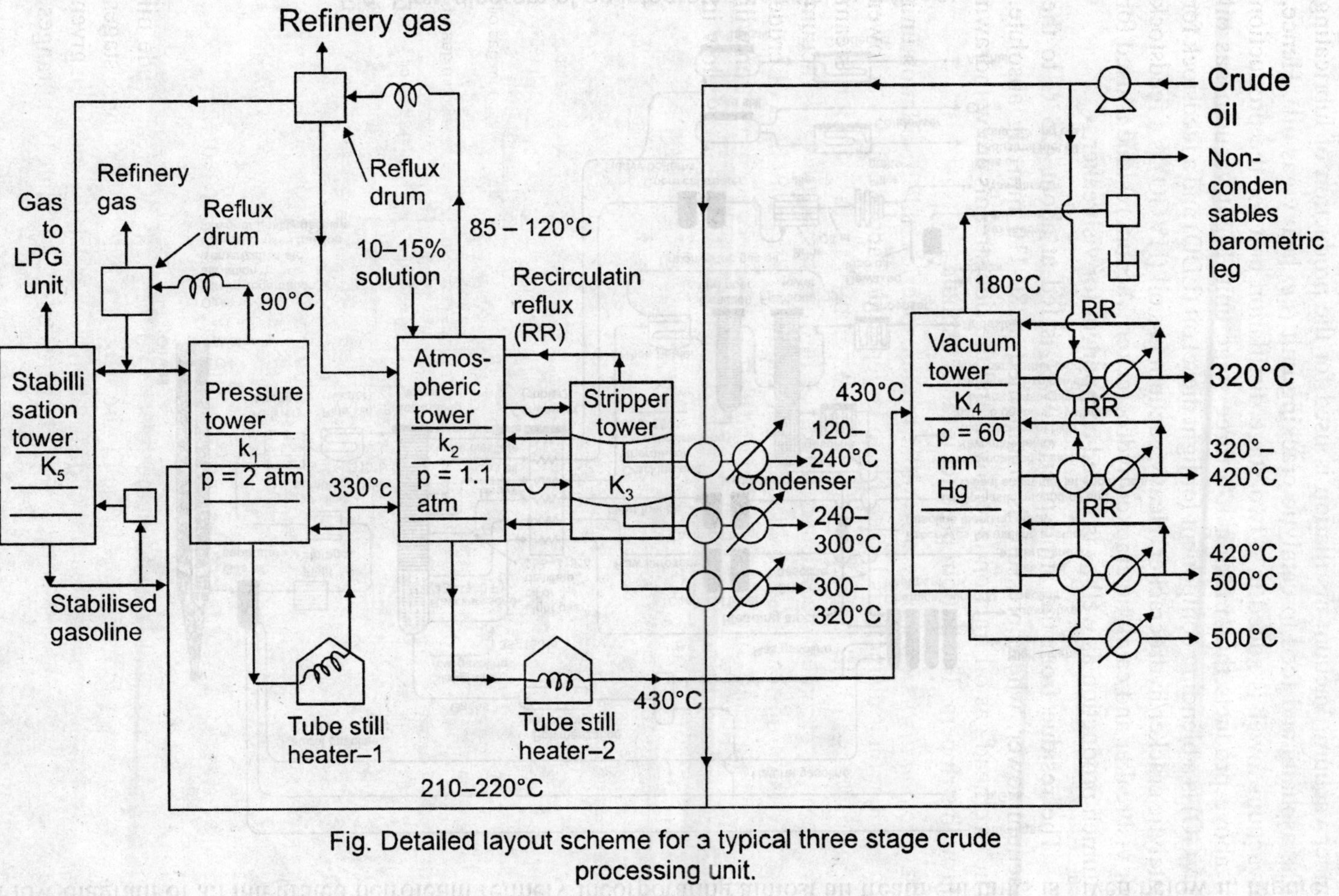

Fig. Detailed layout scheme for a typical three stage crude processing unit.

Flow diagram of an integrated petroleum refinery incorporating almost all treatment units is given below in Figure.

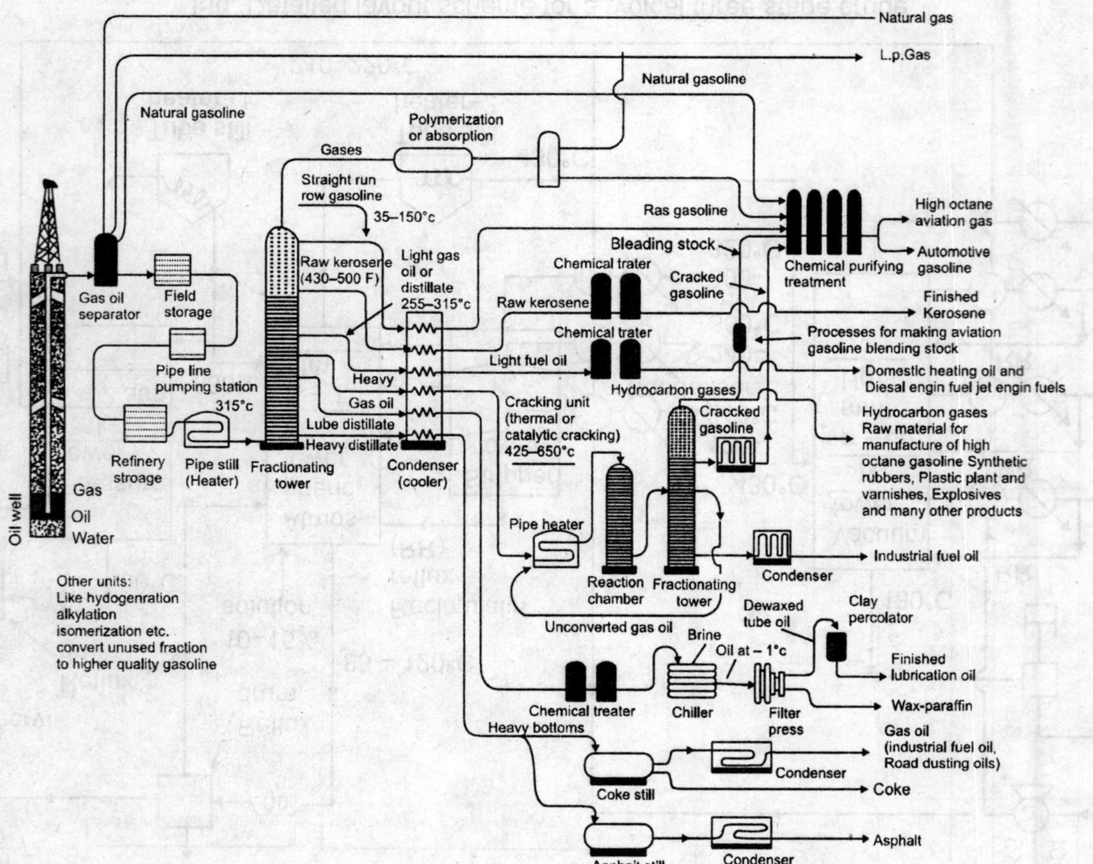

Fig. Flow diagram of an integrated petroleum refinery.

Tower	No. of Plates	Height, metre	Diameter, metre	Boiling point, °C	Products
K1	26	30.3	3.8	120–140	Naphtha
K2	38	30.1	5	240–300	Gas oil
K3	20	22.5	1.6	300–320	Kerosene
K4	–	22.8	8	320	Heavy Kerosene
K5	35	27.8	1.4	320–420	Heavy gas oil
				420–500	Lubricating oil
				> 500	Reduced crude/ tar/asphalt

A schematic simplified flowsheet of various refinning processes generally followed in Petroleum refineries is given below in Figure. Flowsheet of yet another refinery incorporating various unit operations is given below in another Figure. Yet another figure below gives the process flow chart of a modern integrated crude oil refinery incorporating all units.

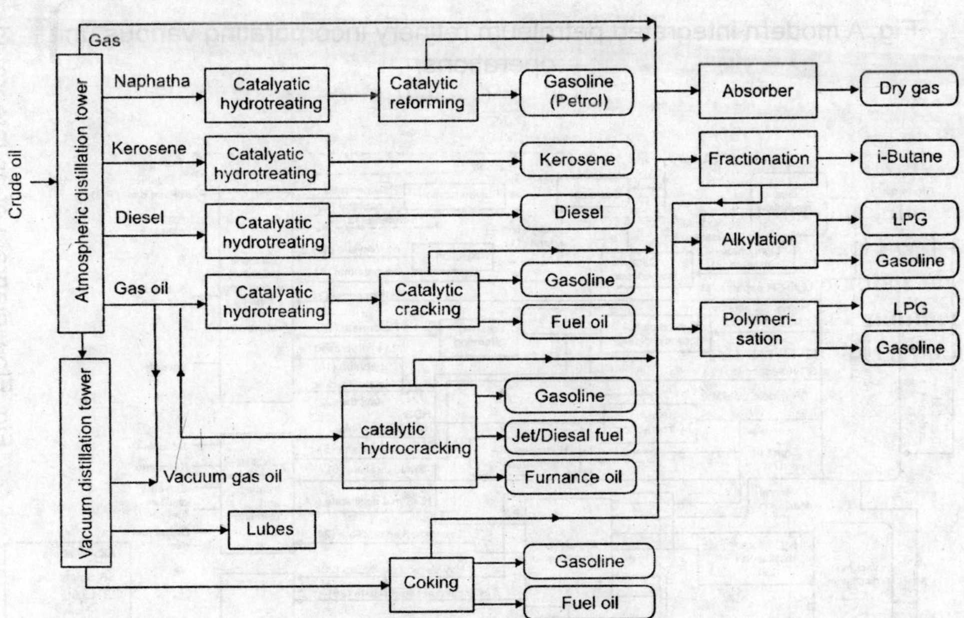

Fig. Flow sheet of Various Petroleum Refinning Processes followed in an oil refinery.

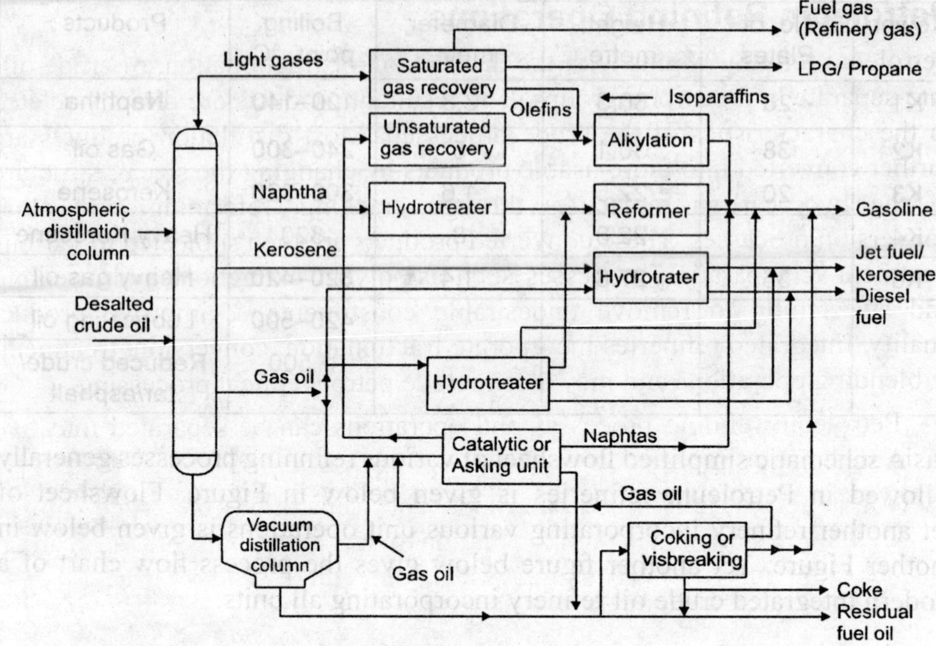

Fig. A modern integrated petroleum refinery incorporating various unit operations.

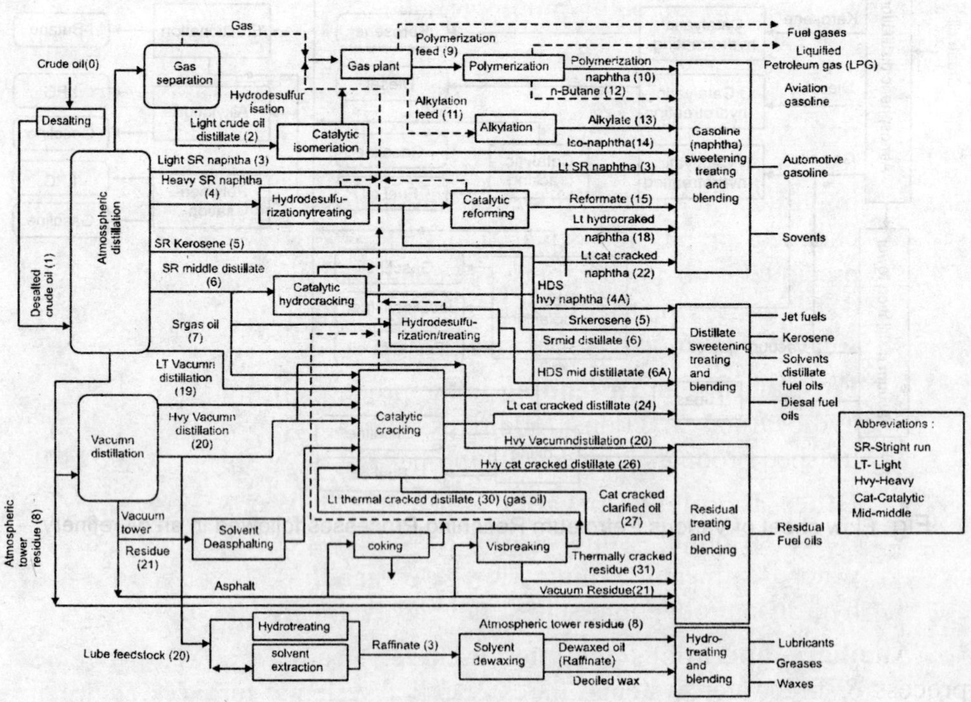

Fig. Process flow chart of a modern integrated crude oil refinery

Petroleum Refining Operations

Petroleum refining begins with the distillation, or fractionation, of crude oils into separate hydrocarbon groups. The resultant products are directly related to the characteristics of the crude processed. Most distillation products are further converted into more usable products by changing the size & structure of the hydrocarbon molecules through cracking, reforming, and other conversion processes. These converted products are then subjected to various treatment & separation processes such as solvent extraction, hydrotreating and sweetening to remove undesirable constituents & improve product quality. Integrated refineries incorporate fractionation, conversion, treatment, & blending operations and may also include petrochemical processing.

Petroleum refining processes and operations can be separated into five basic area:

- **Fractionation** (distillation) is the separation of crude oil in atmospheric & vacuum distillation towers into groups of hydrocarbon compounds of differing boiling point ranges called “fractions" or "cuts."
- **Conversion** processes change the size and/or structure of hydrocarbon molecules. These processes include:
  - Decomposition (dividing) by thermal catalytic cracking;
  - Unification (combining) through alkylation, polymerization; and
  - Alteration (rearranging) with isomerization, catalytic reforming.
- **Treatment** processes prepare hydrocarbon streams for additional processing and to prepare finished products. Treatment may include the removal or separation of aromatics and naphthenes as well as impurities & undesirable contaminants. Treatment may involve chemical or physical separation such as dissolving, absorption, or precipitation using a variety and combination of processes including desalting, drying, hydrodesulfurizing, solvent refining, sweetening, solvent extraction & solvent dewaxing.
- **Formulating and Blending** is the process of mixing & combining hydrocarbon fractions, additives, other components to produce finished products with specific properties.
- **Other Refining Operations** include: light-ends recovery; sour-water stripping; solid waste & wastewater treatment; process-water treatment&cooling; storage & handling; product movement; hydrogen production; acid & tail gas treatment; and sulfur recovery.

Auxiliary operations facilities include: steam & power generation; process & fire water systems; flares & relief systems; furnaces & heaters; pumps &valves; supply of steam, air, nitrogen, &other plant gases; alarms &

sensors; & pollution controls; sampling, testing, &inspecting; and laboratory, control room, maintenance, &administrative facilities.

A flow sheet of crude oil & natural gas processing scheme is shown below in figure:

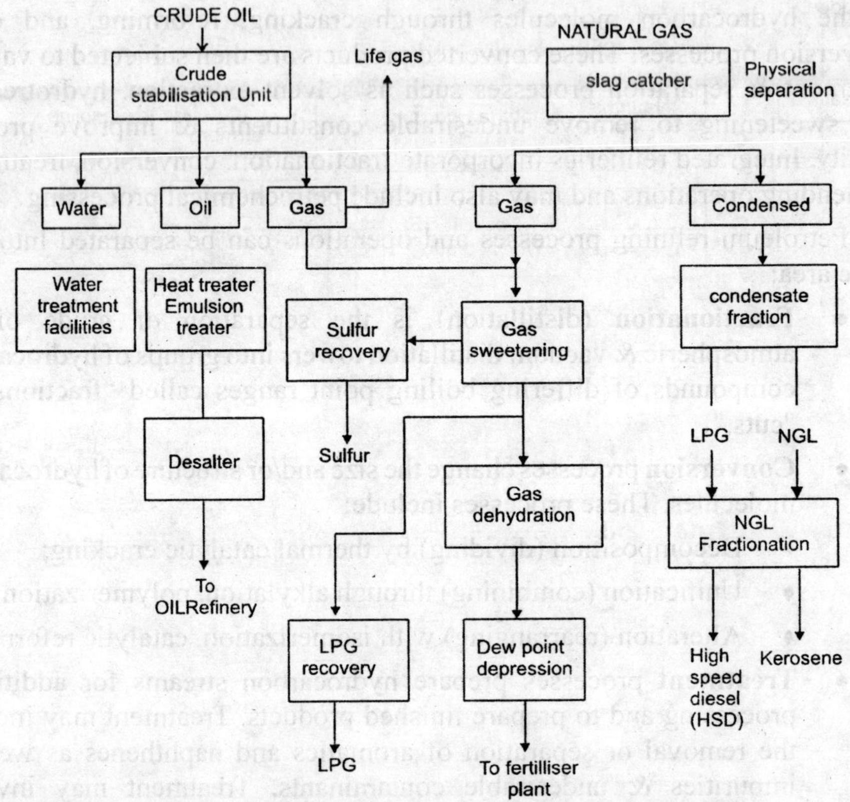

Fig. Crude Oil & Natural Gas processing scheme

Various Physical & Chemical Processes Carried Out in Refineries are Tabulated Below.

Physical	**Chemical**	
	Thermal	**Catalytic**
Distillation Solvent, extraction Propane, deasphalting Solvent, dewaxing Blending	Visbreaking Delayed, coking Flexicoking	Hydrotreating Catalytic reforming Catalytic cracking Hydrocracking Catalytic dewaxing Alkylation Polymerization Isomerization

A typical petroleum refinery processes and products are shown below in figure.

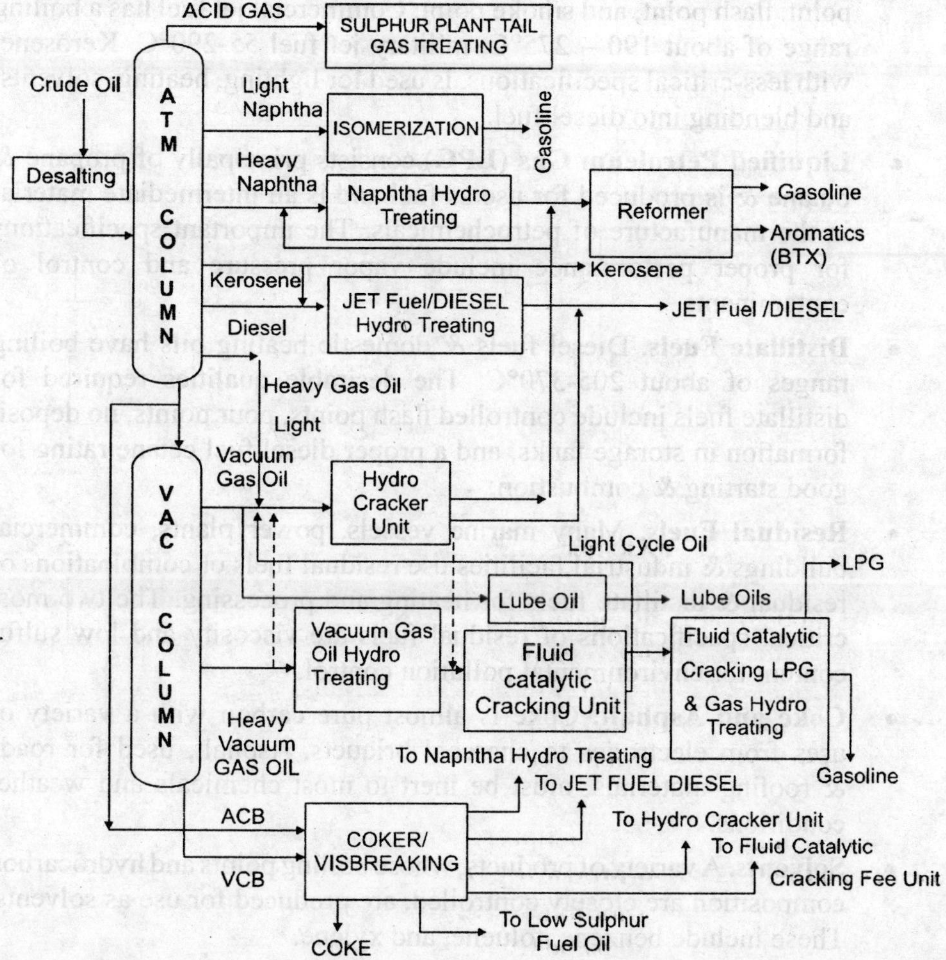

Fig. Typical Refinery Processes & Products (BTX–Benzene, Toluene, Xylene)

Major Petroleaum Refinery Products: Major petroleaum refinery products and common refinery chemical discussed below:

- **Gasoline.** The most important refinery product is motor gasoline, a blend of hydrocarbons with boiling range from ambient temperatures to about 205°C. The important qualities for gasoline are octane number (antiknock), volatility (starting & vapor lock), and vapor pressure (environmental control). Additives are often used to enhance performance and provide protection against oxidation & rust formation.

- **Kerosene.** Kerosene is a refined middle-distillate petroleum product that finds considerable use as a jet fuel and in cooking & space heating. When used as a jet fuel, some of the critical qualities are freezing point, flash point, and smoke point. Commercial jet fuel has a boiling range of about 190 – 275°C, military jet fuel 55-290°C. Kerosene, with less-critical specifications, is used for lighting, heating, solvents, and blending into diesel fuel.
- **Liquified Petroleum Gas (LPG).**consists principally of propane & butane & is produced for use as fuel and is an intermediate material in the manufacture of petrochemicals. The important specifications for proper performance include vapor pressure and control of contaminants.
- **Distillate Fuels.** Diesel fuels & domestic heating oils have boiling ranges of about 205-370°C. The desirable qualities required for distillate fuels include controlled flash points, pour points, no deposit formation in storage tanks, and a proper diesel fuel cetane rating for good starting & combustion.
- **Residual Fuels.** Many marine vessels, power plants, commercial buildings & industrial facilities use residual fuels or combinations of residual & distillate fuels for heating and processing. The two most critical specifications of residual fuels are viscosity and low sulfur content for environmental pollution control.
- **Coke and Asphalt.** Coke is almost pure carbon with a variety of uses from electrodes to charcoal briquets. Asphalt, used for roads & roofing materials, must be inert to most chemicals and weather conditions.
- **Solvents.** A variety of products, whose boiling points and hydrocarbon composition are closely controlled, are produced for use as solvents. These include benzene, toluene, and xylene.
- **Petrochemicals.** Many products derived from crude oil refining, such as ethylene, propylene, butylene, & isobutylene, are primarily intended for use as petrochemical feedstock in the production of plastics, synthetic fibres, synthetic rubbers, and other products / polymers.
- **Lubricants.** Special refining processes produce lubricating oil base stocks. Additives such as demulsifiers, antioxidants, and viscosity improvers are blended into the base stocks to provide the characteristics required for motor oils, industrial greases, lubricants, and cutting oils. The most critical quality for lubricating oil base stock is a high viscosity index, which provides for greater consistency under varying temperatures.

Common Refinery Chemicals

- **Leaded Gasoline Additives.** Tetraethyl lead (TEL) and tetramethyl lead (TML) are additives formerly used to improve gasoline octane ratings but are no longer in common use except in aviation gasoline.
- **Oxygenates.** Ethyl tertiary butyl ether (ETBE), methyl tertiary butyl ether (MTBE), tertiary amyl methyl ether (TAME), and other oxygenates improve gasoline octane ratings and reduce carbon monoxide emissions.
- **Caustics.** Caustics are added to desalting water to neutralize acids and reduce corrosion. They are also added to desalted crude in order to reduce the amount of corrosive chlorides in the tower overheads. They are used in some refinery treating processes to remove contaminants from hydrocarbon streams.
- **Sulfuric Acid & Hydrofluoric Acid.** Sulfuric acid and hydrofluoric acid are used primarily as catalysts in alkylation processes. Sulfuric acid is also used in some treatment processes.

Appendix to Chapter 3

Appendix 3.1:	Overview of petroleum refining processes.
Appendix 3.2:	Main products of crude oil refining and their end uses.
Appendix 3.3:	All about distillation and distillation column used in petroleum refinery.
Appendix 3.4:	Features of lubricants, wax and grease manufacturing processes.

Appendix 3.1: Overview of Petroleum Refining Processes

S-N	Process name	Action	Method	Purpose	Feedstocks	Products
	Fractionation processes					
1	Atmospheric distillation	Separation	Thermal	Separate fractions	Desalted crude oil	Gas, gas oil distillate, residual
.2.	Vacuum distillation	Separation	Thermal	Separate without cracking	Atmospheric tower residual	Gas oil, lube stock, residual
	Conversion processes-Decomposition					
3	Catalytic cracking	Alteration	Catalytic	Upgrade gasoline	Gas oil, coke distillate	Gasoline, petrochemical feedstock
4	Coking	Polymerization	Thermal	Convert vacuum residuals	Residual heavy oil, tar	Naphtha, gas oil, coke
5	Hydrocracking	Hydrogenation	Catalytic	Convert to lighter hydrocarbons	Gas oil, cracked oil residuals	Lighter, higher quality products
6	Hydrogen by steam reforming	Decomposition	Thermal/ catalytic	Produce hydrogen	Desulphurized gas, O_2, steam	Hydrogen, CO, CO_2
7	Steam cracking	Decomposition	Thermal	Crack large molecules	Atmospheric tower heavy fuel/distillate	Cracked naphtha, coke, residuals
8	Visbreaking	Decomposition	Thermal	Reduce viscosity	Atmospheric tower residual	Distillate, tar

S-N	Process name	Action	Method	Purpose	Feedstocks	Products
	Conversion processes-Unification					
9	Alkylation	Combining	Catalytic	Unite olefins and isoparaffins	Tower isobutane/cracker olefin	Iso-octane (alkylate)
10	Grease compounding	Combining	Thermal	Combine soaps and oils	Lube oil, fatty acid, alkyl metal	Lubricating grease
11	Polymerization	Polymerization	Catalytic	Unite two or more olefins	Cracker olefins	High octane naphtha, petrochemical stocks
	Conversion processes-Alteration/rearrangement					
12	Catalytic reforming	Alteration/ dehydrogenation	Catalytic	Upgrade low-octane naphtha	Coker/hydrocracker naphtha	High-octane reformate/ aromatic
13	Isomerization	Rearrangement	Catalytic	Convert straight chain to branch	Butane, pentane, hexane	Isobutane/ pentane/hexane
Treatment processes						
14	Amine treating	Treatment	Absorption	Remove acidic contaminants	Sour gas, hydrocarbons with CO_2 and H_2S	Acid-free gases and liquid hydrocarbons
15	Desalting (pre-treatment)	Dehydration	Absorption	Remove contaminants	Crude oil	Desalted crude oil
16	Drying and sweetening	Treatment	Absorption/ thermal	Remove H_2O and sulphur compounds	Liquid hydrocarbon, LPG, alkylated feedstock	Sweet and dry hydrocarbons

S-N	Process name	Action	Method	Purpose	Feedstocks	Products
17	Furfural extraction	Solvent extraction	Absorption	Upgrade middle distillate and lubes	Cycle oils and lube feedstocks	High-quality diesel and lube oil
18	Hydrodesulphurisation	Treatment	Catalytic	Remove sulphur, contaminants	High-sulphur residual/gas oil	Desulphurized olefins
19	Hydrotreating	Hydrogenation	Catalytic	Remove impurities/ saturate hydrocarbons	Residuals, cracked hydrocarbons	Cracker feed, distillate, lube
20	Phenol extraction	Solvent extraction	Absorption /thermal	Improve lube oil viscosity index, colour	Lube oil base stocks	High-quality lube oils
21	Solvent deasphalting	Treatment	Absorption	Remove asphalt	Vacuum tower residual propane	Heavy' lube oil asphalt
22	Solvent dewaxing	Treatment	Cool/filter	Remove wax from lube stocks	Vacuum tower lube oils	Dewaxed lube base stock
23	Solvent extraction	Solvent extraction	Absorption/ precipitation	Separate unsaturated aromatics	Gas oil reformate, distillate	High-octane gasoline
24	Sweetening	Treatment	Catalytic	RemoveH_2S, convert mercaptan	Untreated distillate/ gasoline	High-quality distillate/gasoline

Note:

- **Basic Refinery Process:** Petroleum refining has evolved continuously in response to changing consumer demand for better and different products. The original requirement was to produce kerosene as a cheaper and better source of light. The development of the internal combustion engine led to the production of gasoline and diesel fuels. The evolution of the airplane created a need first for high-octane aviation gasoline and then for jet fuel (a sophisticated form of the original product, kerosene). Present-day refineries produce a variety of products including many required as feedstock for the petrochemical industry.
- **Distillation Processes.** The first refinery, produced kerosene by simple atmospheric distillation. Its by-products included tar and naphtha. It was soon discovered that high-quality lubricating oils could be produced by distilling petroleum under vacuum. However, for the next 30 years kerosene was the product consumers wanted. Two significant events changed this situation: (1) invention of the electric light decreased the demand for kerosene, and (2) invention of the internal combustion engine created a demand for diesel fuel and gasoline (naphtha).
- **Thermal Cracking Processes.** With the advent of mass production, the number of gasoline powered vehicles increased dramatically and the demand for gasoline grew. accordingly. However, distillation processes produced only a certain amount of gasoline from crude oil. In 1913, the thermal cracking process was developed, which subjected heavy fuels to both pressure & intense heat, physically breaking the large molecules into smaller ones to produce additional gasoline and distillate fuels. Visbreaking, another form of thermal cracking, was developed in the late 1930's to produce more desirable and valuable products.
- **Catalytic Cracking Processes.** Higher-compression gasoline engines required higher-octane gasoline with better antiknock characteristics. The introduction of catalytic cracking and polymerization processes in the late 1930's met the demand by providing improved gasoline yields & higher octane numbers.
- **Alkylation.**another catalytic process developed in the early 1940's, produced more high octane aviation gasoline and petrochemical feedstock for explosives &synthetic rubber. Subsequently, catalytic isomerization was developed to convert hydrocarbons to produce increased quantities of alkylation feedstock. Improved catalysts and process methods such as hydrocracking and reforming were

developed throughout the 1960's to increase gasoline yields & improve antiknock characteristics. These catalytic processes also produced hydrocarbon molecules with a double bond (alkenes) and formed the basis of the modern petrochemical industry.

- **Treatment Processes.** Throughout the history of refining, various treatment methods have been used to remove non hydrocarbons, impurities, and other constituents that adversely affect the properties of finished products or reduce the efficiency of the conversion processes. Treating involve chemical reaction and/or physical separation. Typical examples of treating are chemical sweetening, acid treating, clay contacting, caustic washing, hydrotreating, drying, solvent extraction, and solvent dewaxing. Sweetening compounds & acids desulfurize crude oil before processing and treat products during & after processing.

Following the Second World War, various reforming processes improved gasoline quality & yield and produced higher-quality products. Some of these involved the use of catalysts and/or hydrogen to change molecules and remove sulfur. A number of the more commonly used treating and reforming processes were developed later on.

Appendix 3.2: Main products of crude oil refining and their end uses

S.N	Products /Hydrocarbon gases	Uses of products
1	Liquefied petroleum gases	Cooking and industrial gas Motor fuel gas Illuminating gas Ammonia Synthetic fertilizer Alcohols Solvents and acetone Plasticizers Resins and fibres for plastics and textiles Paints and varnish
2	Chemical industry feedstock	Rubber products
3	Carbon black	Printing inks Rubber industry
	Light distillates	

S.N	Products /Hydrocarbon gases	Uses of products
4	Light naphthas	Olefins Solvents and diluents Extraction solvents Chemical industry feedstocks
5	Intermediate naphthas	Aviation and motor gasoline Dry-cleaning solvents
6	Heavy naphthas	Military jet fuel Jet fuel and kerosene Tractor fuel
7.	Gas oil	Cracking stock Heating oil and diesel fuel Metallurgical fuel Absorber oil-benzene and gasoline recovery
	Heavy distillates	
8	Technical oils	Textile oils Medicinal oils and cosmetics White oil – food industry
9	Lubricating oils	Transformer and spindle oils Motor and engine oils Machine and compressor oils Turbine and hydraulic oils Transmission oils Equipment and cable insulation oils Axle, gear and steam engine oils Metal treating, cutting and grinding oils Quenching and rust inhibitor oils Hear transfer oils Lubricating greases and compound Printing ink oils
10	Paraffin wax	Rubber industry Pharmaceuticals and cosmetics Food and paper industries Candles and matches

S.N	Products /Hydrocarbon gases	Uses of products
11	**Residues**	
	Petrolatum	Petroleum jelly, Cosmetics, Rust inhibitors and lubricants, Cable coating compounds
12	Residual fuel oil	boiler and process fuel oil
13	Asphalts	Paving asphalt Roofing materials Asphaltic lubricants Insulating and foundation protection Waterproof paper products
14	**Refinery by-products**	
	Coke	Electrodes and fuel
15	Sulphonates	Emulsifiers
16	Sulphuric acid	Synthetic fertilizer
17	Sulphur	Chemicals
18	Hydrogen	Hydrocarbon reformation

Note: A number of chemicals are used in, or formed as a result of, oil refining processing. A brief description of those which are specific and pertinent to refining follows:

- **Sulphur Dioxide:** Flue gas from burning high sulphur-content fuels usually contains high levels of sulphur dioxide, which usually is removed by water scrubbing.
- **Caustics:** Caustics are added to desalting water to neutralize acids and reduce corrosion. Caustics are also added to desalted crude in order to reduce the amount of corrosive chlorides in the tower overheads. They are used in refinery treating processes to remove contaminants from hydrocarbon streams.

Nitrogen oxides and carbon monoxide

Flue gas contains up to 200 ppm of nitric oxide, which reacts slowly with oxygen to form nitrogen dioxide. Nitric oxide is not removed by water scrubbing, and nitrogen dioxide can dissolve in water to form nitrous & nitric acid. Flue gas normally contains only a slight amount of carbon monoxide, unless combustion is abnormal.

Hydrogen sulphide

Hydrogen sulphide is found naturally in most crude oils and is also formed during processing by the decomposition of unstable sulphur compounds. Hydrogen sulphide is an extremely toxic, colourless, flammable gas which is heavier than air and soluble in water. It has a rotten egg odour which is discernible at concentrations well below its very low exposure limit. This smell cannot be relied upon to provide adequate warning as the senses are almost immediately desensitized upon exposure. Special detectors are required to alert workers to the presence of hydrogen sulphide, and proper respiratory protection should be used in the presence of the gas. Exposure to low levels of hydrogen sulphide will cause irritation, dizziness & headaches, while exposure to levels in excess of the prescribed limits will cause nervous system depression and eventually death.

Sour water

Sour water is process water which contains hydrogen sulphide, ammonia, phenols, hydrocarbons and low molecular weight sulphur compounds. Sour water is produced by steam stripping hydrocarbon fractions during distillation, regenerating catalyst, or steam stripping hydrogen sulphide during hydrotreating & hydrofinishing. Sour water is also generated by the addition of water to processes to absorb hydrogen sulphide and ammonia.

Sulphuric acid and hydrofluoric acid

Sulphuric acid and hydrofluoric acid are used as catalysts in alkylation processes. Sulphuric acid is also used in some of the treatment processes.

Solid catalysts

A number of different solid catalysts in many forms & shapes, from pellets to granular beads to dusts, made of various materials and having various compositions, are used in refining processes. Extruded pellet catalysts are used in moving &fixed bed units, while fluid bed processes use fine, spherical particulate catalysts. Catalysts used in processes which remove sulphur are impregnated with cobalt, nickel or molybdenum. Cracking units use acid-function catalysts, such as natural clay, silica, alumina and synthetic zeolites. Acid-function catalysts impregnated with platinum or other noble metals are used in isomerization and reforming. Used catalysts require special handling & protection from exposures, as they may contain metals, aromatic oils, carcinogenic poly cyclic aromatic compounds or other hazardous materials, and may also be pyrophoric.

Fuels

The principal fuel products are liquefied petroleum gas, gasoline, kerosene, jet fuel, diesel fuel, heating oil and residual fuel oils.

Liquefied petroleum gas (LPG), which consists of mixtures of paraffinic & olefinic hydrocarbons such as propane & butane, is produced for use as a fuel, and is stored & handled as liquids under pressure. LPG has boiling points ranging from about – 74°C to +38 °C, is colourless, and the vapours are heavier than air & extremely flammable. The important qualities from an occupational health and safety perspective of LPGs are vapour pressure and control of contaminants.

Gasoline: The most important refinery product is motor gasoline, a blend of relatively low-boiling hydrocarbon fractions, including reformate, alkylate, aliphatic naphtha (light straight-run naphtha), aromatic naphtha (thermal & catalytic cracked naphtha) and additives. Gasoline blending stocks have boiling points which range from ambient temperatures to about 204 °C, and a flash point below –40 °C. The critical qualities for gasoline are octane number (anti-knock), volatility (starting &vapour lock) and vapour pressure (environmental control). Additives are used to enhance gasoline performance and provide protection against oxidation and rust formation. Aviation gasoline is a high-octane product, specially blended to perform well at high altitudes.

Tetra ethyl lead (TEL) and tetra methyl lead (TML) are gasoline additives which improve octane ratings and anti-knock performance. In an effort to reduce lead in automotive exhaust emissions, these additives are no longer in common use, except in aviation gasoline.

Ethyl tertiary butyl ether (ETBE), methyl tertiary butyl ether (MTBE), tertiary amyl methyl ether (TAME) & other oxygenated compounds are used in lieu of TEL &TML to improve unleaded gasoline anti-knock performance and reduce carbon monoxide emissions.

Jet fuel and kerosene: Kerosene is a mixture of paraffins and naphthenes with usually less than 20% aromatics. It has a flashpoint above 38 °C and a boiling range of 160 to 290 °C, and is used for lighting, heating, solvents & blending into diesel fuel. Jet fuel is a middle distillate kerosene product whose critical qualities are freezing point, flashpoint and smoke point. Commercial jet fuel has a boiling range of about 190 to 275 °C, and military jet fuel from 55 to 290 °C.

Distillate fuels: Diesel fuels & domestic heating oils are light coloured mixtures of paraffins, naphthenes & aromatics, and may contain moderate quantities of olefins.Distillate fuels have flash points above 60 °C and boiling ranges of about 160 to 370 °C, and are often hydrodesulphurized for improved stability. Distillate fuels are combustible and when heated may emit vapours which can form ignitable mixtures with air. The desirable qualities required for distillate fuels include controlled flash point & pourpoints, clean burning, no deposit formation in storage tanks, and a proper diesel fuel cetane rating for good starting & combustion.

Residual fuels: Many ships and commercial & industrial facilities use residual fuels or combinations of residual & distillate fuels for power, heat and processing. Residual fuels are dark coloured, highly viscous liquid mixtures of large hydrocarbon molecules, with flash points above 121 °C and high boiling points. The critical specifications for residual fuels are viscosity and low sulphur content (for environmental control).

Petrochemical feedstocks

Many products derived from crude-oil refining, such as ethylene, propylene& butadiene, are olefinic hydrocarbons derived from refinery cracking processes, and are intended for use in the petrochemical industry as feedstocks for the production of plastics, ammonia, synthetic rubber, glycol and so on.

Petroleum solvents

A variety of pure compounds, including benzene, toluene, xylene, hexane & heptane, whose boiling points & hydrocarbon composition are closely controlled, are produced for use as solvents. Solvents may be classified as aromatic or non aromatic, depending on their composition. Their use as paint thinners, dry-cleaning fluids, degreasers, industrial& pesticide solvents and so on, is generally determined by their flash points, which vary from well below -18 °C to above 60 °C.

The hazards associated with solvents are similar to those of fuels in that the lower flash point solvents are flammable and their vapours, when mixed with air in the flammable range, are ignitable. Aromatic solvents will usually have more toxicity than non aromatic solvents.

Process oils

Process oils include the high boiling range, straight run atmospheric or vacuum distillate streams and those which are produced by catalytic or thermal cracking. These complex mixtures, which contain large paraffinic, naphthenic & aromatic hydrocarbon molecules with more than 15 carbon atoms, are used as feedstocks for cracking or lubricant manufacturing. Process oils have fairly high viscosities, boiling points ranging from 260 to 540 °C and flash points above 121 °C.

Process oils are irritating to the skin and contain high concentrations of PAHs as well as sulphur, nitrogen &oxygen compounds. Inhalation of vapours& mists should be avoided, and skin exposure should be controlled by the use of personal protective equipments (PPEs) & good hygienic practices.

Lubricants & greases

Lubricating oil base stocks are produced by special refining processes to meet specific consumer requirements. Lubricating base stocks are light to medium coloured, low volatile, medium to high viscous mixtures of paraffinic, naphthenic & aromatic oils, with boiling ranges from 370 to 540 °C. Additives, such as demulsifiers, antioxidants & viscosity improvers, are blended into the lubricating oil base stocks to provide the characteristics required for motor oils, turbine & hydraulic oils, industrial greases, lubricants, gear oils and cutting oils. The most critical quality for lubricating oil base stock is a high viscosity index, providing for less change in viscosity under varying temperatures. This characteristic may be present in the crude oil feed stock or attained through the use of viscosity index improver additives. Detergents are added to keep in suspension any sludge formed during the use of the oil.

Greases are mixtures of lubricating oils and metallic soaps, with the addition of special purpose materials such as asbestos, graphite, molybdenum, silicones and talc to provide insulation or lubricity. Cutting and metal process oils are lubricating oils with special additives such as chlorine, sulphur & fatty acid additives which react under heat to provide lubrication and protection to the cutting tools. Emulsifiers and bacteria prevention agents are added to water soluble cutting oils.

Although lubricating oils by themselves are non irritating and have little toxicity, hazards may be presented by the additives. Users should consult supplier material safety data information to determine the hazards of specific additives, lubricants, cutting oils and greases. The primary lubricant hazard is dermatitis, which can usually be controlled by the use of personal protective equipment together with proper hygienic practices. Occasionally workers may develop a sensitivity to cutting oils or lubricants which will require reassignment to a job where contact cannot occur. There are some concerns about carcinogenic exposure to mists from naphthenic based cutting & light spindle oils, which can be controlled by substitution, engineering controls or personal protection. The hazards of exposure to grease are similar to those of lubricating oil, with the addition of any hazards presented by the grease materials or additives.

Special products

Wax is used for protecting food products; in coatings; as an ingredient in other products such as cosmetics & shoe polish and for candles.

Sulphur is produced as a result of petroleum refining. It is stored either as a heated, molten liquid in closed tanks or as a solid in containers or outdoors.

Coke is almost pure carbon, with a variety of uses from electrodes to charcoal briquettes, depending on its physical, characteristics, which result from the coking process.

Asphalt, which is primarily used for paving roads & roofing materials, should be inert to most chemicals and weather conditions.

Waxes and asphalts are solid at ambient temperatures, and higher temperatures are needed for storage, handling & transportation, with the resulting hazard of bums. Petroleum wax is so highly refined that it usually does not present any hazards. Skin contact with wax can lead to plugging of pores, which can be controlled by proper hygienic practices. Exposure to hydrogen sulphide when asphalt & molten sulphur tanks are opened can be controlled by the use of appropriate engineering controls or respiratory protection. Sulphur is also readily ignitable at elevated temperatures.

Appendix 3.3: All about distillation and distillation columns used in petroleum refinery

Distillation

Distillation is a process in which a liquid or vapor mixture of two or more substances is separated into its component fractions of desired purity, by the application & removal of heat. Distillation is based on the fact that the vapor of a boiling mixture will be richer in the components that have lower boiling points. Thus, when this vapor is cooled & condensed, the condensate will contain the more volatile components. At the same time, the original mixture will contain more of the less volatile components.

Distillation is the most common separation technique and it consumes enormous amounts of energy, both in terms of cooling & heating requirements. Distillation contributes more than 50% of refinery operating costs.

Distillation Columns

Distillation columns are classified by the manner in which they are operated:

***Batch*Distillation** in which the feed to the column is introduced batch wise. The column is charged with a 'batch' and then the distillation process is carried out. When the desired task is achieved, a next batch of feed is introduced.

Continuous distillation columns process-a continuous feed stream.No interruptions occur unless there is a problem with the column or surrounding process units. They are capable of handling high throughputs and are the most common of the two types.

Continuous Distillation Columns are Classified according to:

(a) Nature of the feed that they are processing:

- **binary column** - feed contains only two components;
- **multi-component column** - feed contains more than two components.

(b) Number of product streams they have:

- **multi-product column** - column has more than two product streams.

(c) Where extra feed exits when used to help with the separation:

- **extractive distillation** - where the extra feed appears in the bottom product stream;
- **azeotropic distillation** - where the extra feed appears at the top product stream.

(d) Type of column internals:

- **tray column** - trays of various designs used to hold up the liquid to provide better contact between vapour and liquid;
- **packed column** - packings are used to enhance vapour-liquid contact.

Main Components of Distillation Columns

- A vertical shell where separation of liquid components is done.
- Column internals e.g. trays/plates and/orpackings which are used to enhance component separations.
- A reboiler to provide the necessary vaporization for the distillation process.

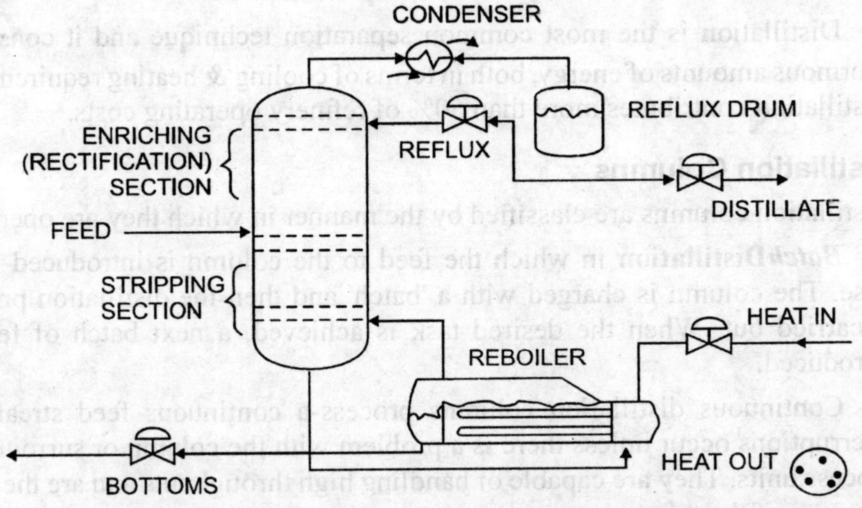

- A condenser to cool and condense the vapour leaving the top of the column.
- A reflux drum to hold the condensed vapour from the top of the column so that liquid (reflux) can be recycled back to the column.

Trays and Plates of Distillation Columns

Bubble cap trays: A riser or chimney is fitted over each hole, and a cap covers the riser. The cap is mounted with a space to allow vapor to rise through the chimney and be directed downward by the cap, finally discharging through slots in the cap, and bubbling through the liquid on the tray.

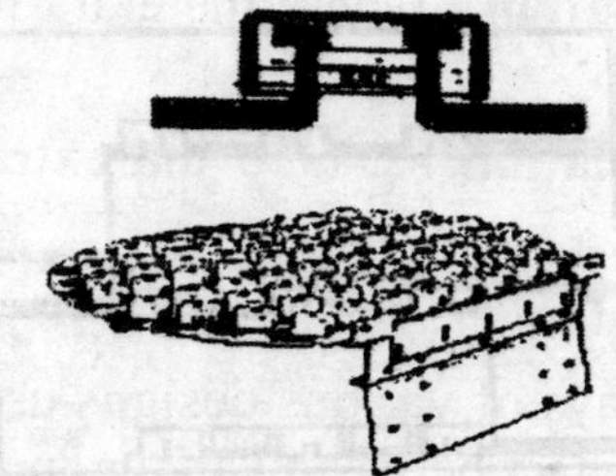

- **Valve trays**: Perforations are covered by caps lifted by vapour, which creates a flow area and directs the vapour horizontally into the liquid.
- **Sieve trays**: Sieve trays are simply metal plates with holes in them. Vapour passes straight upward through the liquid on the plate. The arrangement, number and size of the holes are design parameters.

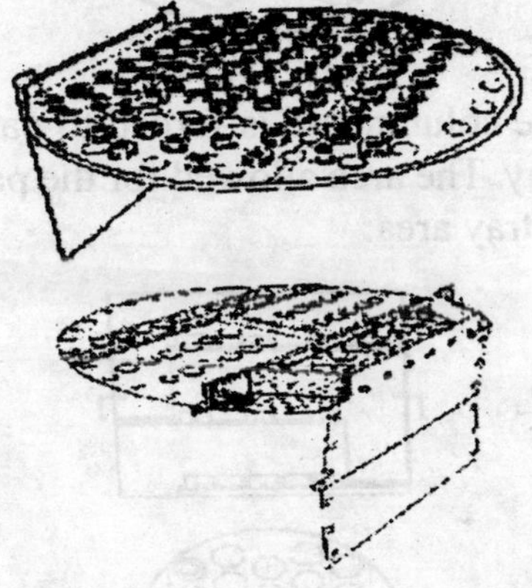

Liquid and Vapour Flows in a Tray Column is Shown Below in Figure.

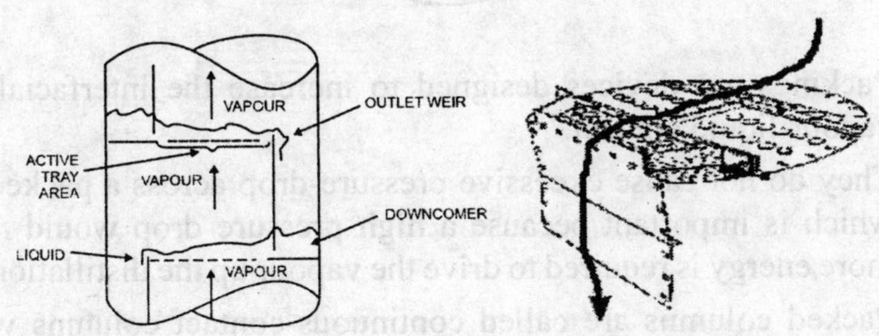

- Each tray has 2 conduits called downcomers: one on each side. Liquid falls by gravity through the downcomers from one tray to the tray below.
- A weir ensures there is always some liquid (holdup) on the tray and is designed such that the holdup is at a suitable height, e.g. such that the bubble caps are covered by liquid.

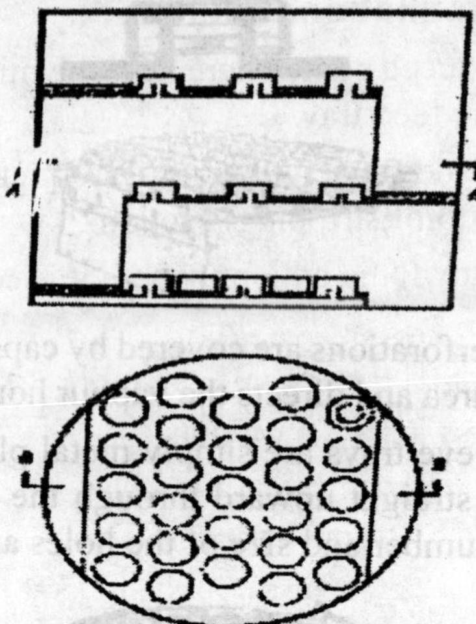

Vapour flows up the column and is forced to pass through the liquid via the openings on each tray. The area allowed for the passage of vapour on each tray is called the active tray area.

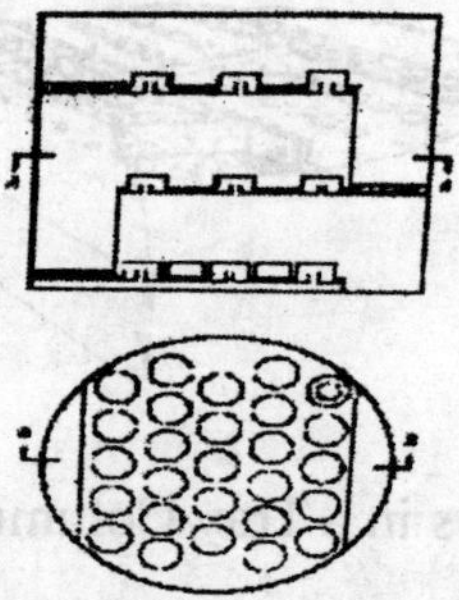

- Packings are devices designed to increase the interfacial area for vapour-liquid contact.
- They do not cause excessive pressure-drop across a packed section, which is important because a high pressure drop would mean that more energy is required to drive the vapour up the distillation column.
- Packed columns are called continuous-contact columns while tray columns are called staged contact columns because of the manner in which vapour and liquid are contacted.

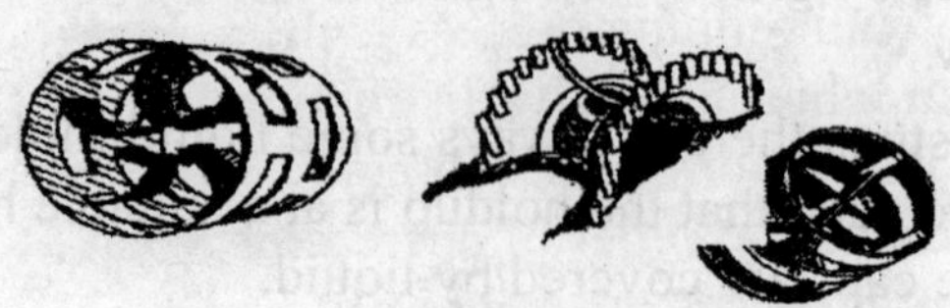

Basic Operation of Distillation Column

- The feed is introduced somewhere near the middle of the column to a tray known as the feed tray.
- The feed tray divides the column into a top (enriching or rectification section and a bottom (stripping) section.
- The feed flows down the column where it is collected in the reboiler.

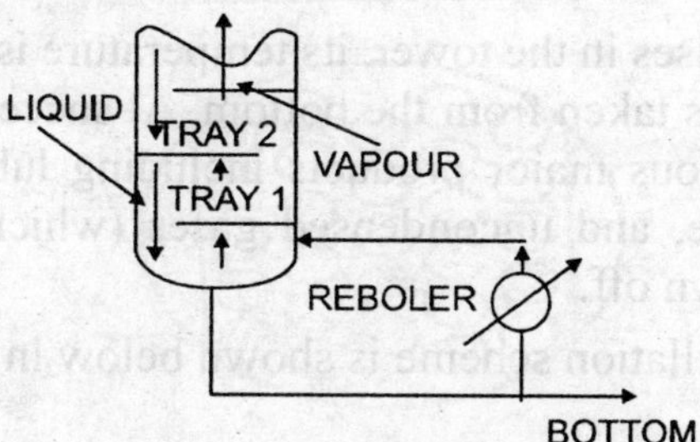

- Heat (usually as steam) is supplied to the reboiler to generate vapour.
- The vapour from the reboiler is reintroduced into the unit at the bottom of the column.
- The liquid removed from the reboiler is known as the bottoms product or simply, bottoms.
- Vapor moves up the column, exits the top, and is cooled in a condenser. The condensed liquid is stored in a holding vessel known as the reflux drum. Some of this liquid is recycled back to the top of the column and this is called the reflux. The condensed liquid that is removed from the system is known as the distillate or top product.
- Thus, there are internal flows of vapour & liquid within the column as well as external flows of feeds & product streams, into & out of the column.

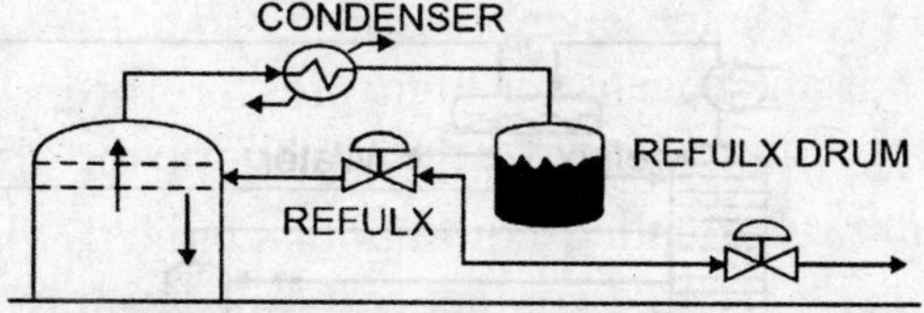

Crude Oil Distillation

First Step in the petroleum refining process is the separation of crude oil into various fractions or straight run cuts by distillation in atmospheric & vacuum towers. The main fractions or "cuts" obtained have specific boiling-point ranges and can be classified in order of decreasing volatility into gases, light distillates, middle distillates, gas oils, and residuum.

Atmospheric distillation

The desalted crude feedstock is preheated using recovered process heat. The feedstock then flows to a direct-fired crude charge heater then into the vertical distillation column just above the bottom, at pressures slightly above atmospheric and at temperatures ranging from 340-370°C (above these temperatures undesirable thermal cracking may occur). All but the heaviest fractions flash into vapor.

As the hot vapor rises in the tower, its temperature is reduced. Heavy fuel oil or asphalt residue is taken from the bottom. At successively higher points on the tower, the various major products including lubricating oil, heating oil, kerosene, gasoline, and uncondensed gases (which condense at lower temperatures) are drawn off.

Atmospheric Distillation scheme is shown below in figure

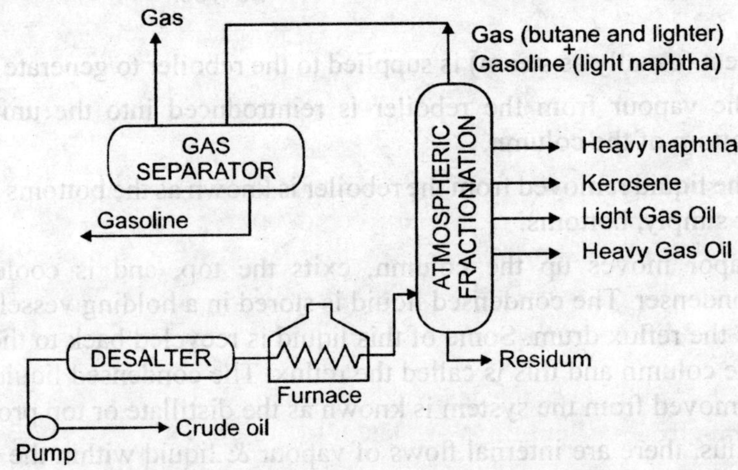

Fig. Atmospheric Distillation Scheme

Simple Crude Oil Distillation Scheme is shown below in figure.

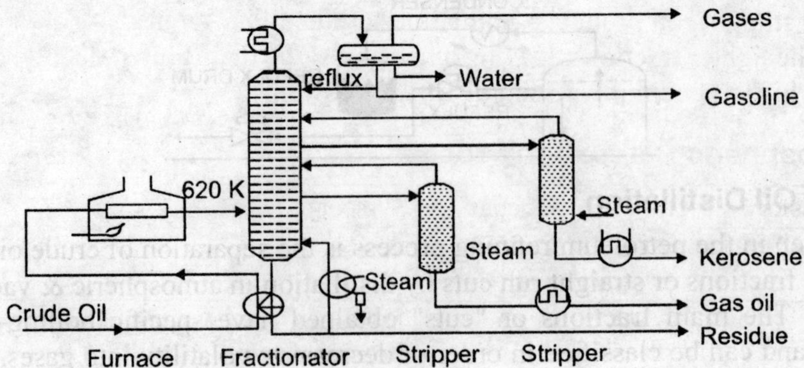

Note: The first step in the petroleum refining process is the separation of crude oil into various fractions or straight run cuts by distillation in atmospheric & vacuum towers. The main fractions or "cuts" obtained have specific boiling point ranges and can be classified in order of decreasing volatility into gases, light distillates, middle distillates, gas oils, & residuum.

Atmospheric Distillation Tower

At the refinery, the desalted crude feedstock is preheated using recovered process heat. The feedstock then flows to a direct fired crude charge heater where it is fed into the vertical distillation column just above the bottom, at pressures slightly above atmospheric and at temperatures ranging from340 to 370°C (heating crude oil above these temperatures may cause undesirable thermal cracking). All but the heaviest fractions flash into vapor. As the hot vapor rises in the tower, its temperature is reduced. Heavy fuel oil or asphalt residue is taken from the bottom. At successively higher points on the tower, the various major products including lubricating oil, heating oil, kerosene, gasoline, and uncondensed gases (which condense at lower temperatures) are drawn off.

The fractionating tower, a steel cylinder about 40 metres high, contains horizontal steel trays for separating and collecting the liquids. At each tray, vapors from below enter perforations and bubble caps. They permit the vapors to bubble through the liquid on the tray, causing some condensation at the temperature of that tray. An overflow pipe drains the condensed liquids from each tray back to the tray below, where the higher temperature causes re-evaporation. The evaporation, condensing, and scrubbing operation is repeated many times until the desired degree of product purity is reached. Then side streams from certain trays are taken off to obtain the desired fractions. Products ranging from uncondensed fixed gases at the top to heavy fuel oils at the bottom can be taken continuously from a fractionating tower. Steam is often used in towers to lower the vapor pressure and create a partial vacuum. The distillation process separates the major constituents of crude oil into so called straight run products. Sometimes crude oil is "topped" by distilling off only the lighter fractions, leaving a heavy residue that is often distilled further under high vacuum in vaccum distillation tower.

Atmospheric Distillation Process

Feedstock	From	Process	Typical products............To
Crude	Desalting	Separation	Gases.............................Atmospheric distillation tower
			Naphthas.....................Reforming or treating

		Kerosene or distillates...Treating
		Gas oil..........................Catalytic cracking
		Residual.....................Vacuum tower or visbreaker

Atmospheric Distillation

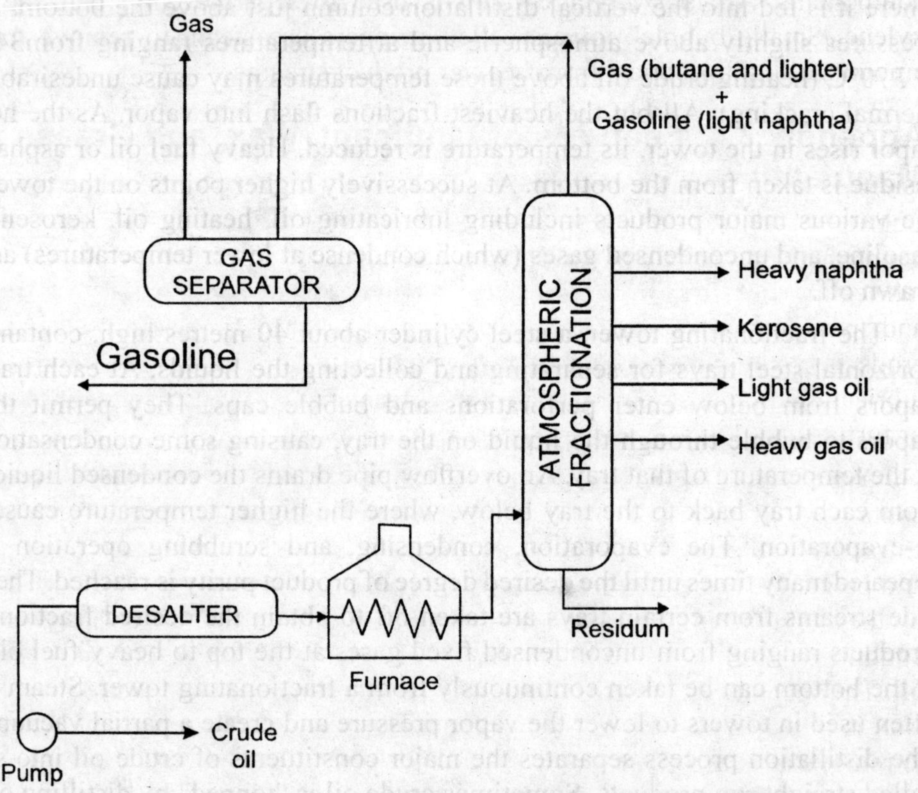

Vacuum Distillation Tower: To further distill the residuum or topped crude from the atmospheric tower at higher temperatures, reduced pressure is required to prevent thermal cracking. The principles of vacuum distillation resemble those of atmospheric fractional distillation and, except that larger diameter columns are used to maintain comparable vapor velocities at the reduced pressures, the equipment is also similar. The internal designs of some vacuum towers are different from atmospheric towers in that random packing and demister pads are used instead of trays. A typical first phase vacuum tower may produce gas oils, lubricating-oil base stocks, and heavy residual for propane deasphalting. A second-phase tower operating at lower vacuum may distill surplus residuum from the atmospheric tower, which is not used for lube-stock processing, and surplus residuum from the first vacuum tower not used for deasphalting. Vacuum towers are typically used to separate catalytic cracking feedstock from surplus residuum.

Other Distillation Towers (Columns): Within refineries there are numerous other, smaller distillation towers called columns, designed to separate specific and unique products. All work on the same principles as the towers described above. For example, a depropanizer is a small column designed to separate propane & lighter gases from butane and heavier components. Another larger column is used to separate ethyl benzene & xylene. Small "bubble" towers called strippers use steam to remove trace amounts of light products from heavier product streams.

Appendix 3.4: Features of Lubricant, Wax, and Grease Manufacturing Processes

Lubricating oils and waxes are refined from the residual fractions of atmospheric & vacuum distillation. The primary objective of the various lubricating oil refinery processes is to remove asphalts, sulfonated aromatics, and paraffinic &isoparaffinic waxes from residual fractions. Reduced crude from the vacuum unit is deasphalted and combined with straight run lubricating oil feedstock, preheated, & solvent extracted (usually with phenol or furfural) to produce raffinate.

Wax Manufacturing Process: Raffinate from the extraction unit contains a considerable amount of wax that must be removed by solvent extraction and crystallization. The raffinate is mixed with a solvent (propane) and precooled in heat exchangers. The crystallization temperature is attained by the evaporation of propane in the chiller and filter feed tanks. The wax is continuously removed by filters and cold solvent washed to recover retained oil. The solvent is recovered from the oil by flashing and steam stripping. The wax is then heated with hot solvent, chilled, filtered, and given a final wash to remove all oil.

Lubricating Oil Manufaturing Process: The dewaxed raffinate is blended with other distillate fractions and further treated for viscosity index, color, stability, carbon residue, sulfur, additive response, and oxidation stability in extremely selective extraction processes using solvents (furfural, phenol, etc.). In a typical phenol unit, the raffinate is mixed with phenol in the treating section at temperatures below 205° F. Phenol is then separated from the treated oil and recycled. The treated lube-oil base stocks are then mixed and/or compounded with additives to meet the required physical and chemical characteristics of motor oils, industrial lubricants, & metal working oils.

Grease Manufaturing Process: Grease is made by blending metallic soaps (salts of long-chained fatty acids) and additives into a lubricating oil medium at temperatures of 205–315°C. Grease may be either batch produced or continuously compounded. The characteristics of the grease depend to a great extent on the metallic element (calcium, sodium, aluminum, lithium, etc.) in the soap and the additives used.

4

Thermal and Catalytic Cracking

Cracking

Cracking means heating of higher boiling petroleum fractions like heavy fuel oil at high temperature (above decomposition temperature) and pressure to produce lower boiling lighter fractions. It is an endothermic reaction.

The main application of cracking is for production of gasoline from gas oil. It is also done to produce olefins (for petrochemicals production) from gas oil & naphtha, lower viscosity furnace oil (visbreaking) and coke (coking). Important chemical reactions in thermal cracking (non catalytic) which is carried out at comparatively higher pressure (1-70 atm.) & temperature (450-750°C) by heating the feed are decomposition, dehydrogenation, isomerisation and polymerisation. First, the higher paraffin (say *n*-decane) decomposes to a lower paraffin and an olefin.

$$\underset{\substack{n-\text{decane}\\ \text{(gas oil component)}}}{CH_3\,(CH_2)_8CH_3} \xrightarrow[\text{T and P}]{\text{cracking}} \underset{\substack{n-\text{hexane}\\ \text{(gasoline component)}}}{CH_3(CH_2)_4CH_3} + \underset{\text{butene}}{CH_3CH_2CH = CH_2}$$

The olefins then undergo isomerisation, dehydrogenation and polymerisation reaction.

$$\text{e.g.,}\quad \underset{\text{butene}}{2CH_3CH_2 - CH = CH_2} \xrightarrow{\text{polymerisation}} \underset{\substack{\text{iso-octene}\\ \text{(gasoline component)}}}{CH_3 - C - CH_3 - CH_2 - CH_3 - \overset{\overset{CH_3}{|}}{C} = CH_2}$$

Gases & solids (coke) are also produced during cracking apart from liquid products. Catalytic cracking compared to the thermal (non catalytic) cracking produces less coke, less gas but more liquid products. Thermal cracking is now mainly used for production of olefins, coking and visbreaking. Due to presence of higher quantity of iso-paraffins and aromatics gasoline produced by catalytic cracking is far more superior.

The catalyst used in cracking are in the form of powder, beads or pellets and include silica, alumina, natural clay, zeolites etc. Use of catalyst minimises the formation of coke during cracking. Catalyst need to be regenerated by burning the deposited coke particles in air as it deactivates the catalyst. Pellets and beads of catalyst are used as fixed bed in a cyclic process(*e.g.* Houdry process) or as a moving bed in a continuous process *(e.g.* TCC process). The powder catalyst (5-500 micron in size) is used either in suspension process or as a fluid bed with the vaporised feed as the fluidising medium(*e.g.* FCC process). Heat requirement of endothermic cracking reaction is met by burning of deposited coke in the catalyst regenerator. In fixed bed process, catalyst is regenerated in the reactor itself. Catalyst movement from reactor to regenerator is more easily achieved in the fluidised cracking process. The demerits of fixed bed processes include variation of temperature & activity of the catalyst along the bed height and with the progress of the cycle causing non uniformity of the yield products quality. While the merits of fluidised bed processes include high heat transfer rate, isothermal reactor conditions and high reaction rate (due to high mass transfer rates between the catalyst surface and gas phase).

Necessity for cracking. An average crude oil on distillation yields 20-25% petrol, 30-45% intermediate oils (kerosene, naphtha, diesel, lube oil etc.) and 25-50% residual fuel oils. Increasing demand for high quality (octane number) petrol exceed the availability of petrol by distillation alone. Hence, cracking of heavier fuel oils (obtained by distillation of crude) is done to produce mainly petrol. Cracking is also done to reduce the viscosity of fuel oil (by visbreaking), to produce coke (by coking) and to produce olefins which is feed for petrochemical industry.

Types of Cracking. There are two types of cracking:thermal cracking and catalytic cracking. When cracking *(i.e.* breaking larger molecules into smaller ones by heating) is done without any catalyst (which increases the rate of reaction), then it is called thermal cracking. Cracking in presence of a catalyst is called catalytic cracking.

Thermal Cracking. Here in this process, the fuel which may be gas oil, fuel oil (heavy or light), atmospheric or vacuum residue is heated upto 450-750°C at pressure ranging from 1-70 atmosphere to produce gas, petrol, diesel cracked residue (coke) etc. Octane number of petrol obtained may be upto 75. The yields & quality of products will depend upon the type of feed, temperature and pressure.

Catalytic Cracking. Here the feed (which may be kerosene, fuel oil, gas oil, lubricating oil; atmospheric, vacuum & cracked residue etc.) is heated in presence of a catalyst (like, platinum, nickel, iron, chromium, silica-alumina, aluminium chloride, sulphuric acid, phosphoric acid etc.) at 350-650°C and 1-15 atm to produce gas, petrol, diesel and residue (coke). Higher temperature gives higher yield of gasoline, whereas higher pressure reduces the octane number of gasoline.

A typical *yield* pattern in cracking is given below:

Feed- residue of vacuum distillation of crude Yields - petrol = 55%, diesel = 20%, fuel oil = 15%, coke = 10%, Octane number of petrol = 92

Advantages of Catalytic Cracking over Thermal Cracking

- Pressure and temperature required for catalytic cracking is lower.
- Yield and octane number of petrol produced is higher.
- Petrol produced has less quantity of gum and gum forming material.
- Sulphur content of the products is low.
- Lead susceptibility of petrol produced is more.

Thermal cracking processes. Depending upon the pressure & temperature employed for the cracking and the characteristics of feed, there are various thermal cracking processes in which the product yields & characteristics are different. Salient features of some of them are given below:

Low temperature and high pressure thermal cracking process. The feed is residue from atmospheric distillation unit. Products are mainly gas, fuel oil and gasoline. Temperature & pressure employed are 500°C & 20 atm. respectively. This process is also called visbreaking when heavy fuel oil is thermally cracked to reduce its viscosity so that it can be properly atomised through the burners. This process is employed for heavy oil cracking.

High temperature and high pressure thermal cracking process

Feed - Gas oil, *Products* - gas and gasoline mainly, *T & P* - 530°C, & 50-70 atm.

This is also called light oil thermal cracking process. When feed is naphtha, it is called *thermal reforming.*

Low pressure and high temperature thermal cracking process:

- **Feed:** residue from atmospheric distillation unit,
- **T & P:** >550°C & 2-5 atm.
- **Products**: gasoline and gas rich in unsaturated hydrocarbons (a feed for petrochemicals).

High temperature and low pressure thermal cracking process. This is also called pyrolysis. Pressure & temperature employed is 1 atm. & 700°C. Product is mainly gas rich in aromatic and unsaturated hydrocarbons.

Coking. Feed is vacuum residue. Pressure & temperature are 2 atm. & 400-450°C respectively.

The products are gas, coker gasoline, coke for electrode manufacture (main product) and coke distillate.

Thermal Cracking Reactions

Few important thermal cracking reactions are:

- Decomposition and destructive condensation.
- Hydrogenation and dehydrogenation.
- Polymerisation.
- Cyclisation.

Octane number increases in the order paraffin → Olefin → Naphthene → iso-paraffin → aromatic. That means aromatics have the highest octane number and paraffins have the lowest. Hence, cracking reactions should aim for producing hydrocarbons with higher octane number, if production of gasoline is the main product.

Decomposition and destructive condensation of olefins to produce high octane number aromatics is shown below:

$$\underset{\text{(olefin)}}{\underset{\text{ethylene}}{2C_2H_4}} \xrightarrow{\text{decomposition}} \underset{\text{Butadiene}}{C_4H_6} + H_2$$

$$C_4H_6 + C_2H_4 \xrightarrow{\text{condensation}} \underset{\text{Benzene (aromatic)}}{C_6H_6 + H_2}$$

Hydrogenation(*i.e.* hydrocracking) of higher boiling paraffin to lower boiling paraffin is shown below:

$$CH_3(CH_2)_6CH_3 + H_2 \xrightarrow{\text{hydrogenation}} CH_3CH_2CH_3 + CH_3(CH_2)_3CH_3$$

Octane (Higher boiling paraffind) → Propane, Pentane (Lower boiling paraffins)

Since, lower boiling paraffins have higher octane number, hydrogenation (*i.e.* hydrocracking) improves octane rating *i. e.*, quality of gasoline (petrol).

Dehydrogenation of naphthenes to aromatics also helps in increasing the octane number of cracking products.

$$C_6H_{12} \xrightarrow{\text{dehydrogenation}} C_6H_6 + 3H_2$$

cyclo-hexane (naphthene) → Benzene (aromatic)

Polymerisation. This reaction Converts C_3 and C_4 Olefins to higher diolefins *e.g.*

$$2\,(CH_3)_2C=CH_2 \xrightarrow{\text{Polymerisation}} CH_3-C(CH_3)_2-Ch_2-C(CH_3)=Ch_2$$

iso -butylene → di-isobutylene

Cyclisation. This reaction converts /i-paraffins to aromatics which have higher octane number. This reaction is also called dehydrocyclisation because hydrogen is a co-product.

$$C_7H_{16} \xrightarrow{\text{cyclisation}} C_6H_5CH_3 + 4H_2$$

n–heptane (*n*–paraffin) → toluene (aromatic)

Gas-Oil Cracking. For the cracking of gas oil, the reaction pattern is shown below. Paraffins crack to olefins & smaller paraffins in the gas, gasoline, and gas-oil boiling ranges. Side chains of alkyl naphthenes & alkyl aromatics crack to yield olefins plus naphthenes and aromatics with shorter side chains.

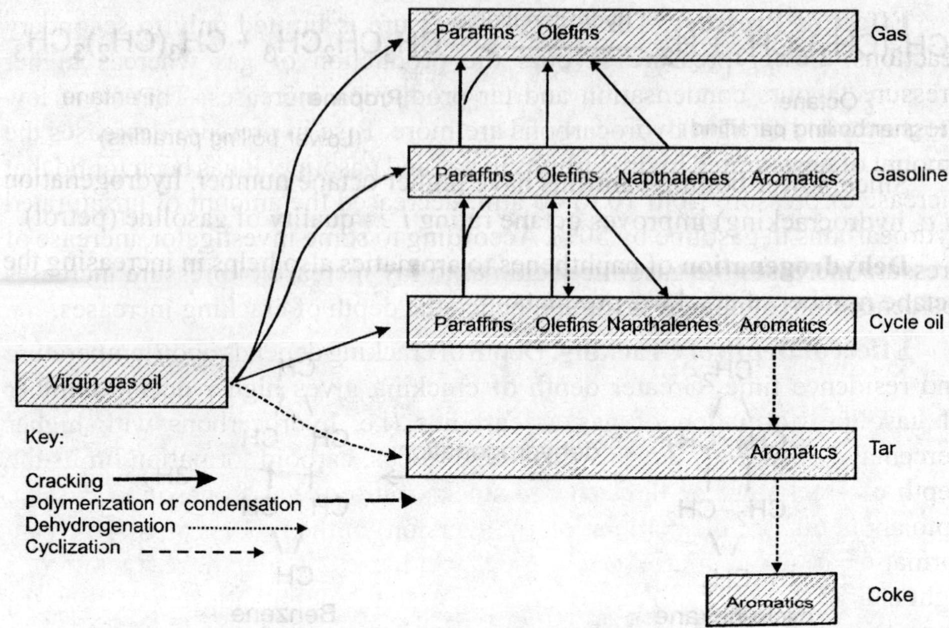

Fig. Reactions in thermal cracking of gas oil.

Aromatics condense to large molecules. All the primary products except the aromatics crack further as the reaction proceeds. Also, olefins polymerize to larger molecules and cyclize to naphthenes, including condensed ring naphthenes. Naphthenes dehydrogenate to aromatics, and condensed ring aromatics condense to larger molecules and eventually coke. The *overall effects* may be *summarized* as follows:

- Cracking of paraffins, alkyl naphthenes, & alkyl aromatics to gas, gasoline, and cracked gas oil.
- Cracking of gasoline to gas.
- Reaction of olefins to produce larger molecules, including aromatics heavy ends ("tar") and eventually coke.
- Further cracking of cracked gas oil to produce gasoline and gas.

Influence of Various Parameters in Cracking

Effect of Temperature. Higher temperature decreases the residence time of charge stock. Small change in temperature can change the depth of cracking to a considerable extent, hence, it is necessary to obtain the temperature very accurately. Between 450 to 550°C there is no change in product quality at the same depth of cracking.

Effect of Residence Time. Residence time has a great effect on quality of products as increased time leads to more severity in cracking.

Effect of Pressure. The effect of pressure is limited only to secondary reactions. Lower pressure favours the production of gas whereas higher pressure favours condensation and tar production increases. That is, at low pressure yield of light hydrocarbons are more. Rise in pressure decreases the amount of unsaturated hydrocarbons in gas and gasoline. It has been found that increase of pressure from 10 to 40 atm. decreased the amount of unsaturated hydrocarbons in gasoline by 50%. According to some investigator, increase of pressure increases the gasoline yield *i.e.* to say increase of pressure increases residence time of charge stock and hence the depth of cracking increases.

Effect of Depth of Cracking. Depth of cracking depends upon temperature and residence time. Greater depth of cracking gives higher octane number of gasoline, formation of gas & carboids (i.e. hydrocarbons with higher percentage of carbon). For heavy charge stocks, carboid formation limits the depth of cracking. For light charge stocks, ratio of gas to gasoline is same upto about 60% of the completion (conversion) of the process after which the formation of gas is more. This critical yield fixes the depth of cracking for light charge stock. The depth of cracking is so fixed that the yield of carboids in no case exceeds 0.1%.

Chemical Composition of Feed. This parameter effects the process only in the beginning. The greater the depth of cracking, the less the influence of chemical composition. But composition of feed effects the yield of products. More paraffins in feed stock gives higher yield of gasoline whereas naphthenic feed gives less quantity of gasoline.

Commercial Thermal Cracking Processes

Mainly there are four commercial processes employed for thermal cracking in oil refineries. They are:

- Dubbs thermal cracking process (old process followed in old refineries).
- Pyrolysis
- Visbreaking
- Coking.

Dubbs Thermal Cracking Process

It is an old thermal cracking process used in oil refineries upto about 1939 on gas oil and having fuel oil stocks.Heavy oil or topped crude (residue of atmospheric distillation of crude oil) is pumped into a bubble cap fractionator where it is mixed with heavy cracked oil from the process. The mixed feed at pressure upto 70 kg/cm^2 is passed through a tube furnace where it is heated maximum upto 540°C before passing to the soaking chamber. This permits a time dwell during which a controlled amount of cracking can take place.

Pressure is reduced before passing into a flash chamber where heavy cracked fuel oil separates and then to the bubble cap fractionator (distillation column) from which the cracked gas and gasoline passes out at the top and furnace distillate from side.

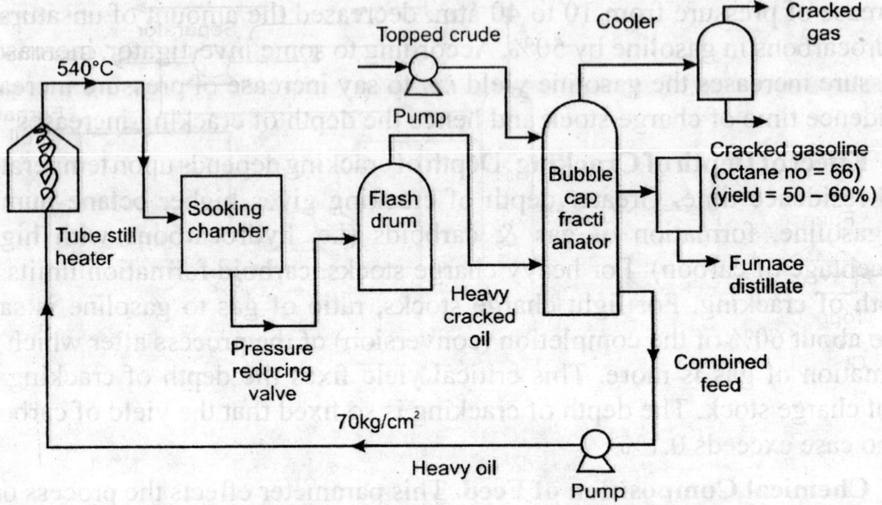

Fig. Flow sheet of Dubbs thermal cracking process

Flow sheet of Dubbs thermal cracking is given above in Figure.

Pyrolysis. Pyrolysis (mild thermal cracking) is done mainly for the production of lighter products mainly unsaturates like olefin (ethylene) & naphthene polymers, diolefins, benzenes and toluenes etc. It is carried out at high temperature (650-700°C) and low pressure. Initially this process was used for the production of toluene. The feed to pyrolysis is raw crude oil, kerosene and natural gasoline. A higher yield of gas, benzene and toluene are obtained by this process.

The feed is heated in special furnaces. The products of the pyrolysis are cooled and delivered to the fractionating column. The pyrolysis gas is drawn from the top of the fractionator. The aromatics (benzene and toluene) are taken from middle section of the column. The residue (aromatic tar, carbon etc.) which is resinous in nature is discharged from its bottom.

The pyrolysis process is used mainly for producing ethylene (by pyrolysis of propane- propylene & low octane gasoline fractions), benzene and toluene. Optimum temperature in pyrolysis is 650 to 700°C respectively for benzene and toluene to be the principal product.

Visbreaking. It literally means viscosity breaking (*i.e.* reduction). It is done to reduce the viscosity and pour point of heavy fuel oil without coke formation. Feed is residue from atmospheric distillation unit *i.e.* topped crude charge (having gravity ÅPI = 20.3 and Conradson carbon from 10.5 to 13%). The main product is fuel oil of lower viscosity though gasoline and gas oil are also produced depending upon the intensity of operation.

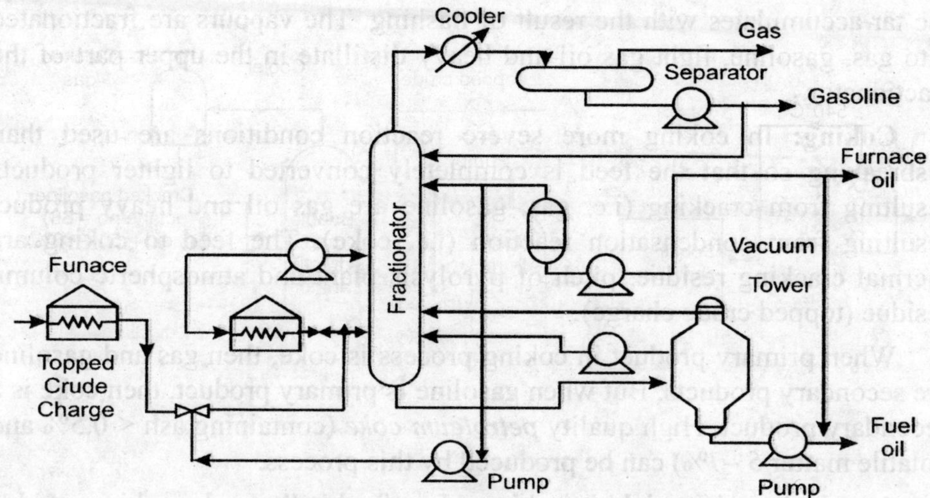

Fig. Visbreaking of heavy fuel oil.

Lower temperature is employed in visbreaking than in thermal cracking because residual stocks have greater coking tendency than gas oils. Typical reaction condition in visbreaking is 510°C and 17 kg/cm^2. Typical *product yield* is (volume basis) gasoline = 7%, gas oil = 12.4% and fuel oil = 80.6%.

To minimise coking during visbreaking, rapid & even heating of feed and high vapour velocity is desirable. To increase the vapour velocity, steam/water is injected in the feed. Coking tendency is reduced if vacuum distillation is employed to produce very heavy charge stock.

High viscosity heavy fuel oil is also subjected to visbreaking to reduce its viscosity so that it can be pumped economically (at lower pressure drop in pipeline) and atomised smoothly in burners.

The ***visbreaking process*** is described below and is illustrated by a flow-sheet above in Figure.

The heavy gas oil is withdrawn from the tower and charged to heavy gas oil cracking furnace. The effluent material is quenched and fed to the evaporators which is flashed with the admixture of visbreaking furnace effluent. The tower bottoms are flashed in vacuum and distillate material is returned to the fractionator where it aids in making up the recycle charge to the heavy gas oil cracking furnace. The vacuum oil is blended with light gas oil to meet the desired specifications, having a net production of furnace oil.

The topped crude charge is pumped to a visbreaking furnace where it is heated and thermally cracked slightly. The effluent product from visbreaking furnace is quenched with light gas oil. Now the quenched product is fed to the evaporator section of fractionator where it is flashed. In the base of the tower the tar accumulates with the result of flashing. The vapours are fractionated into gas, gasoline, light gas oil and heavy distillate in the upper part of the fractionator.

Coking: In coking more severe reaction conditions are used than visbreaking so that the feed is completely converted to lighter products resulting from cracking (i.e. gas, gasoline are gas oil and heavy product resulting from condensation reaction (i.e. coke). The feed to coking are thermal cracking residue, pitch of pyrolysis plant and atmospheric column residue (topped crude charge).

When primary product in coking process is coke, then gas and gasoline are secondary products. But when gasoline is primary product, then coke is a secondary product. High quality *petroleum coke* (containing ash < 0.5% and volatile matter 5 –7%) can be produced by this process.

A typical product yield in coking when feed is thermal cracking residue is given below:

Products	Yield, weight %
Gas	13
Gasoline	7
Gas oil	49
Coke	31
Octane number of gasoline	65

Gasoline production goes upto 16%, if feed is atmospheric column residue and the gas yield is reduced. Gasoline produced by coking is called coker gasoline and is rich in aromatics (20-30%) & unsaturates (40-60%).

Quality of product from a coking unit depends upon the feed. The yield of gas and coker gasoline increases with increasing temperature but the yield of gas oil decreases. The yield of gas oil increases by coking at lower pressure. At higher pressure more gas & coke are produced and the liquid product contains more gasoline. The Octane number of coker gasoline increases with temperature. However coker gasoline produced at higher temperature are unstable and require finishing operations such as clay treating or mild hydrogenation. The gases produced at higher temperature (say 525°C) are more olefinic (upto 50% olefins).

There are batch, semi continuous (also called delayed coking process) and continuous processes for coking. Coke produced is called petroleum coke which is used in the manufacture of carbon electrodes (used mainly in ahiminium industry), motor brushes, dry cells etc.

Delayed Coking Process. In this process coke is produced from the feed by increasing the time of liquid phase cracking. The furnace type coking unit is called delayed coker because it heats up the reduced crude feed to coking temperature (500°C) so rapidly that little reaction occurs while the charge is in the furnace. Effluent from the furnace is discharged at about 500°C into a large coke drum operating at 2-4 atmosphere pressure where it remains until it cracks and passes off as a vapour or condenses into coke. When the drum is filled up, it is taken out of service and steamed to remove volatile matter in the coke. The coke is then cut out with water jet at a pressure of about 80 atmosphere and removed from the drum. The cleaned drum is returned to service. Two or more drums are used in parallel and is typically on stream from 24 hours and out for cleaning for 24 hours.

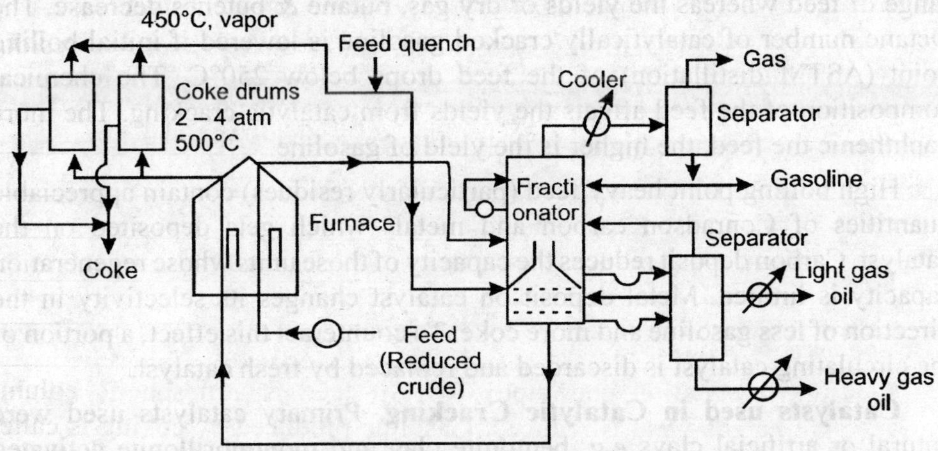

Fig. Delayed coking process.

A flow-sheet of delayed coking process is given above in Figure.

Catalytic Cracking. Cracking of feedstock in presence of a catalyst is called catalytic cracking which gives greater yield and higher octane number of gasoline than is possible by thermal cracking or reforming. Moreover, it employs reduced severity of operation as the temperature & pressure is low *i.e.* 450–550°C & 0.5–1.5 atm. (gage).

Feed for Catalytic Cracking. Feed ranges from light gas oil to reduced crude. For producing aviation gasoline, the boiling range of feed should be 350–420°C. Whereas for motor gasoline the same should be about 350–500°C. *Feed stocks* are generally:

(a) Crude oil/primary products of crude oil.

(b) Kerosene and gas oil from atmospheric distillation unit.

(c) Coker distillate - This promotes coke deposition on catalyst.

(d) Light lubricating oil (from lube oil reprocessing plant) - Its cracking produces aromatics which decompose to give deposits on catalyst.

(e) Deasphalted crude - It may contain deleterious metal component.

However, *straight run distillates are the best feed for catalytic cracking.* Performance of feed depends upon the boiling range, distribution of hydrocarbon types, the concentration of sulphur & nitrogen, the metallic constituents and Conradson carbon.

At constant conversion, gasoline and coke yield increases with boiling range of feed whereas the yields of dry gas, butane & butenes decrease. The Octane number of catalytically cracked gasoline is lowered if initial boiling point (ASTM distillation) of the feed drops below 250°C. The chemical composition of the feed affects the yields from catalytic cracking. The more naphthenic the feed, the higher is the yield of gasoline.

High boiling point heavy feed (particularly residues) contain appreciable quantities of Conradson carbon and metals which gets deposited on the catalyst. Carbon deposit reduces the capacity of those units whose regeneration capacity is limited. Metal deposit on catalyst changes its selectivity in the direction of less gasoline and more coke. To counteract this effect, a portion of the circulating catalyst is discarded and replaced by fresh catalyst.

Catalysts used in Catalytic Cracking. Primary catalysts used were natural or artificial clays *e.g* .bentonite clay and montmorillonite activated by sulphuric acid. Metallic catalysts being used presently are platinum, chromium, iron and nickel. Synthetic catalysts being used presently include silica-alumina and silica-magnesia.

Desirable properties of the catalysts are:

- high reactivity
- good selectivity
- sufficient strength and hardness
- absence of sulphur, nitrogen & metallic constituents

- easy regenerability
- high surface area
- high porosity (about 50%).

Size of the fresh catalyst in moving bed catalytic cracking process is 3-5 mm and that of regenerated catalyst is 3-4 mm. In fluidised bed catalytic cracking process the size of the catalyst is about 20-40 microns.

In fluidised bed catalytic cracking, Silica-magnesia catalyst gives the highest gasoline yield whereas silica-alumina gives the highest octane number of gasoline. Bentonite clays are intermediate in both respects.

Metallic contamination in feed lowers the gasoline yield and increases the coke production. Salt such as sodium chloride if present in feed deactivates the catalyst. Nitrogen compounds present in feed are adsorbed on catalyst and deactivate it. However, burning of nitrogen during regeneration of catalyst restores its activity completely. Catalysts also gets deactivated when exposed to steam at higher temperatures. Catalyst consumption is 0.1-0.4% of charge stock including losses in cyclone separator.

Catalytic cracking reactions. Cracking reactions are endothermic whereas the regeneration of catalysts is exothermic. The reaction occurring during thermal cracking are accelerated in catalytic cracking.

- Paraffin crack faster to lower paraffins and olefins, n-paraffins crack 5-60 times faster than corresponding thermal process.
- i-paraffins crack faster than n-paraffins.
- Dehydrocyclisation reaction takes place resulting in the formation of cyclics (*e.g.* aromatics) from paraffins.

paraffins dehydrocyclisation

—C—C—C—C—C—C ⟶ (benzene ring)

- Olefins (most reactive) crack 1000-10000 times faster than thermal process.
- Olefins undergoes isomerisation to give iso-olefins and coke (with high C/H ratio).
- Olefins undergo polymerisation.
- Cracking of naphthenes or alkenes.
- Aromatics are not responsive to cracking catalytic conditions as it is stable under the conditions unlike thermal cracking. Under severe conditions, it gives carbon deposits.

- However, alkyl aromatics (*e.g.* alkyl benzene) with side chain of carbon atom greater than 2 gives aromatic and olefins on catalytic cracking.
- The ease of cracking of alkyl aromatics increases with the length of the side chain at least upto a length of four carbon atoms.

Catalytic Cracking Processes. Depending upon the physical condition of the catalyst bed there are three main processes for catalytic cracking namely:

- Hourdry's fixed bed catalytic cracking process.
- T.C.C. (Thermofer catalytic cracking) moving bed process.
- Fluidised bed catalytic cracking (F.C.C.) Process.

Houdry's fixed bed process. This catalytic cracking process was first used in 1936 has become obsolete now by the development of moving bed and fluidised bed process. In this process, catalyst (silica-alumina) in the form of compressed pellets is contained in reactors (catalyst cases) arranged normally in a group of three. The reactors are tubular heat exchangers which supply heat necessary to break down the heavier molecules of feed and to remove the heat evolved when burning of the carbon on the catalyst during the regeneration cycle.

Operation of the process is cyclic occurring in four stages and consuming a total time of 30 minutes *e.g.*

- reaction time = 10 minutes at about 2 kg/cm^2 and 480°C.
- purging time = 5 minutes.
- regeneration of catalyst = 10 minutes at 510°C.
- purging = 5 minutes.

The temperature of the catalyst bed is controlled by circulating a molten mixture of alkali metal salts through vertical tubes within the reactor core. Life of catalyst is about 18 months but the yield of gasoline falls from 52 to 40% during this period. Octane number of gasoline produced is about 73.

Several reactors (generally in a multiple of 3 *i.e.* 3, 6, 9 or 12) are provided in commercial unit so that feed may be switched cyclically from a deactivated reactor to an active one. Provision is made to heat the catalyst during cracking reaction (which is endothermic) and cool the catalyst during regeneration (which is exothermic).

A flow-sheet of the Houdry's fixed bed catalytic cracking process is given below in Figure.

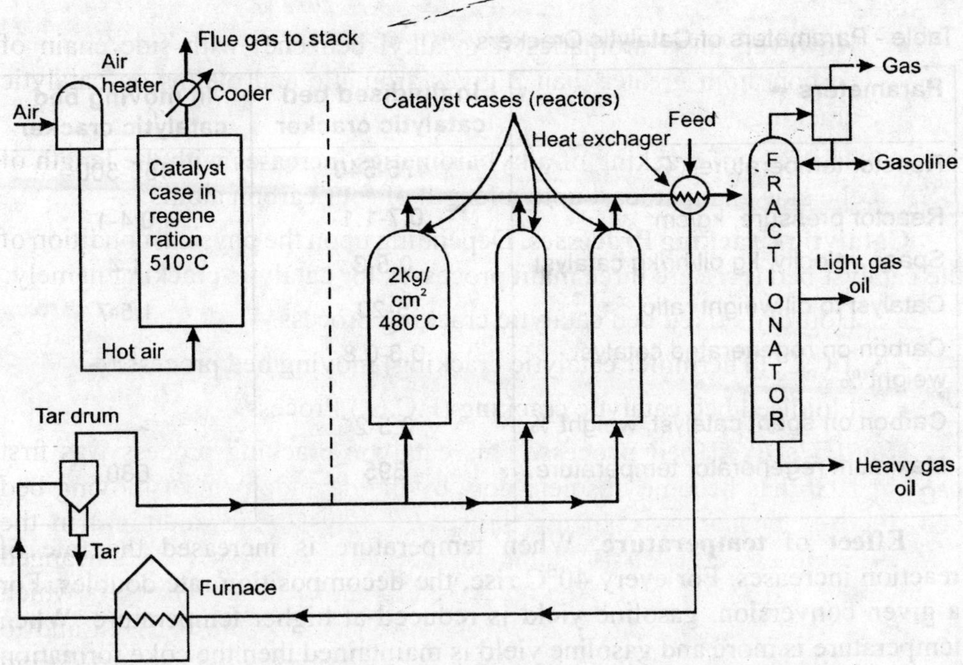

Fig. Houdry's fixed bed catalytic cracking process.

Material balance for a catalytic cracker:

For a charge stock having:

(a) Boiling range = 320-500°C,

(b) Sulphur = 1.9%

(c) Specific gravity = 0.9128,

(d) Pour point = 30°C

When subjected to catalytic cracking the above charge stock at 450°C and 0.5 atm. (gage) gave the following product distribution:

Gas (including C_4 fractions) = 13% by weight

C_5 →Gasoline (upto final boiling point of 200°C) = 24%

Heavy gas oil (H.G.O.) = 26.9%

Light gas oil (L.G.O.) (200-320°C) = 24%

Coke = 5%, Losses = 2.3%.

Effect of reaction conditions in catalytic cracking. What happens in a moving bed or fluidised bed catalytic cracking reactor depends upon the temperature, pressure, space velocity, ratio of catalyst to oil passing through the reactor, concentration of coke on the incoming & outgoing catalyst and the catalyst activity. The normal range of each variable is given below in Table.

Table - Parameters of Catalytic Crackers

Parameters	In fluidised bed catalytic cracker	In moving bed catalytic cracker
Reactor temperature, °C	475-540	455-500
Reactor pressure, kg/cm^2	0.7-1.1	0.4-1
Space velocity, kg oil/hr/kg catalyst	0.5-3	1-4
Catalyst to oil weight ratio	5-20	1.5-7
Carbon on regenerated catalyst, weight %	0.3-0.8	0.1
Carbon on spent catalyst, weight %	0.5-2	—
Maximum regenerator temperature, °C	595	680

Effect of temperature. When temperature is increased the rate of reaction increases. For every 40°C rise, the decomposition rate doubles. For a given conversion, gasoline yield is reduced at higher temperature. When temperature is more and gasoline yield is maintained then the coke formation rate decreases when residence time is less. This is achieved in two stage operation. Higher temperature results in gasoline of slightly higher sulphur content.

Effect of pressure. When reactor pressure is more, then the residence time is also more which results in more unsaturates in the product and less octane number of gasoline. At a given conversion, increasing the pressure increases the production of coke.

Effect of catalyst to oil ratio. Conversion increases with increase in catalyst to oil ratio as it shortens the time required for the catalyst to pass through the reactor and thereby reduces the extent of its deactivation from coke production. At a given conversion and temperature, coke yield is higher at higher catalyst to oil ratio.

Effect of coke concentration on catalyst. Lesser is the concentration of coke on catalyst, better is the effective activity of the catalyst. Lower coke concentration is obtained by increasing catalyst to oil ratio. On silica-alumina catalyst, higher coke concentration catalyses the production of more coke.

Effect of catalyst activity. With decrease in catalyst activity, the conversion declines. Product yield is poorer at low activity hence catalyst is sometimes deliberately discarded so that more fresh catalyst can be added to increase activity and improve yields.

Product characteristics. Gas produced comprises mainly of propane and propylene (used in polymerisation).

- Gasoline has more aromatics and iso-paraffin; hence has a high octane number. One third of total gasoline fraction is olefins. About one fourth of C_4 is associated with gasoline fraction. Gasoline has a final boiling point of 200°C. 40% of it evaporates within 100°C.
- Light gas oil (having boiling range 200-320°C) produced has a cetane number of 25-35, if the feedstock is rich in naphthenes and aromatics. However, if feed is of paraffinic base oil then cetane number is 45-55. It contains about 30-55% aromatics and 6% olefins.
- Heavy gas oil is the last fraction from cracking unit. Its viscosity and pour point is similar to the feedstock. Its specific gravity is about 0.89-0.96. This H.G.O. is used as feed to the coking or conventional thermal cracking unit or as a fuel oil.

Thermofer Catalytic Cracking (TCC) Moving Bed Process

In this process, the regenerated catalyst beads (similar to that in Houdry's process) goes down from an elevated hopper to the reactor where it comes in contact with partially vapourised feed. Catalyst & oil can flow concurrently into a section where the cracked products are separated and removed. Additional oil is often obtained by steam stripping of catalyst (i.e. by steaming the catalyst at the base of the reactor). The spent catalyst (hot, carbon coated) then descends by gravity into the regenerator where a counter current low oxygen content flue gas is used to burn off the deposited coke. The catalyst containing 2-4% by weight coke at a temperature of about 450°C is regenerated to a coke per cent of 0.5 and leaves at 550°C.

A cooler at the bottom of the regenerator reduces catalyst temperature to the level desired for entry in lift line where a low pressure of 0.15-0.2 kg/cm^2 is used for the lifting of catalyst to the catalyst hopper. A high velocity of 25-30 metres/second is employed in catalyst lift line.

A flowsheet of T.C.C. moving bed process is shown below in Figure.

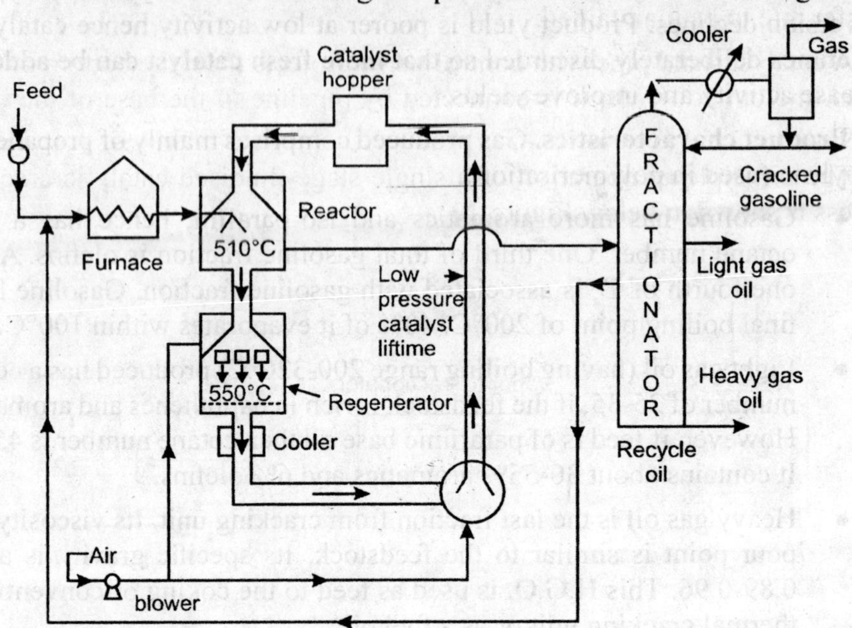

A catalyst/oil ratio = 2/5 and recycle ratio = 0-2 is maintained in T.C.C. moving bed process.

A typical result shows:

Gasoline yield	43-56%
Light gas oil	15-27%
Heavy gas oil	17-19%
Coke	5%
Octane number of gasoline	95
Catalyst consumption	0.2 lb/barrel oil = 0.57 kg/m^3 oil.

Fluidised Bed Fluidised Catalytic Cracking (F.C.C.) Process

Fluidised bed catalytic cracking is the most widely used process in oil refineries. The catalyst is in powdered form and is recirculated between two vertical vessels operating as reactor and regenerator.

Fluidised bed of powder is obtained by forming the vapour through columns of powder until the minimum fluidised velocity is reached at which the particles of catalyst are arranged as a loose packing in the gas stream so that they will flow as a fluid. A slight increase in gas velocity causes vapour to flow through a zone of low solids concentration in which high mass transfer rate between catalyst surface and gas enables high reaction rates and high heat transfer rate to be maintained.

A catalyst stand pipe with control valve projects from the bottom of each vessel, the outlet of which is connected by pipeline to the base of the other vessel.

The principle of operation of a single stage fluidised catalytic cracking process is shown in below Figure.

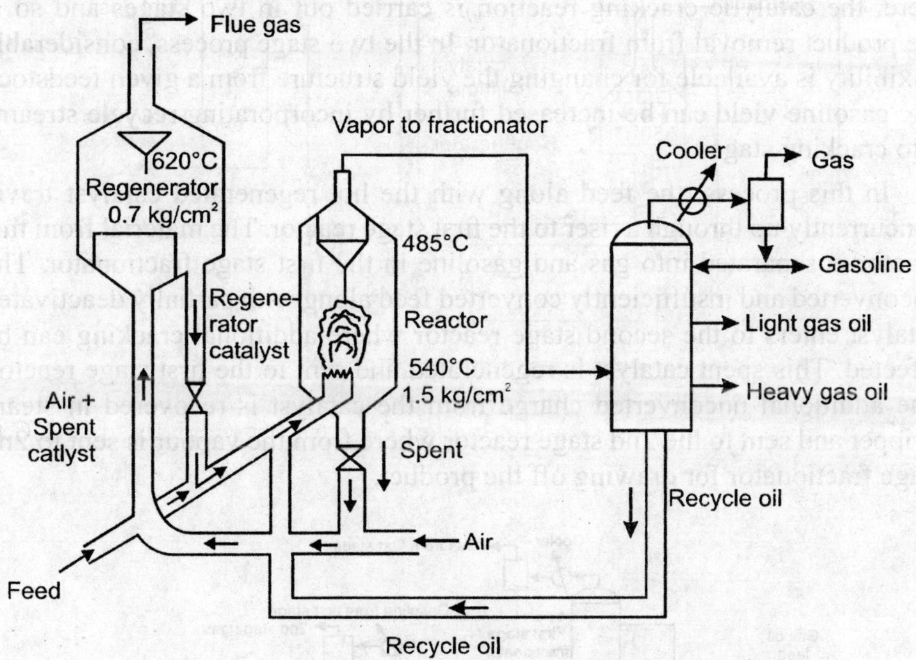

Fig. Fluidised bed catalytic cracking (single stage).

The sequence of events in the fluidised catalytic cracking process as evident from the above figure is as follows :

(a) The feed is heated and vapourised by a stream of hot catalyst passing into the reactor at about 540°C. The catalyst separates from the vapour and sinks to the base of reactor from which it flows, with air into the regenerator.

(b) In the regenerator, the temperature rises to about 680°C due to combustion of carbon on the catalyst. The regenerated catalyst flows down the stand pipe into the fresh oil feed.

(c) The cracked vapour pass from top of the reactor to the fractionating column from which the heavier fractions are recycled through the reactor and the products are drawn off.

The catalyst to oil ratio is 5-15 and the reactor space velocity is 1-3 kg oil/hr/kg catalyst. A typical result from a single stage F.C.C. process shows :

Gasoline yield = 53-59%	Heavy gas oil = 2-3%
Light gas oil = 18-25%	Coke = 5-8%
Octane number of gasoline = 92-94 (unleaded)	

Two Stage Fluidised Catalytic Cracking (F.C.C.) Process

Here, the catalytic cracking reaction is carried out in two stages and so is the product removal from fractionator. In the two stage process, considerable flexibility is available for changing the yield structure from a given feedstock *e.g.* gasoline yield can be increased further by incorporating recycle streams into cracking stages.

In this process, the feed along with the hot regenerated catalyst travel concurrently up through a riser to the first stage reactor. The material from this reactor is separated into gas and gasoline in the first stage fractionator. The unconverted and insufficiently converted feed along with partially deactivated catalyst enters to the second stage reactor where additional cracking can be effected. This spent catalyst is regenerated and sent to the first stage reactor. The additional unconverted charge from the catalyst is recovered in steam stripper and sent to the 2nd stage reactor where from the vapour is sent to 2nd stage fractionator for drawing off the product.

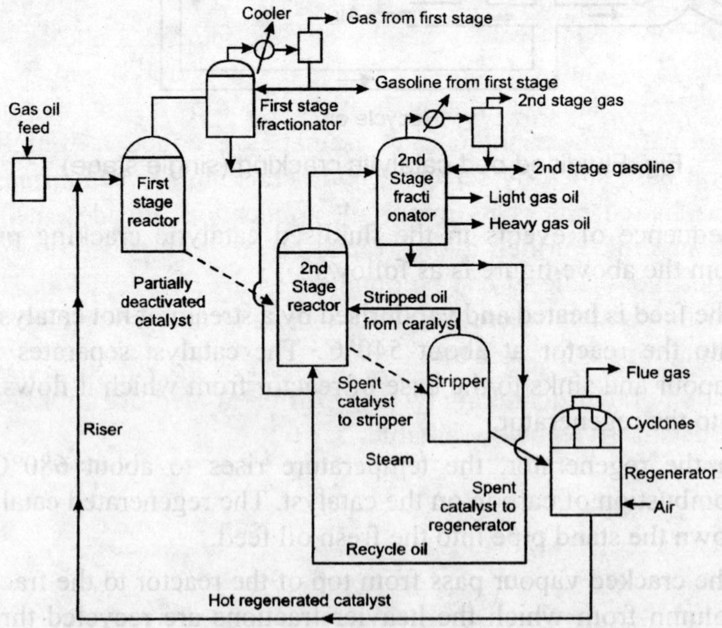

Fig. Two-stage fluidised bed catalytic cracking (F.C.C.) process.

Reaction conditions- 450-520°C & 8-15 kg/cm^2.

Regeneration condition - 550-600°C & 1.3-1.6 kg/cm^2.

Catalyst / oil ratio = 5-15

Average consumption of catalyst = 0.6 lb/bbl of feed = 1.7 kg/m^3 feed.

Octane number of gasoline produced from this process is about 97.

Hydrocracking

Cracking in presence of hydrogen is called hydrocracking. It is also called destructive hydrogenation and leads to the formation of saturated compounds of lower molecular weight by the cracking of heavy fractions and residues under pressure of hydrogen. Hydrogen suppresses the formation of tar & coke and also raises the crackability of polycyclic aromatic compounds.

Hydrocracking achieves high conversion into gasoline and produces high quality diesel fuels. Mild hydrocracking of petroleum residues supply feed for 'other refinery processes. Important catalyst used are tungsten sulphide for vapour phase hydrocracking and an iron catalyst or bauxite for liquid phase hydrocracking. Hydrocracking of residues and heavy fractions boiling over 330°C is done in the liquid phase. Reaction conditions in hydrocracking are more severe *viz.*400-550°C and 70-700 kg/cm^2.

Hydrocracking and other hydrogenation processes in refinery has been possible mainly because of the availability of surplus hydrogen rich gases from the catalytic reforming of naphthas. Hydrocracking produces desired products with a lower molecular weight than the feedstock and any fraction from naphtha to asphalt can be processed. It completely removes sulphur, nitrogen & oxygen besides saturating the olefins thereby giving stable products comprising of paraffins, naphthenes & aromatics. Production of high quality lubricating oil base stock by selective hydrocracking of vacuum gas oils & propane deasphalted oils is its important application. Besides, it increases the yield of lube oil. Further, avoids range of stable distillate products like gasoline or jet fuel or diesel can be obtained in the hydrocracker. However, compared to fluid catalytic cracker, hydrocracking unit is 1.6 times costlier and has a higher operating cost too.

Application of hydrocracking by using different feed stock for getting the desired products is given below in table.

Table - Hydrocracking applications to get desired products

Feedstock charged to hydrocracker	Desired Products obtained
Naphtha	Butane & Propane (LPG)
Condensate from natural gas	Naphtha
Kerosene	Naphtha
Atmospheric gas oils	Naphtha, jet fuel and/or distillates
Vacuum gas oils	Naphtha, jet fuel, distillates, lube oils
Deasphalted oils	Naphtha, jet fuels, distillate, lube oils
Reduced crude oil	Distillate &LSHS fuel oil
Vacuum residue	Naphtha, distillate, Vacuum gas oil &LSHS fuel oil
Light & heavy cycle oils (from FCC)	Naphtha and/or distillates
Coker distillates & heavy gas oils	Naphtha and/or distillates

Hydrotreating. It is a comparatively mild operation carried out at 300 - 400°C & 15-17 Kg/cm^2, aimed at removal of impurities like sulphur, oxygen, nitrogen, halides & trace metals from products of all boiling ranges from naphtha to lubricating oils. Hydrotreating is also used for stabilisation of petroleum products by saturating olefins present in them. Catalyst used is cobalt-molybdena. Hydrogenation processes like hydrocracking, hydrotreating and aromatic saturation has been possible due to availability of surplus hydrogen rich gases from catalytic reforming of naphtha.

Catalytic Hydrotreating

Catalytic hydrotreating is a hydrogenation process used to remove about 90% of contaminants such as nitrogen, sulfur, oxygen, and metals from liquid petroleum fractions. These contaminants, if not removed from the petroleum fractions as they travel through the refinery processing units, can have detrimental effects on the equipment, the catalysts, & the quality of the finished product. Typically, hydrotreating is done prior to processes such as catalytic reforming so that the catalyst is not contaminated by untreated feedstock. Hydrotreating is also used prior to catalytic cracking to reduce sulfur, improve product yields, and to upgrade middle-distillate petroleum fractions into finished kerosene, diesel fuel, & heating fuel oils. In addition, hydrotreating converts olefins and aromatics to saturated compounds.

Catalytic Hydrodesulfurization Process

Hydrotreating for sulfur removal is called hydrodesulfurization. In a typical catalytic hydrodesulfurization unit, the feedstock is deaerated and mixed with hydrogen, preheated in a fired heater (315 – 425°C) and then charged under pressure (up to 70kg/cm^2) through a fixed bed catalytic reactor. In the reactor, the sulfur & nitrogen compounds in the feedstock are converted into H_2S & NH_3. The reaction products leave the reactor and after cooling to a low temperature enter a liquid/gas separator. The hydrogen-rich gas from the high-pressure separation is recycled to combine with the feedstock, and the low-pressure gas stream rich in H_2S is sent to a gas treating unit where H_2S is removed.

The clean gas is then suitable as fuel for the refinery furnaces. The liquid stream is the product from hydrotreating and is normally sent to a stripping column for removal of H_2S and other undesirable components. In cases where steam is used for stripping, the product is sent to a vacuum drier for removal of water. Hydrodesulfurized products are blended or used as catalytic reforming feedstock.

Other Hydrotreating Processes

- Hydrotreating processes differ depending upon the feedstock available and catalysts used. Hydrotreating can be used to improve the burning characteristics of distillates such as kerosene. Hydrotreatment of a kerosene fraction can convert aromatics into naphthenes, which are cleaner-burning compounds.
- Lube-oil hydrotreating uses catalytic treatment of the oil with hydrogen to improve product quality. The objectives in mild lube hydrotreating include saturation of olefins and improvements in color, odor, & acid nature of the oil. Mild lube hydrotreating also may be used following solvent processing. Operating temperatures are usually below 350°C and operating pressures below 55 kg/cm^2. Severe lube hydrotreating, at temperatures in the 315-400°C range and hydrogen pressures up to 210kg/cm^2 is capable saturating aromatic rings, along with sulfur and nitrogen removal, to impart specific properties not achieved at mild conditions.
- Hydrotreating also can be employed to improve the quality of pyrolysis gasoline (pygas) a by-product from the manufacture of ethylene. Traditionally, the outlet for pygas has been motor gasoline blending, a suitable route in view of its high octane number. However, only small portions can be blended untreated owing to the unacceptable odor, color, & gum-forming tendencies of this material. The quality of pygas, which is high in diolefin content, can be satisfactorily improved by hydrotreating, whereby conversion of diolefins into mono-olefins provides an acceptable product for motor gas blending.

Appendix 4.1: Comparative features of thermal, catalytic & hydro cracking processes

Thermal Cracking

- Because the simple distillation of crude oil produces amounts & types of products that are not consistent with those required by the marketplace, subsequent refinery processes change the product mix by altering the molecular structure of the hydrocarbons. One of the ways of accomplishing this change is through "cracking," process that break/cracks the heavier, higher boiling-point petroleum fractions into more valuable products such as gasoline, fuel oil, and gas oils. The two basic types of cracking are thermal cracking, using heat &pressure, and catalytic cracking.
- The first thermal cracking process was developed in 1913. Distillate fuels and heavy oils were heated under pressure in large drums until they cracked into smaller molecules with better antiknock characteristics. However, this method produced large amounts of solid, unwanted coke. This early process has evolved into the following applications of thermal cracking: visbreaking, steam cracking, & coking.

Visbreaking Process. Visbreaking, a mild form of thermal cracking, significantly lowers the viscosity of heavy crude oil residue without affecting the boiling point range. Residual from the atmospheric distillation tower is heated (430–510°C) at atmospheric pressure and mildly cracked in a heater. It is then quenched with cool gas oil to control overcracking, and flashed in a distillation tower. Visbreaking is used to reduce the pour point of waxy residues and reduce the viscosity of residues used for blending with lighter fuel oils. Middle distillates may also be produced, depending on product demand. The thermally cracked residue tar, which accumulates in the bottom of the fractionation tower, is vacuum flashed in a stripper and the distillate is recycled.

Steam Cracking Process. Steam cracking is a petrochemical process sometimes used in refineries to produce olefinic raw materials (e.g., ethylene) from various feedstock for petrochemicals manufacture. The feedstock range from ethane to vacuum gas oil, with heavier feeds giving higher yields of by-products such as naphtha. The most common feeds are ethane, butane, & naphtha. Steam cracking is carried out at temperatures of (815-870°C) and at pressures slightly above atmospheric. Naphtha produced from steam cracking contains benzene, which is extracted prior to hydrotreating. Residual from steam cracking is sometimes blended into heavy fuels.

Coking Processes. Coking is a severe method of thermal cracking used to upgrade heavy residuals into lighter products or distillates. Coking produces straight-run gasoline (coker naphtha) and various middle distillate fractions used as catalytic cracking feedstock. The process so completely reduces hydrogen that the residue is a form of carbon called "coke." The two most common processes are delayed coking and continuous (contact or fluid) coking. Three typical types of coke are obtained (sponge coke, honeycomb coke, and needle coke) depending upon the reaction mechanism, time, temperature, & the crude feedstock.

Delayed Coking. In delayed coking process the heated charge (typically residuum from atmospheric distillation towers) is transferred to large coke drums which provide the long residence time needed to allow the cracking reactions to proceed to completion. Initially the heavy feedstock is fed to a furnace which heats the residuum to high temperatures (430-510^{0}C) at low pressures (1-2 kg/cm^2) and is designed & controlled to prevent premature coking in the heater tubes. The mixture is passed from the heater to one or more coker drums where the hot material is held approximately for 24 hours (delayed) at pressures, of 1.8-5.5 kg/cm^2, until it cracks into lighter products. Vapors from the drums are returned to a fractionators where gas, naphtha, and gas oils are separated out. The heavier hydrocarbons produced in the fractionator are recycled through the furnace.

After the coke reaches a predetermined level in one drum, the flow is diverted to another drum to maintain continuous operation. The full drum is steamed to strip out uncracked hydrocarbons, cooled by water injection, and decoked by mechanical or hydraulic methods. The coke is mechanically removed by an auger rising from the bottom of the drum. Hydraulic decoking consists of fracturing the coke bed with high-pressure water ejected from a rotating cutter.

Continuous Coking. Continuous (contact or fluid) coking is a moving-bed process that operates at temperatures higher than delayed coking. In continuous coking, thermal cracking occurs by using heat transferred from hot, recycled coke particles to feedstock in a radial mixer, called a reactor, at pressure of 3.5 kg/cm^2. Gases &vapors are taken from the reactor, quenched to stop any further reaction, and fractionated. The reacted coke enters a surge drum and is lifted to a feeder and classifier where the larger coke particles are removed as product. The remaining coke is dropped into the preheater for recycling with feedstock. Coking occurs both in the reactor and in the surge drum. The process is automatic in that there is a continuous flow of coke and feedstock.

Catalytic Cracking

- Catalytic cracking breaks complex hydrocarbons into simpler molecules in order to increase the quality & quantity of lighter, more desirable products and decrease the amount of residuals. This process rearranges the molecular structure of hydrocarbon compounds to convert heavy hydrocarbon feedstock into lighter fractions such as kerosene, gasoline, LPG, heating oil, and petrochemical feedstock.
- Catalytic cracking is similar to thermal cracking except that catalysts facilitate the conversion of the heavier molecules into lighter products. Use of a catalyst (a material that assists a chemical reaction but does not take part in it) in the cracking reaction increases the yield of improved quality products under much less severe operating conditions than in thermal cracking. Typical temperatures are from 455-510^{0}C and much lower pressures of 0.7-1.4 kg/cm^2. The catalysts used in refinery cracking units are typically solid materials (zeolite, aluminum hydrosilicate, treated bentonite clay, fuller's earth, bauxite, and silica- alumina) that come in the form of powders, beads, pellets or shaped materials called extrudites.
- There are three basic functions occuring the catalytic cracking process:
  - **Reaction**: Feedstock reacts with catalyst and cracks into different hydrocarbons;
  - **Regeneration**: Catalyst is reactivated by burning off coke; and
  - **Fractionation**: Cracked hydrocarbon stream is separated into various products.
- The three types of catalytic cracking processes are fluid catalytic cracking (FCC), moving-bed catalytic cracking, and Thermofor catalytic cracking (TCC). The catalytic cracking process is very flexible, and operating parameters can be adjusted to meet changing product demand. In addition to cracking, catalytic activities include dehydrogenation, hydrogenation, and isomerization.

Fluid Catalytic Cracking

The most common process is FCC, in which the oil is cracked in the presence of a finely divided catalyst which is maintained in an aerated or fluidized state by the oil vapors. The fluid cracker consists of a catalyst section and a fractionating section that operate together as an integrated processing unit. The catalyst section contains the reactor and regenerator, which, with the stand pipe and riser, forms the catalyst circulation unit. The fluid catalyst is continuously circulated between the reactor and the regenerator using air, oil vapors, & steam as the conveying media.

A typical FCC process involves mixing a preheated hydrocarbon charge with hot, regenerated catalyst as it enters the riser leading to the reactor. The charge is combined with a recycle stream within the riser, vaporized, and raised to "reactor temperature (430-540° C) by the hot catalyst. As the mixture travels up the riser, the charge is cracked at 0.7-2.1kg/cm^2. In the more modern FCC units, all cracking takes place in the riser. The "reactor" no longer functions as a reactor; it merely serves as a holding vessel for the cyclones. This cracking continues until the oil vapors are separated from the catalyst in the reactor cyclones. The resultant product stream (cracked product) is then charged to a fractionating column where it is separated into fractions, and some of the heavy oil is recycled to the riser.

Spent catalyst is regenerated to get rid of coke that collects on the catalyst during the process. Spent catalyst flows through the catalyst stripper to the regenerator, where most of the coke deposits burn off at the bottom where preheated air and spent catalyst are mixed. Fresh catalyst is added and worn-out catalyst removed to optimize the cracking process.

Moving Bed Catalytic Cracking. The moving-bed catalytic cracking process is similar to the FCC process. The catalyst is in the form of pellets that are moved continuously to the top of the unit by conveyor or pneumatic lift tubes to a storage hopper, then flow downward by gravity through the reactor, and finally to a regenerator. The regenerator and hopper are isolated from the reactor by steam seals. The cracked product is separated into recycle gas, oil, clarified oil, distillate, naphtha, & wet gas.

Thermofor Catalytic Cracking. In a typical thermofor catalytic cracking unit, the preheated feedstock flows by gravity through the catalytic reactor bed. The vapors are separated from the catalyst and sent to a fractionating tower. The spent catalyst is regenerated, cooled, and recycled. The flue gas from regeneration is sent to a carbon-monoxide boiler for heat recovery.

Hydrocracking

Hydrocracking is a two-stage process combining catalytic cracking and hydrogenation, wherein heavier feedstocks are cracked in the presence of hydrogen to produce more desirable products. The process employs high pressure, high temperature, a catalyst, and hydrogen. Hydrocracking is used for feedstocks that are difficult to process by either catalytic cracking or reforming, since these feedstocks are characterized usually by a high polycyclic aromatic content and/or high concentrations of the two principal catalyst poisons, sulfur & nitrogen compounds.

The hydrocracking process largely depends on the nature of the feedstock and the relative rates of the two competing reactions, hydrogenation and cracking. Heavy aromatic feedstock is converted into lighter products under a very high pressures (70-140 kg/cm^2) and fairly high temperatures (400-815°C) in the presence of hydrogen and special catalysts. When the feedstock has a high paraffinic content, the primary function of hydrogen is to prevent the formation of polycyclic aromatic compounds. Another important role of hydrogen in the hydrocracking process is to reduce tar formation and prevent buildup of coke on the catalyst. Hydrogenation also serves to convert sulfur & nitrogen compounds present in the feedstock to hydrogen sulfide & ammonia.

Hydrocracking produces relatively large amounts of isobutane for alkylation feedstock. Hydrocracking also performs isomerization for pour point control and smoke point control, both of which are important in high-quality jet fuel.

Hydrocracking Process

In the first stage, preheated feedstock is mixed with recycled hydrogen and sent to the first-stage reactor, where catalysts convert sulfur & nitrogen compounds to hydrogen sulfide & ammonia. Limited hydrocracking also occurs. After the hydrocarbon leaves the first stage, it is cooled and liquefied and run through a hydrocarbon separator. The hydrogen is recycled to the feedstock. The liquid is charged to a fractionator. Depending on the products desired (gasoline components, jet fuel, and gas oil), the fractionator is run to cut out some portion of the first stage reactor out turn. Kerosene range material can be taken as a separate side draw product or included in the fractionator bottoms with the gas oil.

The fractionator bottoms are again mixed with a hydrogen stream and charged to the second stage. Since this material has already been subjected to some hydrogenation, cracking, & reforming in the first stage, the operations of the second stage are more severe (higher temperatures & pressures). Like the out turn of the first stage, the second stage product is separated from the hydrogen and charged to the fractionator.

5

Thermal and Catalytic Reforming

Reforming means rearrangement of molecules without much affecting the average molecular weight of feed which is generally naphtha of gasoline boiling range. Reforming is carried out to produce high quality (octane number) gasoline by heating with or without catalyst the naphtha. Low octane number thermally cracked gasoline (is also a feed for reforming) is also subjected to reforming to improve its quality (octane number).

Reforming can be thermal or catalytic as in the case of cracking. Catalyst apart from accelerating the process also enhances the yield & quality (octane number) of gasoline. The gasoline produced by reforming is called reformed gasoline or reformate. Thermal reforming has been almost completely replaced by catalytic reforming.

Feed for reforming. The feed is generally naphtha produced during straight run distillation of crude oil, catalytic cracking and coking process. The best result is obtained with naphtha (produced during straight run distillation) having high naphthene content.

Reforming products. Besides the main product *i.e.* reformate or reformed gasoline, reforming also produces lighter hydrocarbons (gases), hydrogen and traces of very high boiling materials. The hydrogen rich gaseous product is recycled with the naphtha feed. The reformed gasoline has a wider boiling range than the feed and its volatility is increased.

Chemical Reactions in catalytic reforming. Since, aromatics have the highest octane number and iso-paraffins are next to them, most of the desirable reactions during reforming tend to produce high octane number hydrocarbons from lower ones. Octane number increases in the order paraffin —>olefin —>naphthene —>iso-paraffin —> aromatic.

The main reaction in the reforming process is dehydrogenation of naphthene to produce aromatics. Other reactions occurring during reforming are dehydrocyclisation of paraffins, isomerisation, hydrocracking, dehydroisomerisation, dehydrogenation etc.

The various *reforming reactions* are illustrated below.

Dehydrogenation of naphthene to aromatic:

CH3
H2C CH2
| | — CH3
H2C CH2
CH2
methyl cyclohexane
(naphthene)

—dehydrogenation—> Ch3 (ring) + CH2
Toluene
(aromatic)

Dehydroisomerisation: In a typical reaction methyl cyclopentane isomerises to cyclohexane which dehydrogenates to yield benzene

CH2
CH2 Ch — Ch3
CH2 — CH2
methyl cyclohexane
(naphthene)

—isomerisation—> CH2
H2C CH2
H2C CH2
CH2
cyclohexane

—dehydrogenation—> + 3H2
Benzene
(aromatic)

Dehydrocyclisation of paraffins to aromatic:

C_7H_{16}
n–pentane
(paraffin)

—cyclisation—> CH2
H2C CH2 — CH3
H2C CH2
CH2
methyl
cyclohexane

—dehydrogenation—> CH3 + 4H2
Toluene
(aromatic)

Hydrocracking of higher paraffins to lower paraffins:

$$C_8H_{18} + H_2 \xrightarrow{\text{Hydrocracking}} C_3H_8 + C_5H_{12}$$

octane (higher boiling point paraffin) → propane pentane (lower boiling point paraffin)

$$C_{10}H_{22} + H_2 \xrightarrow{\text{hydrocracking}} n\text{–}C_5H_{12} + i - C_5H_{12}$$

Decane → *n*–pentane, *iso*–pentane

Lower boiling point paraffins have higher octane number.

Isomerisation of paraffins: Here, n-paraffins are converted into iso-paraffins which have higher octane number.

C — C — C — C — C — C → C — C(—C) — C — C — C

n–paraffins → *i*–paraffins

Dehydrogenation of n-paraffins to olefins:

$$C_nH_{2n+2} \xrightarrow{\text{dehyrogenation}} C_nH_{2n} + H_2$$

n–paraffin → olefin

Though olefins have higher octane number than corresponding n-paraffins, still this reaction is undesirable because lower olefins polymerise to produce gummy materials which is unwanted.

Thermal reforming: This is carried out in absence of catalyst and is similar to high temperature, low pressure thermal cracking. Feed is usually naphtha and the products are mainly gas (13-14%) and gasoline (75-80%) of octane number 81-86 ; rest being polymer (unwanted gum) and losses. A flowsheet for thermal reforming is given below in Figure.

Catalytic Reforming: Reforming in the presence of a catalyst is called catalytic reforming. Two types of catalyst are used *viz.* non-precious metal oxide type *(e.g.* molybdena or chromia supported on alumina base) and precious metal oxide type *(e.g.* platinum on a silica-alumina or alumina base). Platinum is more active and selective catalyst, however, it is easily poisoned by compounds containing sulphur, nitrogen, arsenic etc. and needs pretreatment of the feed for the removal of these impurities. Catalyst may be in the form of powder, bead or pellet.

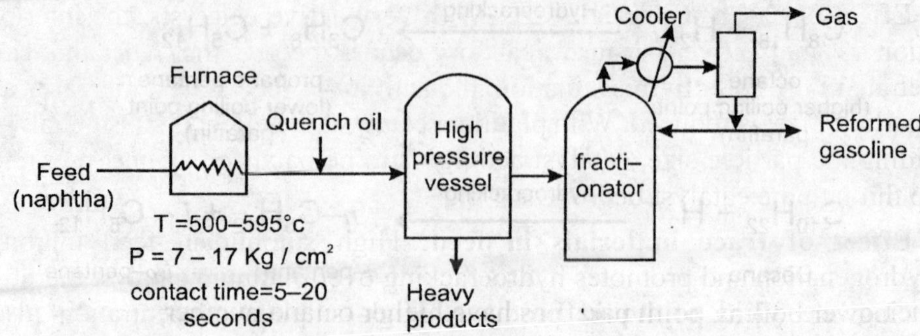

Fig. Flowsheet for thermal reforming

Fixed bed, moving bed and fluidised bed processes are in use. Various commercial catalytic reforming processes in use in oil refineries are Platforming, Catforming, Hydroforming, Fluidised Hydroforming and Thermofer Catalytic Reforming (TCR).

Reforming is normally a once through process. The reactor conditions are normally in the range of $T = 450–550°C$ and $P = 15–45$ kg/cm^2. Fluidised bed and moving bed processes are run in high temperature ranges at about 450–550°C. High pressure upto 45 kg/cm^2 are for the fixed bed process.

Effect of variables in catalytic reforming: Catalytic reforming is carried out over a wide range of reaction conditions, the most important of which is pressure. Other variables are recycle gas rate, temperature, space velocity and the catalyst particle size.

Effect of Temperature. Temperature in catalytic reforming is in the range of 450–550°C. If temperature is less than 450°C then catalyst is inactive whereas if temperature exceeds 550°C then hydrocracking predominates. Cyclisation reaction occurs at 470-490°C whereas isomerisation reaction occurs at 40-50°C above that for cyclisation. Dehydrogenation reaction occurs upto 500-550°C. At 700-800°C, olefin formation starts.

Effect of Pressure: Dehydrogenation of naphthene which is desirable in reforming reaction and is favoured by low pressure and high temperature; however, this condition also favours the formation of coke which deactivates the catalyst. Deposition of coke can be reduced by operating in an atmosphere of hydrogen; the higher the pressure of hydrogen, the slower is the coke deposition. Increasing the pressure of hydrogen does not severely retards the dehydrogenation of naphthenes which produce high octane number aromatics. Hence, catalytic reforming is carried out under substantial pressure with the hydrogen product recycled to the reaction zone. As pressure and hydrogen recycle ratio is increased, gasoline yield (at a given octane level) reduces. Lower pressure are permissible with lower boiling feed which has a lower tendency to form coke.

Effect of catalyst particle size: With very active catalysts, the interior portion is ineffective because reaction occurs before the reactants can penetrate to the core. In this situation, the activity of the catalyst increases as the particle size, is reduced. With platinum catalyst, the effect is substantial. In deciding the particle size, catalyst activity is to be balanced against pressure drop through the catalyst bed.

Effect of trace materials in feed: High sulphur in feed inhibits dehydrogenation and promotes hydrocracking over platinum catalyst thereby reducing the yield of liquid products. High sulphur concentration also accelerates deactivation of catalyst especially at low hydrogen pressure.

Presence of nitrogen in feed to platinum catalyst reforming unit inhibits dehydrocyclisation of paraffin and hydrocracking.

Halides promote the activity of a platinum on alumina catalyst and its some concentration in the catalyst is necessary for good operation. However, if halide concentration becomes too high, excessive hydrocracking results. Halides are not necessary when silica-alumina base is used for the catalyst.

Metals such as lead and arsenic poison the catalyst. Tetraethyl lead is also a catalyst poison.

Yield in catalytic reforming depends upon the charge stock, catalyst and operating conditions. If feed has 15–25% naphthenes then the reformate will have 11–18% benzene, 18–20% toluene and 22–24% xylene. With increase in severity of operation, yield increases at the cost of quality *(i.e.* octane number reduces).

In a typical catalytic reforming of naphtha of boiling range 85-180°C, the yield was as follow:

gasoline component	= 84%	unsaturates	= 4%
hydrogen	= 11%	losses	= 1%

Commercial Catalytic Reforming Processes

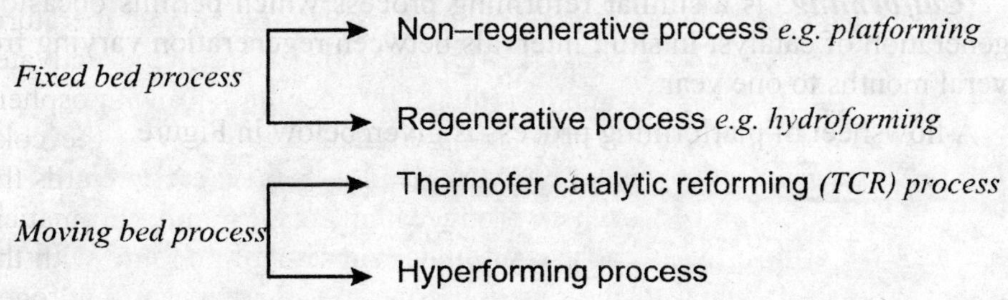

Fluidised bed process -Fluidised hydroforming process

Some of the important commercial processes are described below.

Non-Regenerative Fixed Bed Platforming Process

One of the most widely used catalytic reforming process is the non-regenerative platforming process which employs normally a three to five reactor system. The catalyst is platinum and halogen (normally platinic chloride) on alumina base. The catalyst is not regenerated in situ but is replaced when operating temperature cannot be increased further to maintain the octane number of gasoline produced. A catalyst life of 200 barrels *(i.e.* 7000 imperial gallon) of feed per pound of catalyst has been reported.

Preheated naphtha with an end point of 190°C is mixed with hydrogen rich recycle gas and partially preheated by indirect heat. Final preheat is obtained in a preheat furnace and the feed to reactor is at a temperature between 455 to 530°C. Pressures of 33–47 kg/cm^2 is used. Reactor space velocity varies from 1.5 to 3 volume of feed per hour per volume of catalyst. Hydrogen rich recycle gas is dehydrated and freed from hydrogen sulphide by scrubbing with monoethanolamine (MEA).

Product after passing through the water cooled condensors enters a high pressure separator where the unstabilised reformate is recovered. The major portion of the gas is recycled while the tail gas is sent to the absorber system for removal of C_4 and higher hydrocarbons. The two most important reactions are aromatisation (endothermic) which takets place mainly in the first reactor, and the hydrocracking (exothermic) which takes place in the later reactors.

Gasoline yield and its octane number varies with the reforming severity *(i.e.* temperature and pressure mainly). Reforming of a sample of naphtha gave the following result.

Low severity - gasoline yield = 91.5% and octane number = 85 (unleaded)
High severity - gasoline yield = 85.2% and octane number = 95 (unleaded)

'Catforming' is a similar reforming process which permits occasional regeneration of catalyst in situ ; intervals between regeneration varying from several months to one year.

A flowsheet of platforming process is given below in Figure.

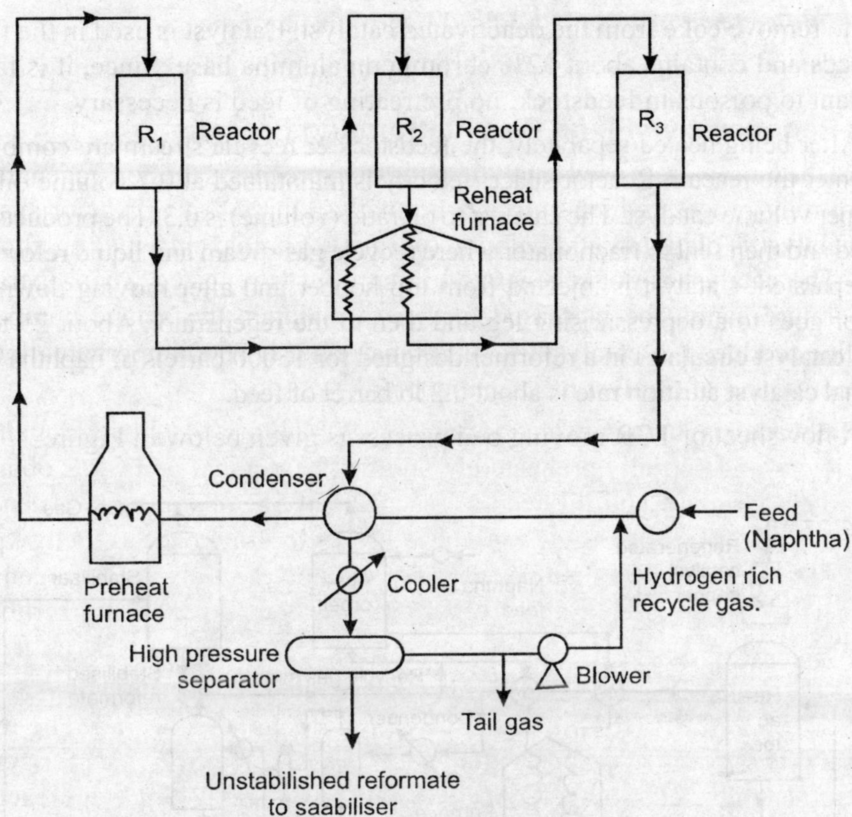

Fig. Flowsheet of Platforming process

Regenerative Fixed Bed Hydroforming Process

This process uses a catalyst containing 8-10% of molybdena on an alumina support.

Temperatures of 480–540°C and pressures from 14 to 20 kg/cm^2 are used. The catalyst beads are regenerated after 4 to 8 hours operation. Both upflow and downflow regeneration is done to hold the temperature between 565 to 595°C. A high partial pressure of hydrogen is maintained during hydroformingprocess by recycling the hydrogen produced by the dehydrogenation of naphthene and cyclisation of paraffins. Desul- phurisation also takes place, most of the sulphur present being removed as hydrogen sulphide. Gasoline yield is more than 75% with an octane number of 80-85 (unleaded).

Thermofer Catalytic Reforming (TCR) Moving Bed Process

TCR process uses a single reactor operating at 12 kg/cm^2 and 510–540°C. A catalyst regenerating kiln operating at atmospheric pressure and 540–570°C is used to remove coke from the deactivated catalyst. Catalyst is used in the form of beads and contains about 32% chromia on alumina base. Since, it is fairly resistant to poisons in feedstock, no pretreating of feed is necessary.

After being heated separately, the feedstock & recycle stream are combined and enter the reactor. Reactor space velocity is maintained at 0.7 volume oil per hour per volume catalyst. The catalyst to oil ratio (volume) is 0.3. The products are cooled and then sent to fractionator where recycle gas stream and liquid reformate are separated. Catalyst is injected from top hopper and after moving down the reactor goes to a depressurising leg and then to the regenerator. About 25 tons/hr. of catalyst circulates in a reformer designed for 15000 barrels of naphtha/day. Normal catalyst attrition rate is about 0.2 lb/barrel of feed.

A flowsheet of TCR moving bed process is given below in Figure.

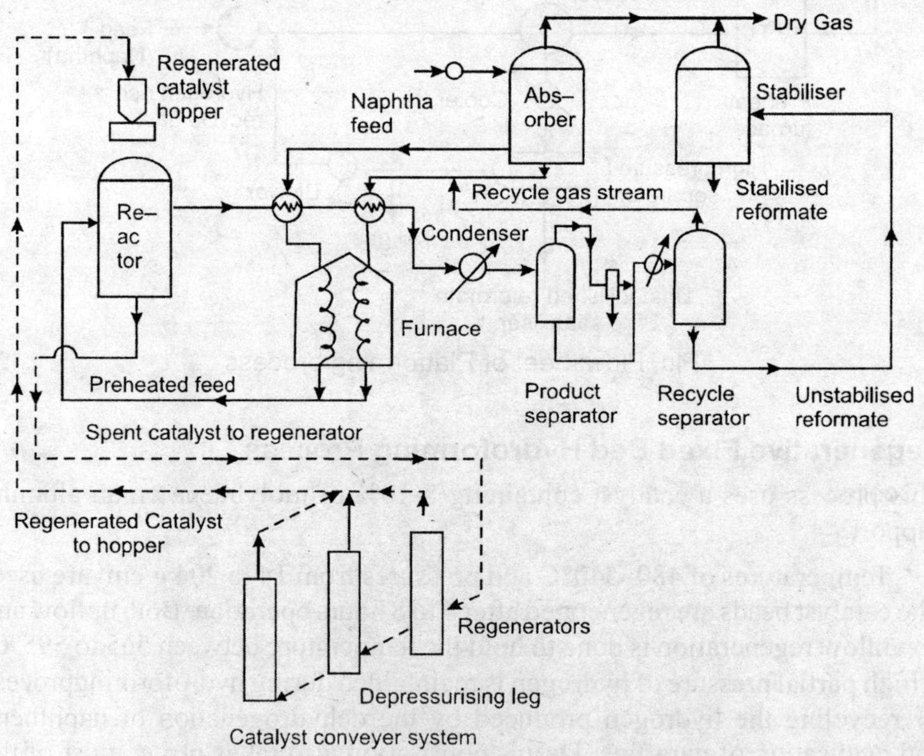

Flowsheet of TCR moving Bed Process

Fluidised Bed Hydroforming Process

In this process, a powdered catalyst containing 10% of molybdena,and 90% alumina is circulated between a reactor and a regenerator by the fluidised technique. Fluid hydroformers operate at 480°C and 14–20 kg/cm^2 (same as fixed bed hydroforming unit). To maintain this reaction temperature while supplying the endothermic heat of reaction, the recycle gas is heated to 650°C. The feed temperature is kept below 540°C to minimise thermal cracking.

Fluidisation enables the temperature to be controlled in the reactor very closely preventing the excessive reforming with production of gases that take place in fixed bed units. Similarly, regeneration temperature is more closely controlled by the use of cooling coils in the regeneration unit.

Catalyst circulation rate is about 1 kg catalyst per kg naphtha charge. A gasoline yield of about 80% with octane number of 95 (unleaded) can be obtained by this process.

A flowsheet of fluidised bed hydroforming unit is given below in Figure.

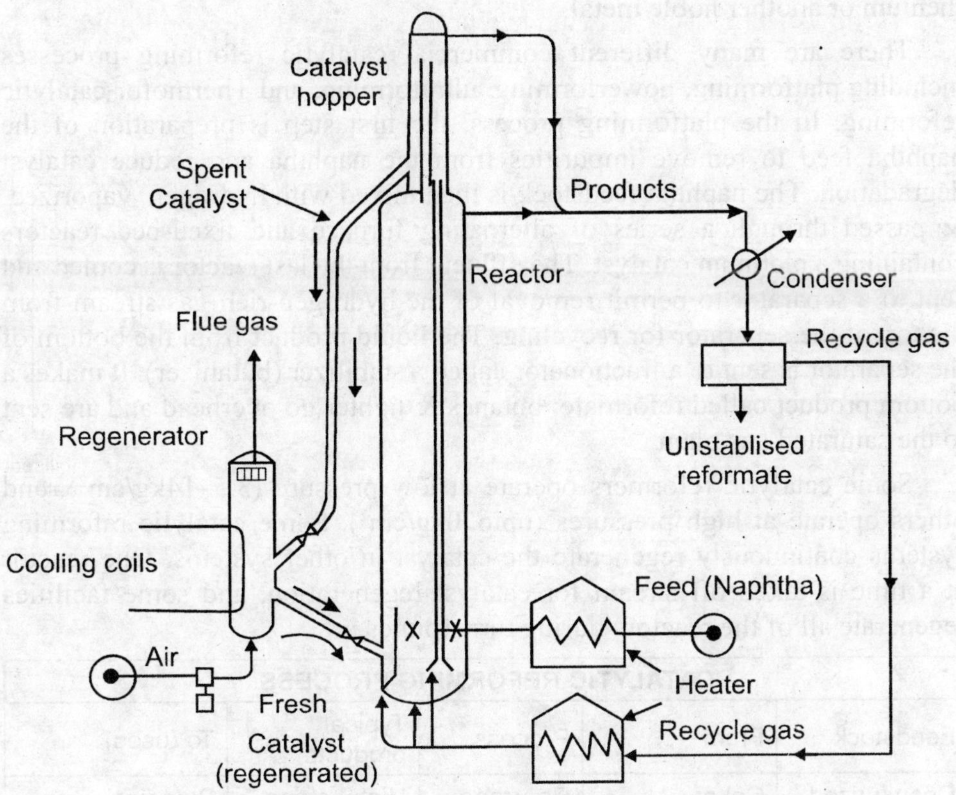

Fig. Flow sheet of Fluidised Bed Hydroforming Unit.

Appendix 5.1: Feature of Catalytic Reforming Process

Catalytic reforming is an important process used to convert low-octane naphthas into high-octane gasoline blending components called reformates. Reforming represents the total effect of numerous reactions such as cracking, polymerization, dehydrogenation, and isomerization taking place simultaneously. Depending on the properties of the naphtha feedstock (as measured by the paraffin, olefin, naphthene, & aromatic content) and catalysts used, reformates can be produced with very high concentrations of toluene, benzene, xylene, & other aromatics useful in gasoline blending and petrochemical processing. Hydrogen, a significant by-product, is separated from the reformate for recycling and use in other processes.

A catalytic reformer comprises a reactor section and a product-recovery section. More or less standard is a feed preparation section in which, by combination of hydrotreatment and distillation, the feedstock is prepared to specification. Most processes use platinum as the active catalyst. Sometimes platinum is combined with a second catalyst (bimetallic catalyst) such as rhenium or another noble metal.

There are many different commercial catalytic reforming processes including platforming, powerforming, ultraforming, and Thermofor catalytic reforming. In the platforming process, the first step is preparation of the naphtha feed to remove impurities from the naphtha and reduce catalyst degradation. The naphtha feedstock is then mixed with hydrogen, vaporized, & passed through a series of alternating furnace and fixed-bed reactors containing a platinum catalyst. The effluent from the last reactor is cooled and sent to a separator to permit removal of the hydrogen-rich gas stream from the top of the separator for recycling. The liquid product from the bottom of the separator is sent to a fractionator called a stabilizer (butanizer). It makes a bottom product called reformate; butanes & lighter go overhead and are sent to the saturated gas plant.

Some catalytic reformers operate at low pressure (3.5–14kg/cm^2), and others operate at high pressures (upto70kg/cm^2). Some catalytic reforming systems continuously regenerate the catalyst in other systems. One reactor at a time is taken off stream for catalyst regeneration, and some facilities regenerate all of the reactors during turnarounds.

CATALYTIC REFORMING PROCESS				
Feedstock	From	Process	Typical products	To (uses)
Desuifurized naphtha Naphthene-rich fractions Straight-run naphtha	Coker hydrocracker, hydrodesulfur Atmospheric fractionator	Rearrange dehydr-ogenate	High octane gasoline Aromatics . . . Hydrogen . . . Gas	Blending Petrochemical Recycle, hydrotreaty etc. Gas plant

6

Polymerisation, Alkylation and Isomerisation Processes

All these three processes namely *polymerisation, alkylation and isomerisation* are complimentary to catalytic cracking & reforming and are meant for production of superior quality higher octane number gasoline. Olefin rich cracker gases produced in refineries undergo polymerisation in presence of catalyst (like H_2SO_4 or H_3PO_4) at appropriate pressure and temperature to give *polymer gasoline* (rich in branched-chain hydrocarbons), catalyst

$$\underset{\substack{\text{iso–butane} \\ \text{(olefin)}}}{2CH_3C(CH_3) = CH_2} \xrightarrow[\text{P and T}]{\text{catalyst}} \underset{\substack{\text{di–isobutane} \\ \text{(branched–chain hydrocarbon)}}}{CH_3C(CH_3)_2CH_2.C(CH_3){=}CH_2}$$

In alkylation, a gaseous iso-paraffin (*e.g.* iso-butane) is reacted with an olefin (*e.g.* propene or butene) in presence of a catalyst (*e.g.* H_2SO_4 or H_3PO_4) at suitable pressure and temperature to produce superior alkylated gasoline (having larger iso-paraffin)

$$\underset{\substack{\text{iso–butane} \\ \text{(iso–paraffin)}}}{C_4H_{10}} + \underset{\substack{n\text{–butane} \\ \text{(olefin)}}}{C_4H_8} \xrightarrow[\text{T and P}]{\text{catalyst}} \underset{\substack{\text{iso–octane} \\ \text{(larger iso–paraffin)}}}{C_8H_{18}}$$

In isomerisation, n-paraffin (*e.g.* butane) is converted into iso-paraffin (*i*-butane) in presence of a catalyst (*e.g.* $AlCl_3$ or bauxite). Iso-butane is then used for production of superior gasoline by alkylation process. Pentane and hexane are also isomerised to produce high volatile gasoline components, iso-pentane & iso-hexane.

Pure petrochemicals are also produced using the processes like polymerisation and isomerisation.

Polymerisation

The gases produced in cracking are rich in C_2 and C_4 olefins (*i.e.iso*-butylene, iso-propylene etc.). These gases undergo polymerisation in presence of a catalyst (*e.g.* phosphoric acid or sulphuric acid) at suitable temperature and pressure to give products of gasoline boiling range rich in branched-chain hydrocarbons.

$$2\ \underset{\text{isobutylene}}{(CH_3)_2C = CH_2} \xrightarrow{\text{Polymerisation}} \underset{\text{di-isobutylene}}{CH_3-\overset{\overset{CH_3}{|}}{\underset{\underset{CH_3}{|}}{C}}-CH_2-\overset{\overset{CH_3}{|}}{C} = CH_2}$$

The higher liquid hydrocarbons produced (by polymerisation of lower gaseous olefinic hydrocarbon) in the gasoline boiling range is called polymer gasoline. Hence, polymerisation process is mainly for the production of superior gasoline and is complimentary to catalytic cracking and reforming.

Polymerisation is of two types namely thermal polymerisation and catalytic polymerisation. In thermal polymerisation (which is carried out without a catalyst) category, there are two processes namely Unitary thermal polymerisation and Multiple coil thermal polymerisation.

Thermal Polymerisation

In Unitary thermal polymerisation, the gases (containing mainly propane, propylene, butane & butylene) from cracker unit are subjected to polymerisation at 510–595°C & 70–310 kg/cm^2. The liquid products produced are polymer gasoline and gas oil. In multiple coil thermal polymerisation, the feed (olefinic gases + paraffinic gases) are subjected to polymerisation at 480–705°C & 3–55 kg/cm^2. This polymerisation takes place in three stages. In the first stage, the olefinic hydrocarbons polymerise. In the second stage, the polymerized olefinic hydrocarbons and the residue gas containing paraffinic gas after polymerisation are cracked. In the third stage, the olefinic hydrocarbons obtained from cracking in the second stage are polymerised. The products obtained are gasoline, fuel oil and tar.

Though polymer gasoline can be produced from any hydrocarbon (whether olefinic or paraffinic), very severe conditions are required for paraffinic gases lighter than butane. It is not necessary to convert all butane and butene produced in refinery because they can be blended into gasoline upto 10%.

Catalytic Polymerisation

The first catalytic polymerisation process employed sulphuric acid as a catalyst. There are two processes under this heading namely cold acid process and hot acid process. The other process uses phosphoric acid as catalyst.

Cold Sulphuric Acid Polymerisation Process

In this process *i*-butene is polymerised selectively to di-isobutylene & tri-isobutylene. The cold acid process consists of two steps *i.e.* extraction of *i*-butene followed by polymerisation. The extraction of *i*-butene is done by 65% sulphuric acid at 21 to 38°C in two stages. The extract phase is separated and heated to about 100°C for about 1 minute to effect polymerisation. It is then cooled to 38°C and the product is separated from acid catalyst by settling. Conversion of i-butene is about 90%. The product is generally 75% dimer *(i.e.* di-isobutylene) plus 25% trimer *(i.e.* tri-isobutylene).

A flowsheet of cold sulfuric acid process is given below in Figure.

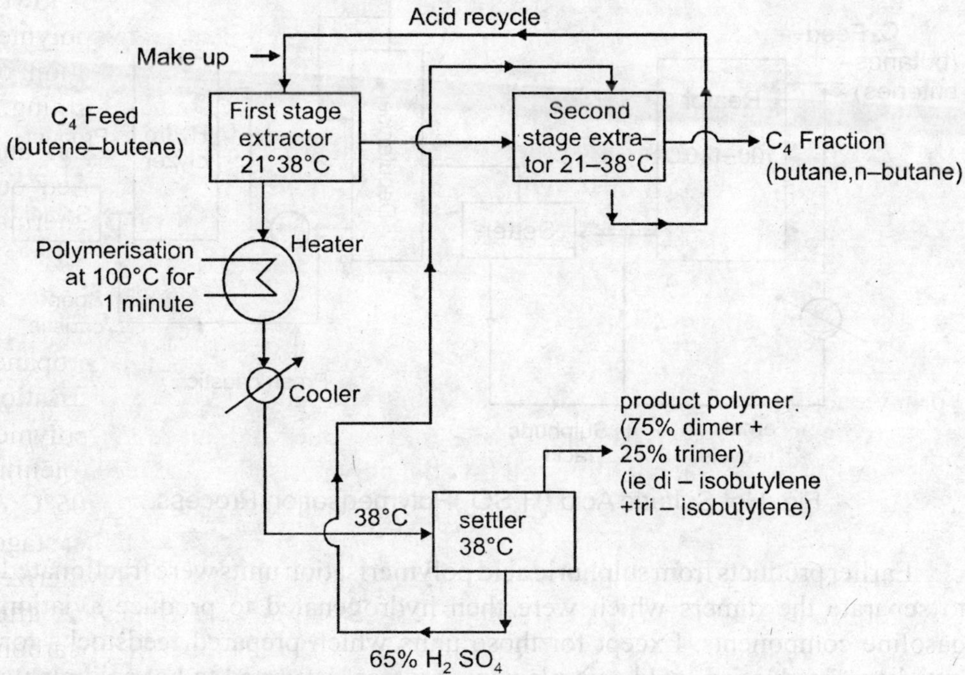

Figure-Cold Sulfuric Acid Polymerisation Process.

Cold acid process is also important in preparing feedstock for the manufacture of butadiene, *iso*- *Iso*-butene which interferes with the production of butadiene is removed from butanes-butenes *via* cold acid route.

Hot Sulpharic Acid Polymerisation Process

In hot acid process, all the butenes is polymerised (unlike cold acid process which polymerises only *i*-butene). In this process, extraction and polymerisation are carried out in single step at about 100–140°C. A hydrocarbon-acid emulsion is circulated through the reactor to reduce the concentration of iso-butene in the feed. Circulation promotes copolymerisation of more reactive *i*-butene with the less reactive *n*-butenes and thereby improves yields. Reactor effluent is settled to separate product from the catalyst, which is recycled. The product is fractionated to eliminate butenes and then hydrolysed with caustic to breakdown any acidic substances. At contact time of 10-15 minutes, all the *i*-butene and an equal amount of *n*-butenes are converted. The product contains 90–95% octenes.

A flow sheet of hot sulfuric acid polymerisation process is given below in Figure.

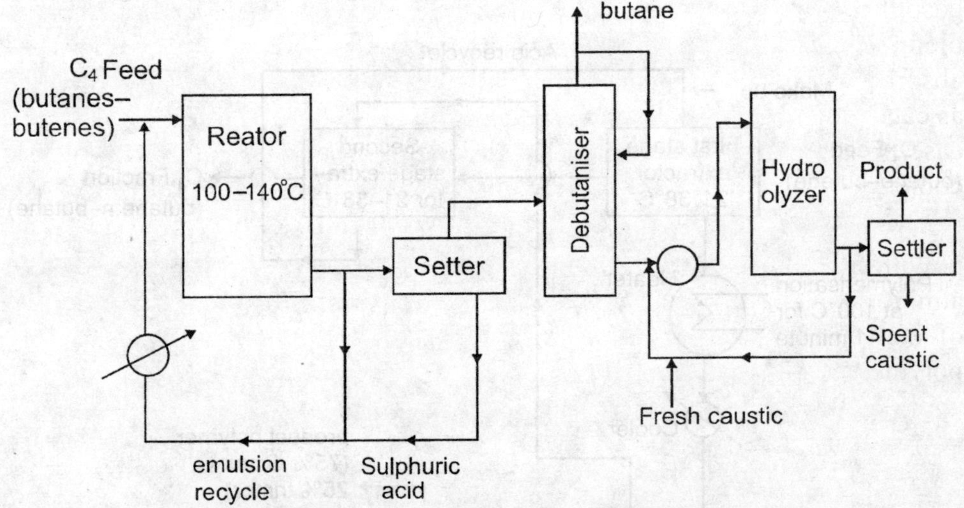

Fig. Hot Sulfuric Acid (H_2SO_4)Polymerisation Process.

Earlier products from sulphuric acid polymerisation units were fractionated to separate the dimers which were then hydrogenated to produce aviation gasoline components. Except for those units which prepared feedstocks for butadiene production, cold acid plants were later converted to hot acid plants due to higher production rate of the latter.

Phosphoric Acid (H_3PO_4)Polymerisation Process

In this process of catalytic polymerisation, H_3P0_4 catalyst used are of four types—three of them use solid catalyst and the fourth employs liquid phosphoric acid. Commercial solid catalysts include (a) phosphoric acid on kieselguhr or on silica and (b) copper pyrophosphate on charcoal. However, the latest trend is the use of liquid phosphoric acid as polymerisation catalyst. Liquid phosphoric acid is always used with 2% steam to prevent meta & para phosphoric acid formation which are inactive.

Feed is pretreated to remove hydrogen sulphide & mercaptans which would otherwise enter the product and reduce its octane rating. Feed is washed with caustic to remove H_2S & mercaptans and is then washed with water to remove organic bases & any carry over of caustic which poisons the catalyst. Both the feedstock & wash water should be free of dissolved oxygen as it promotes the deposition of tarry material on the catalyst. Since polymerisation units are operated at high pressure, coke deposition on beaded solid catalyst is greatly reduced, hence catalyst regeneration facility is not provided. Most units employ solid catalyst.

After pretreatment of feed, it is passed over a bed of catalyst at a temperature of 150–225°C & a pressure of 10–80 atm. Olefin conversion to gasoline are 85% or more. To remove the large exothermic heat of reaction, olefin lean stream(*i.e.* C_3–C_4 paraffin quench) is recycled to act as heat absorber to control the reaction temperature. When the liquid H_3PO_4 is used as catalyst, only a reaction tank will substitute the packed bed reactor. In this case, temperature is controlled by circulating the liquid catalyst through a cooler.

Typical process Conditions in the reactor having beaded solid catalyst for polymerisation are 230°C & 60-70 atm. for C_3 rich feed and 200°C & 40–45 atm. for C_4 rich feed. Quench stream is C_3–C_4 paraffins for cooling and control of olefin concentration in reactor to 25-30%. Beyond this concentration, polymerisation is vigorous. Product has an octane number (motor) of 80–81.

A flow sheet of phosphoric acid polymerisation unit is given below in Figure.

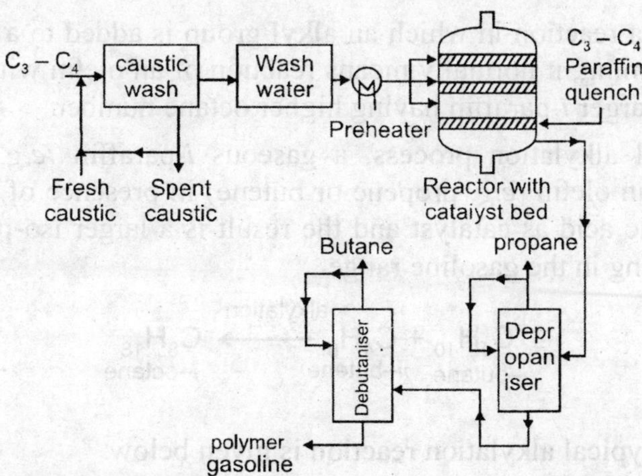

Fig. Flow sheet of phosphoric acid polymerisation process

Effect of reaction conditions in polymerization

Conversion of olefins to polymers depend upon temperature, pressure, reaction time, catalyst activity and the feed composition. Reaction rates increase with temperature, however higher temperature also increases rate of accumulation of tarry deposits on catalyst thereby shortening its life.

Increasing the pressure increases conversion by increasing contact time. Higher pressure permits the use of lower temperatures and thus reduce the formation of heavy polymers and coke.

Reaction time must be long enough to give satisfactory conversions, but not so long as to cause the production of polymer chains too long to be included in gasoline.

Reaction rate differs for the several olefins,Iso-butene is the most reactive which accelerates the polymerisation of *n*-butene which in turn accelerates the polymerisation of propene.

Polymerisation product quality

Polymer gasolines are chemically stable, clean burning fuel of high octane quality. They are usually mixture of polymers, mainly dimers and trimers. With a propene feed, trimers predominate while with a butene feed, dimers are major constituents.

Catalyst life

Catalyst life ranges from 100 to 200 gallons of polymer per pound for phosphoric acid on kieselguhr and between 100 to150 gallon per pound for copper pyrophosphate. Catalyst life in a liquid phosphoric acid operation exceeds 45 gallons of polymer per pound of acid.

Alkylation

Alkylation is a reaction in which an alkyl group is added to a compound. In petroleum refining, it normally means reaction of an olefin with an *i*-paraffin to produce a larger *i*-paraffin having higher octane number.

In typical alkylation process, a gaseous *i*-paraffm (*e.g.* iso-butane) is reacted with an olefin (*e.g.* propene or butene) in presence of sulphuric acid or hydrofluoric acid as catalyst and the result is a larger iso-paraffin (called alkylate) boiling in the gasoline range

$$\underset{i\text{–butane}}{C_4H_{10}} + \underset{n\text{–butene}}{C_4H_8} \xrightarrow{\text{alkylation}} \underset{i\text{–octane}}{C_8H_{18}}$$

Another typical alkylation reaction is given below.

$$i\text{-butane} + \text{propene} \xrightarrow{\text{alkylation}} i\text{-heptane}$$

Alkylation reactions. The reactions occurring during alkylation are :

(a) *iso* – paraffin + olefin ⟶ higher *iso* – paraffin. This is the main reaction.

(b) olefins $\xrightarrow{\text{Polymerisation}}$ dimer + *iso* – butane → *iso* – paraffin

(c) Hydrogen transfer : $\underset{iso\text{–butane}}{2i - C_4H_{10}} + \underset{n\text{–butene}}{nC_4H_8} \longrightarrow \underset{\text{iso–octane}}{C_8H_{18}} + \underset{n\text{–butane}}{C_4H_{10}}$

Alkylation processes.

There are two processes namely thermal & catalytic alkylation. In thermal alkylation, iso-paraffins are alkylated without catalyst but higher temperature (550°C) & pressure (250-350 atm) are required. As for example, neohexane (2-2 dimethyl butane) having an octane number of 94 can be produced by thermal alkylation by the reaction as shown below :

$$iso\text{ - butane} + C_2\text{ olefin} \rightarrow \text{neohexane}$$

Normally, reactivity of iso-butane decreases with increase in olefin carbon atom.

Catalytic alkylation process

In this process, catalyst like sulphuric acid, hydrogen fluoride and aluminium chloride are used for alkylation.

Sulphuric acid is used as an alkylation catalyst with propene and higher boiling olefins. The acid concentration is maintained between 90 to 98%. Less than 90% acid concentration promotes polymerisation whereas more than 98% acid concentration promotes cracking reaction. Acid catalyst deactivation rate depends upon the type of feed and the *i*-butane charge rate. Acid consumption ranges from 0.3 to 2 pounds per gallon of alkylated product.

Hydrogen fluoride having an acidity *(i.e.* concentration) of 85–92% is also used as catalyst for alkylation of *i*-butane with propene and higher boiling olefins. High acidity is used to maintain high conversion and high antiknock quality of alkylated gasoline. Hydrogen fluoride consumption ranges from 0.2 to 0.8 pound per barrel of product.

Aluminium chloride (with HCl as promoter), is used as a catalyst to produce di-isopropyl (for aviation gasoline) from *i*-butane and ethylene by alkylation. The optimum aluminium chloride concentration in the catalyst is 63-84% and its consumption is 1.4 to 3 pounds per barrel of product.

Alkylation process variables

Important process variables in alkylation are *i*-butane/olefin charge ratio, *i*-butane content in reactor effluent, temperature, olefin space velocity and contact time. Their effects are discussed below :

***i*-butane/olefin ratio.** iso-butane/olefin ratio is maintained at 5 to 10 in commercial alkylation plants. Higher ratio produces product with higher octane number'and having lower final boiling point besides reducing catalyst consumption rate.

***i*-butane in effluent.** Increasing the *i*-butane concentration in the reactor effluent has the same effect as increasing the *i*-butane/olefin ratio and it can be done by reducing the amount of *n*-butane charged to the reactors in the fresh feed and in the *i*-butane recycle stream.

Temperature. Low temperatures are used in all alkylation processes.

Operating temperature in various processes are :

(a) H_2S0_4 alkylation process — 5-10°C (This requires refrigeration).

(b) HF alkylation process — 25-38°C

(c) $AICI_3$ alkylation process — 45-60°C.

If temperature is less, then the octane number of final alkylate is more and yield of polymer is low. If temperature is too low then alkylate yield is reduced whereas at very high temperature, acid attacks olefins.

Space velocity of olefins. It means volume of olefins charged per hour per unit volume of acid in the reactor. Lower olefin space velocity improves the octane number of product, reduces the production of heavy materials and lowers the consumption of acid catalyst.

Contact time. It is defined as the residence time of the fresh feed and externally recycled *i*-butane in the reactor. Contact time for H_2SO_4 alkylation process is 5–45 minutes and for HF alkylation process it is 5–25 minutes. Long contact time in the absence of *i*-butane is undesirable because olefins undergo side reaction thereby lowering the product quality.

Feed stocks for alkylation. Any olefin containing hydrocarbon can be used to alkylate *i*-butane. The chief sources of olefins are catalytic cracking and dehydrogenation of paraffins (mainly butane). Butane is also isomerised to *i*-butane which is cracked thermally to produce an alkylation feed. Iso-butane is obtained from crude oils, from cracking operations, from catalytic reformers and from natural gas.

Alkylation feed should be free from hydrogen sulphide and mercaptans which makes the product corrosive, lowers leaded octane rating of products and increases the catalyst consumption rate.

Sulphuric acid alkylation process

Feed is given caustic wash to remove sulphur compound and then contacted with recycle *i*-butane. It is cooled to the required low temperature and contacted with the acid catalyst. Catalyst acid is settled from the hydrocarbon product and recycled. Caustic wash is then applied to remove traces of acid from the product before fractionating recycle *i*-butane stream from it. To prevent propane from building up in the recycle *i*-butane, a depropaniser is also provided. Also, a debutaniser is usually included to separate *n*-butane from the product and a rerun tower is required to remove heavy ends if the product is to be blended into aviation gasoline.

A flow sheet of the sulfuric acid alkylation process is given below in Figure.

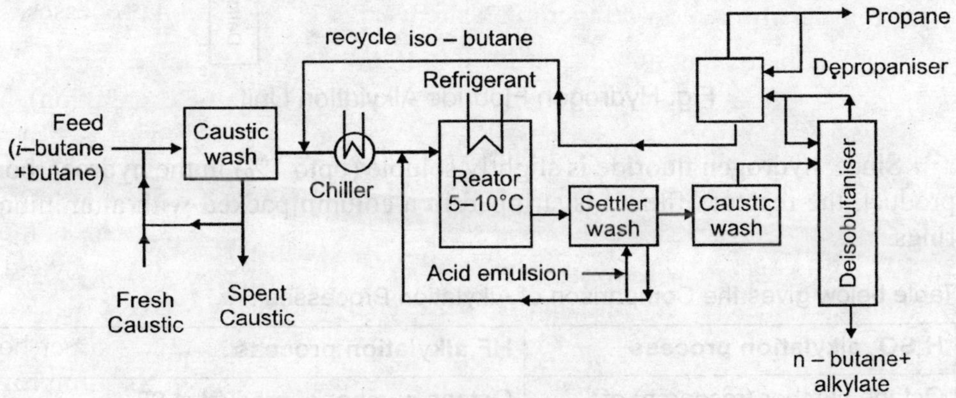

Fig. Sulphuric Acid Alkylation Process.

Sulphuric acid alkylation unit needs a refrigeration system to cool the reactants to 5–10°C.

Hydrogen Fluoride Alkylation Process. The process is almost same as H_2SO_4 alkylation process. This process does not require refrigeration equipment as the reactor temperature is 25-38°C. However, it requires, additional equipment to dry the feed, to strip HF from product, to rerun a slip stream of the circulating catalyst to reject water & tar and to remove traces of flourinated compounds from the product.HF is corrosive at working concentration, hence costly alloy monel is used in the equipment.

Acid/hydrocarbon ratio varies from 0.2 to 1 and optimum $i\ C_4$/olefin ratio is about 7 in this process. A flow sheet of the process is given below in Figure.

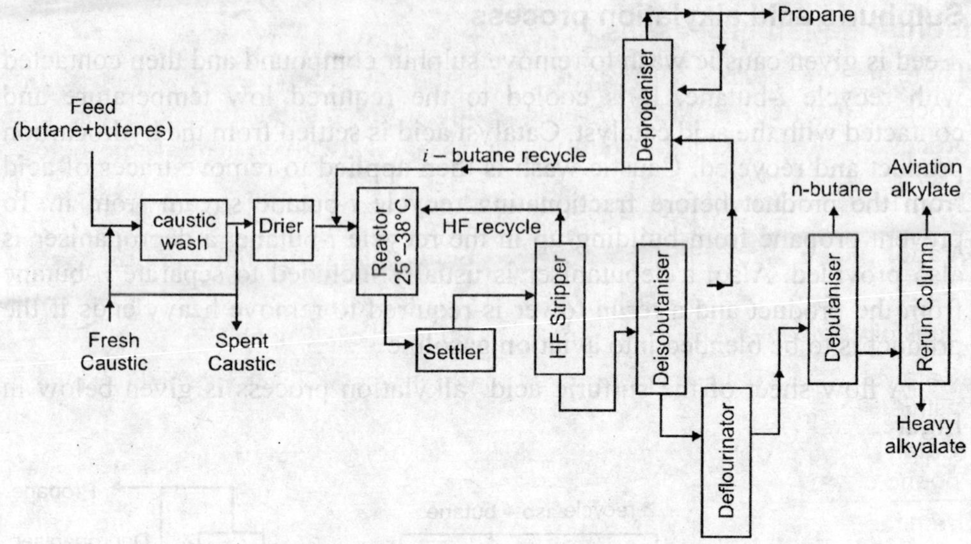

Fig. Hydrogen Flouride Alkylation Unit.

Since, hydrogen fluoride is slightly soluble (upto 1%) in the hydrocarbon product, the reactor effluent is stripped in a column packed with aluminium rings.

Table below gives the Comparison of Alkylation Processes :

H_2SO_4 alkylation process	HF alkylation process
Octane number (research) of product is 95.	Octane number of product is 93.
Plant running cost is more as refrigeration unit is there.	Initial investment is more as costly monel clad equipments needed.
Acid consumption is more.	Acid consumption is less.
Spent H_2SO_4 disposal is a problem.	HF poses safety hazard, hence protective clothing required for handling.

Isomerisation

Isomerisation process converts *n*-paraffins to *i*so-paraffms (which has a higher octane number) in presence of $A1C1_3$ as catalyst. Commercially, it is mostly practised to obtain *i*so-butane from *n*-butane which is then used in the production of good quality gasoline by alkylation process. Pentane & hexane are also isomerised to produce *i*-pentane & *i*-hexane which are high volatile gasoline components.

Isomerisation processes

There are two processes for isomerisation of paraffins namely low temperature process and high temperature process, the former being the more common process. Features of the low temperature process are given in the Table below.

Table: Features of Low Temperature Isomerisation Process (for butane isomerisation)

Vapor phase batch process	**Liquid phase continuous process**
This process has been developed by shell oil company.	This process has been developed by standard oil company.
Catalyst used is bauxite impregnated with $AlCl_3$.	Catalyst is $AlCl_3$with HCl as promoter.
Catalyst life i.e. gallon of i-butane per pound of $AlCl_3$ used is 200.	Catalyst life is 50-100.
Temperature & pressure for isomerisation is 100-150°C & 14 kg/cm^2 for C4 and 20 kg/cm^2 for C_3 - isomerisation..	Temperature and pressure is 100°C and 25 kg/cm^2.
Conversion is 40%.	Conversion is 38%.
HCl concentration (weight %) is 2-14%.	HCl concentration is 4%.

Low Temperature Isomerisation Process

Feedstock, dried and preheated to reaction temperature is combined with a recycle stream, mixed with HCl and passed through a reactor & an $AlCl_3$ recovery section. Reactor effluent is cooled and flashed to discharge any light gases through an absorber that recover HCl carried off in the gases. Liquid from the flash drum is stripped to recover HCl and is caustic washed to remove the traces of acid. Recycle paraffin stream is fractionated from the product isomer and then recycled back to the reactor.

Flow sheet of liquid phase, low temperature continuous isomerisation is given below in Figure.

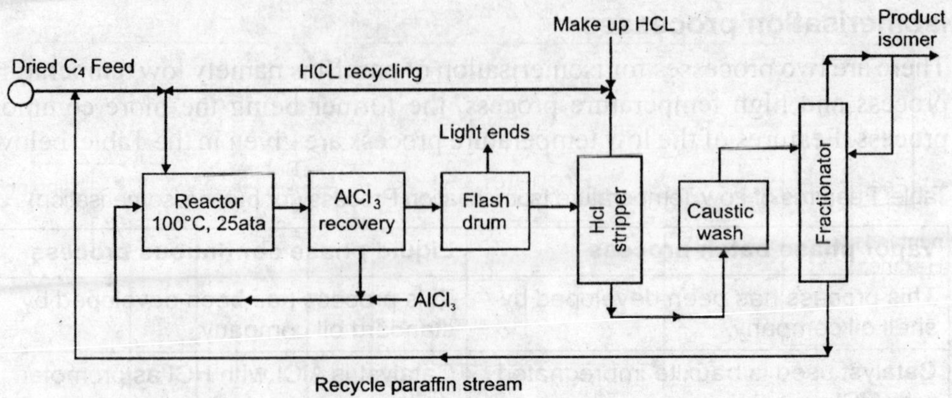

Fig. Low Temperature Liquid Phase Butane Isomerisation Process.

The HCl stripping column is operated at high enough pressure so that the stripped HCl can be returned directly to the reactor.

In the vapour phase butane isomerisation process, $AlCl_3$ is adsorbed on refined bauxite to prepare the catalyst. In vapour phase batch process, the plant has to be shutdown periodically to change the catalyst.

For pentane and hexane isomerisation, the process conditions are same as that for butane except that inhibitor *(e.g.* hydrogen at high pressure) are used to suppress the side reactions.

High temperature isomerisation process

In this process, catalyst used in the reactor is platinum on silica-alumina and the reaction conditions are 370-480°C & 20–45 kg/cm^2. Typically, a pentane -hexane feed combined with pentane recycle stream is fractionated to take *i*-pentane overhead. The fractionator bottoms are mixed with hydrogen, preheated and charged to the reactor. Reactor effluent is cooled & flashed to separate recycle hydrogen from the product which is stabilised and depentanised. The pentanes are recycled to the deisopentaniser. *iso*-pentane and the iso-hexane bottoms from the depentaniser constitute the products.

Hexanes isomerise more readily than pentanes. However, with a nickel catalyst, equal rates, are obtained when hexanes are processed at a temperature about 20°C lower than pentanes. For minimizing cracking reactions hexanes are processed at lower temperature than pentanes. Desulphurisation of feed and removal of nitrogen is done to improve product quality and life of catalyst.

Product yield varies from 94 to 98% (by volume) and the octane number ranges from 66 to 72.

A flowsheet of high temperature isomerisation process is given below in Figure.

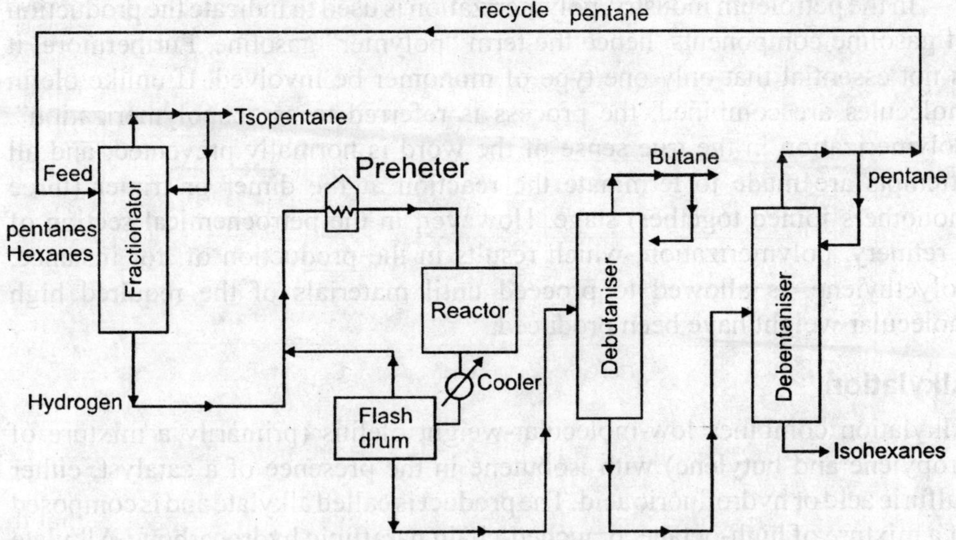

Fig. High Temperature Isomerisation Process.

Low temperature isomerisation processes are more common for butane isomerisation. High temperature isomerisation process, however, has the advantage of avoiding the handling of $AlCl_3$ and HCl. Balanced against this advantage are the less favourable paraffin equilibriums at higher operating temperature.

Appendix 6.1: Comparative features of polymerisation, alkylation and isomerisation process

Polymerization in the petroleum industry is the process of converting light olefin gases including ethylene, propylene, and butylene into hydrocarbons of higher molecular weight &higher octane number that can be used as gasoline blending stocks. Polymerization combines two or more identical olefin molecules to form a single molecule with the same elements in the same proportions as the original molecules. Polymerization may be accomplished thermally or in the presence of a catalyst at lower temperatures.

The olefin feedstock is pretreated to remove sulfur and other undesirable compounds. In the catalytic process the feedstock is either passed over a solid phosphoric acid catalyst or comes in contact with liquid phosphoric acid, where an exothermic polymeric reaction occurs. This reaction requires cooling water and the injection of cold feedstock into the reactor to control temperatures between 150-230 ºC at pressures from 15-80 kg/cm^2. The reaction products leaving the reactor are sent to stabilization and/or fractionator systems to separate saturated and unreacted gases from the polymer gasoline product.

In the petroleum industry, polymerization is used to indicate the production of gasoline components, hence the term "polymer" gasoline. Furthermore, it is not essential that only one type of monomer be involved. If unlike olefin molecules are combined, the process is referred to as "copolymerization". Polymerization in the true sense of the word is normally prevented, and all attempts are made to terminate the reaction at the dimer or trimer (three monomers joined together) stage. However, in the petrochemical section of a refinery, polymerization, which results in the production of, for instance, polyethylene, is allowed to proceed until materials of the required high molecular weight have been produced.

Alkylation

Alkylation combines low-molecular-weight olefins (primarily a mixture of propylene and butylene) with isobutene in the presence of a catalyst, either sulfuric acid or hydrofluoric acid. The product is called alkylate and is composed of a mixture of high-octane, branched-chain paraffinic hydrocarbons. Alkylate is a premium blending stock because it has exceptional antiknock properties and is clean burning. The octane number of the alkylate depends mainly upon the kind of olefins used and upon operating conditions.

Sulfuric Acid Alkylation Process

In cascade type sulfuric acid (H_2SO_4) alkylation units, the feedstock (propylene, butylene, amylene, and fresh isobutane) enters the reactor and contacts the concentrated sulfuric acid catalyst (in concentrations of 85% to 95% for good operation and to minimize corrosion). The reactor is divided into zones, with olefins fed through distributors to each zone, and the sulfuric acid and isobutanes flowing over baffles from zone to zone.

The reactor effluent is separated into hydrocarbon and acid phases in a settler, and the acid is returned to the reactor. The hydrocarbon phase is hot-water washed with caustic for pH control before being successively depropanized, deisobutanized, and debutanized. The alkylate obtained from the deisobutanizer can then go directly to motor-fuel blending or be rerun to produce aviation-grade blending stock. The isobutane is recycled to the feed.

Hydrofluoric Acid Alklation Process

Phillips and UOP are the two common types of hydrofluoric acid alkylation processes in use. In the Phillips process, olefin and isobutane feedstock are dried and fed to a combination reactor/settler system. Upon leaving the reaction zone, the reactor effluent flows to a settler (separating vessel) where the acid separates from the hydrocarbons. The acid layer at the bottom of the separating vessel is recycled. The top layer of hydrocarbons (hydrocarbon phase), consisting of propane, normal butane, alkylate, and excess (recycle) isobutane, is charged to the main fractionator, the bottom product of which is motor alkylate. The main fractionator overhead, consisting mainly of propane, isobutane, and HF, goes to a depropanizer. Propane with trace amount of HF goes to an HF stripper for HF removal and is then catalytically defluorinated, treated, and sent to storage. Isobutane is withdrawn from the main fractionator and recycled to the reactor/settler, and alkylate from the bottom of the main fractionator is sent to product blending.

The UOP process uses two reactors with separate settlers. Half of the dried feedstock is charged to the first reactor, along with recycle and makeup isobutane. The reactor effluent then goes to its settler, where the acid is recycled and the hydrocarbon charged to the second reactor. The other half of the feedstock also goes to the second reactor, with the settler acid being recycled and the hydrocarbons charged to the main fractionator. Subsequent processing is similar to the Phillips process. Overhead from the main fractionator goes to a depropanizer. Isobutane is recycled to the reaction zone and alkylate is sent to product blending.

Isomerization converts n-butane, n-pentane and n-hexane into their respective isoparaffins of substantially higher octane number. The straight-chain paraffins are converted to their branched-chain counterparts whose component atoms are the same but are arranged in a different geometric structure. Isomerization is important for the conversion of n-butane into isobutane, to provide additional feedstock for alkylation units, and the conversion of normal pentanes & hexanes into higher branched isomers for gasoline blending. Isomerization is similar to catalytic reforming in that the hydrocarbon molecules are rearranged, but unlike catalytic reforming, isomerization just converts normal paraffins to isoparaffins.

There are two distinct isomerization processes, butane (C_4) and pentane/hexane (C_5/C_6). Butane isomerization produces feedstock for alkylation. Aluminum chloride catalyst plus hydrogen chloride are universally used for the low-temperature processes. Platinum or another metal catalyst is used for the higher-temperature processes. In a typical low-temperature process, the feed to the isomerization plant is n-butane or butanes mixed with hydrogen (to inhibit olefin formation)which is passed to the reactor at 110–170°C & 15–22 kg/cm^2 Hydrogen is flashed off in a high-pressure separator and the hydrogen chloride is removed in a stripper column. The resultant butane mixture is sent to a fractionators (deisobutanizer) to separate n-butane from the isobutane product.

Pentane/hexane isomerization increases the octane number of the light gasoline components n-pentane and hexane, which are found in abundance in straight-run gasoline. In a typical C_5/C_6 isomerization process, dried & desulfurized feedstock is mixed with a small amount of organic chloride &

recycled hydrogen, and then heated to reactor temperature. It is then passed over supported-metal catalyst in the first reactor where benzene and olefins are hydrogenated. The feed next goes to the isomerization reactor where the paraffins are catalytically isomerized to isoparaffins. The reactor effluent is then cooled and subsequently separated in the product separator into two streams: a liquid product (isomerate) and a recycle hydrogen gas stream. The isomerate is washed (caustic and water), acid stripped, and stabilized before going to storage.

7

Purification of Petroleum Products

Impurities and undesirable components in petroleum products come from the original crude or may develop during its processing. Purification processes are aimed at :

- **Minimizing**
  - (a) acid and corrosion
  - (b) carbon residue
  - (c) catalyst deactivation
  - (d) gum formation.
- **Improving**
  - (a) burning characteristics of kerosene, diesel fuels etc.
  - (b) lead susceptibility and octane number of petrol.
  - (c) pour point of diesel.
  - (d) colour and odour.
- **recovering**
  - (a) hydrogen sulphide and mercaptans
  - (b) alkyl phenols
  - (c) naphthenic acids
  - (d) sulphonates and nitrogen bases.

Sweetening Processes

Removal of sulphur and its compounds like hydrogen sulphide, mercaptans etc. from petroleum products is called its sweetening. Presence of these compounds makes the oil sour.

Petroleum products are termed as sweet if they pass the doctor test. In 'doctor test', oil sample is shaken with an equal volume of sodium plumbite solution and a small amount of sulphur. The mercaptans present in oil first reacts with sodium plumbite to form oil soluble lead mercaptides which on further reaction with elemental sulphur form sweet smelling oil soluble disulphide and give a black precipitate of lead sulphide. Hydrogen sulphide also gives black precipitate with the plumbite solution. If this test does not produce dark colour at the interface, the oil sample is sweet.

Sour stocks have bad odour. The processes by which the foul smelling sour petroleum., products are converted into sweet smelling or odourless products are called sweetening processes.

Important sweetening processes are ;

- Doctor's sweetening process
- Copper chloride sweetening process
- Solutizer process
- Catalytic desulphurisation process
- Hydrofining process.

Doctor's Sweetening Process

In this process, mercaptans present in the petroleum products are converted into disulphide by treatment with doctor's solution (sodium plumbite Solution) and sulphur. Lead sulphide is converted back (regeneration) to sodium plumbite by air blowing the solution.

Reactions : $\underset{\text{mercaptans}}{2RSH} + NaPbO_2 \longrightarrow Pb(RS)_2 + 2NaOH$

$$Pb(RS)_2 + S \longrightarrow PbS + RSSR$$

Regeneration : $PbS + 2O_2 \longrightarrow PbSO_4$

$$PbSO_4 + 4NaOH \longrightarrow Na_2PbO_2 + Na_2SO_4 + 2H_2O$$

Because the presence of sodium sulphate reduces the solubility of lead sulphate, a portion of the sodium is discarded periodically to control the build up of sodium sulphate.

Discarded doctor solution may be regenerated by settling to separate an emulsion of caustic, oil and lead sulphide from a top oil phase & a bottom sodium plumbite phase; breaking the emulsion by heating it to 65°C and separating a slurry of lead sulphide and air blowing it at 80°C for about 43 hours at atmospheric pressure.

Doctor's sweetening process is carried out at *65°C*. 20 parts of doctors solution are circulated per 100 parts of lead. Feed & doctor solution are contacted for 30 seconds before sulphur solution is added and the total mixture is then stirred for 5-15 minutes.Sulphur addition should be controlled as excess sulphur produces polysulphides [$Pb(RS)_2 + 2S \longrightarrow PbS + RSSR$] which reduces octane rating of gasoline and lower the burning qualities of distillate fuels.

Regeneration of doctor's solution is carried out at 80-110°C & 1-7 atmosphere pressure.

Flowsheet of the Doctor sweetening process is given below in Figure.

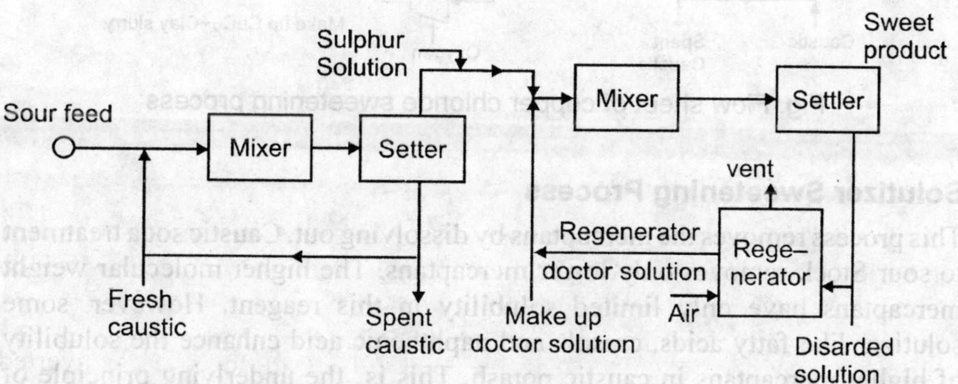

Fig. Doctor's Sweetening Process.

Copper Chloride Sweetening Process

In this process, no extra sulphur is added. Catalyst is a slurry of $CuCl_2$.

Reactions : $4RSH + CuCl_2 \longrightarrow RSSR + HCl + Cu_2Cl_2$

$$Cu_2Cl_2 + 4HCl + O_2 \longrightarrow 4CuCl_2 + 2H_2O$$

This process is used for the sweetening butane, gasoline, fuel oils and kerosene. After caustic washing & warming, the feed is mixed with air or oxygen and then with a slurry of copper chloride (one part) & clay (four parts). The mixture is sent to a reactor-separator from which catalyst is recycled to the feed. Treated product is water washed before withdrawing from the system. Reaction temperature is between 25 to 50°C. At lower temperature, the reactions are slow, whereas at higher temperature there are undesirable side reactions.

A flowsheet of the copper chloride sweetening process is given below in Figure.

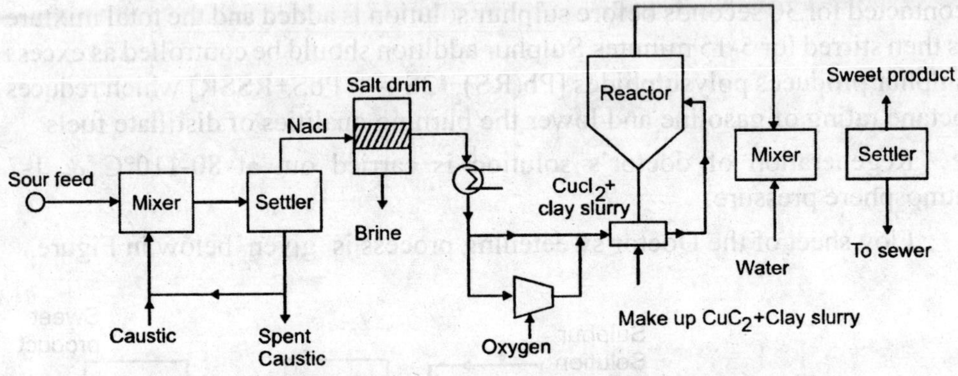

Fig. Flow sheet of copper chloride sweetening process

Solutizer Sweetening Process

This process removes the mercaptans by dissolving out. Caustic soda treatment to sour Stock removes only lower mercaptans. The higher molecular weight mercaptans have only limited solubility in this reagent. However, some solutizer like fatty acids, cresols and naphthenic acid enhance the solubility of higher mercaptans in caustic potash. This is the underlying principle of solutizer process.

When methanol is used as solutizer solution then the process is known as *Unisol Process.* When naphthenic acid is used as *solutizer,* then it is called *'Mercapsol Process'.*

In the solutizer process, the sour charge is caustic washed and then mixed with solutizer solution in a rotating disc contactor (RDC) maintained at 30°C & 3-7 atm. pressure. The spent solution is regenerated and recycled with make up solutizer solution. This is a costlier process and hence is used only when mercaptans in feed is high (> 300 ppm).

A flowsheet of the Solutizer sweetening process is given below in Figure.

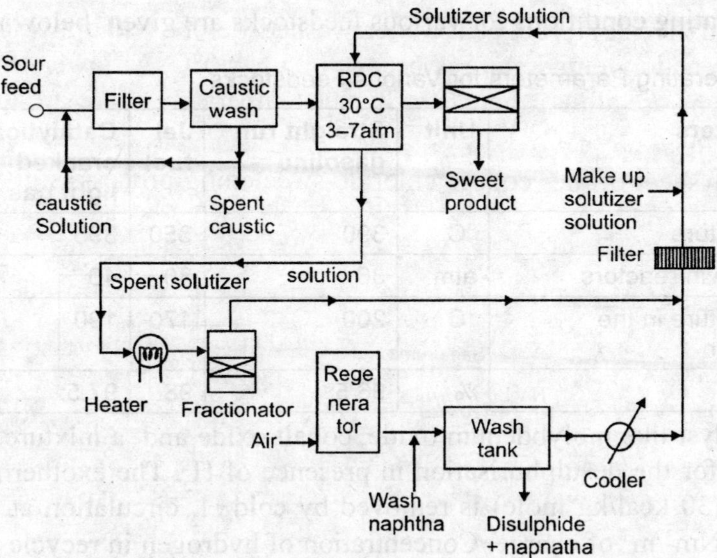

Fig. Flow sheet of Solutizer sweetening process

Catalytic Desulphurisation Process

Catalysts like bauxite, fuller's earth, activated clay etc. catalyse the conversion of organic sulphides & mercaptans to hydrogen sulphide & hydrocarbons at 350-400°C.

e.g.,

$$\underset{\text{mercaptans}}{C_9H_{19}SH} = C_9H_8 + H_2S$$

$$2C_9H_{19}SH = C_9H_{19}SC_9H_{19} + H_2S$$

$$C_9H_{19}SC_9H_9 = C_9H_{19}SH + C_9H_{18}$$

It is used for sweetening (desulphurisation) of straight run gasoline and kerosene. Removal of sulphur from 0.425 to 0.011% and octane number improvement from 45 to 48 can be achieved.

Hydrofining Desulphurisation Process

This process is most recent and most effective desulphurisation process. In this process, catalytic desulphurisation is carried out in presence of hydrogen recirculation. Hydrogen enhances the process of desulphurisation and minimises coke formation. Reaction conditions are 350-450°C & 20-50 atmosphere depending upon the nature of feedstocks.

Operating conditions for various feedstocks are given below in table.

Table: Operating Parameters for Various Feedstocks

Parameters	Unit	Straight run gasoline	Jet fuel	Catalytically cracked light gas oil (LGO)
Temperature	°C	390	350	395
Pressure in reactors	atm.	36	30	40
Temperature in the separator	°C	200	170	190
Yield	%	96.5	98	97.5

Catalyst like molybdenum oxide, cobalt oxide and a mixture of the two are used for the desulphurisation in presence of H_2. The exothermic heat of reaction (30 kcal/kg mole) is removed by cold H_2 circulation at the rate of 300-600 Nm^3/m^3 of charge. Concentration of hydrogen in recycle gas stream is maintained at 70-90%.

A flowsheet of the hydrofining sweetening process is given below in Figure.

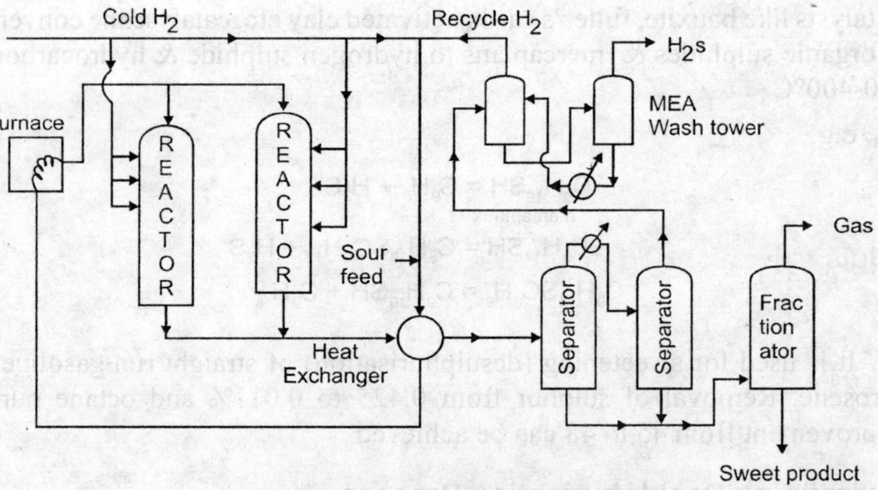

Fig. Hydrofining sweetening process.

Dewaxing of Petroleum Products

Removal of wax from petroleum products (mainly the lubricating oil) is called dewaxing. Petroleum waxes are solid hydrocarbons (melting point 35-95°C) and are soluble in crude oils & their products. Waxes can't be separated by distillation as their boiling points overlap with the petroleum products. At lower temperatures, waxes crystallise and precipitate out from fusel fuels and lubricating oils. Hence, to maintain their fluidity they must be dewaxed.

There are three methods of dewaxing:

(a) Chilling and pressing process

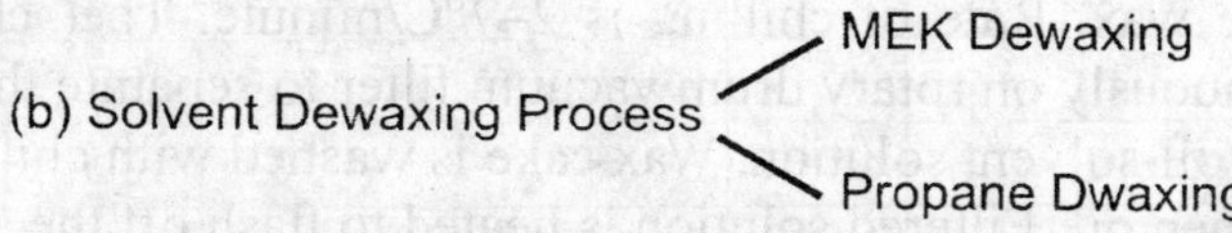

(c) Urea dewaxing Process.

Dewaxing raises the viscosity of the oil (*i.e.* the viscosity index is lowered) because the viscosity of dissolved wax is lower than that of oil with which it is associated.

Chilling and Pressing Dewaxing Process

This is the oldest method in which wax bearing oil is chilled by refrigeration in double pipe heat exchangers to crystallise the waxes which are then separated by filtration,centrifuging or settling. Since the viscosity of oil increases after chilling, filtration becomes difficult, time consuming & costly. Hence solvent dewaxing method has been developed.

Solvent Dewaxing Process

In this process, wax bearing oil is first dissolved in the solvent, and the solution is chilled. The crystallised wax is then more conveniently separated from the solution in vacuum filters. Methyl Ethyl Ketone (MEK) and propane are the two solvents widely used in the process.

Requisites of a good dewaxing solvent:

- It should have complete miscibility with oil at high temperature.
- It should have negligible solubility of wax.
- Presence of wax in solvent should be in filterable form.
- Its boiling temperature should be low so as to facilitate easy separation for easy recovery without much heat expense.
- It should have low solvent to feed oil ratio.
- It should be thermally stable and non-corrosive.

MEK (Solvent) Dewaxing Process

Normally, solvent methyl ethyl ketone (MEK) is mixed with aromatics like benzene & toluene in following proportion : MEK = 25–50%, Benzene = 40–60%, Toluene = 12–15%.

The MEK causes wax to solidify in a crystalline, easily filterable form while the aromatics increases the capacity of the solvent for dissolving oil.

The wax bearing lube oil is mixed with the solvent (by heating) and the mixture is chilled first by heat exchange with cold filtrate from filtration and then with a refrigerant (ammonia or propane refrigeration to –20°C) to crystallise the wax. Rate of chilling is 2–7°C/minute. The(chilled feed is filtered continuously on rotary drum vacuum filter to separate the crystallised wax from the oil-solvent solution. Wax cake is washed with chilled solvent to displace retained oil. Filtered solution is heated to flash off the solvent which is condensed and recycled. Oil from the flashing operation still contains traces of solvent which are stripped out by steam.

The wax cake contains the original water content as ice & also some solvent. The cake is melted and sent to a settling drum where melted wax &water separate ; solvent distributing between them. Dilute solvent from oil stripper is also charged to the settling drum, some of its solvent passing to the melted wax. The aqueous phase from the drum is fractionated to reject water as bottoms, the overhead being recycled to the drum to recover its solvent content by transfer to the melted wax. Melted wax from the drum is sent to an evaporator and a steal stripper. Solvent from the evaporator is returned to the warm solvent stream. Dilute solvent from the wax stripper is sent back to the wax settling drum.

A flowsheet of the MEK dewaxing process is given below in Figure.

Solvent to feed oil ratio in MEK Dewaxing process generally ranges from 1 : 1 to 4 : 1 .Heavier feedstocks require larger ratio. Yield of dewaxed oil is 89 to 92% while that of wax is 8 to 11%.

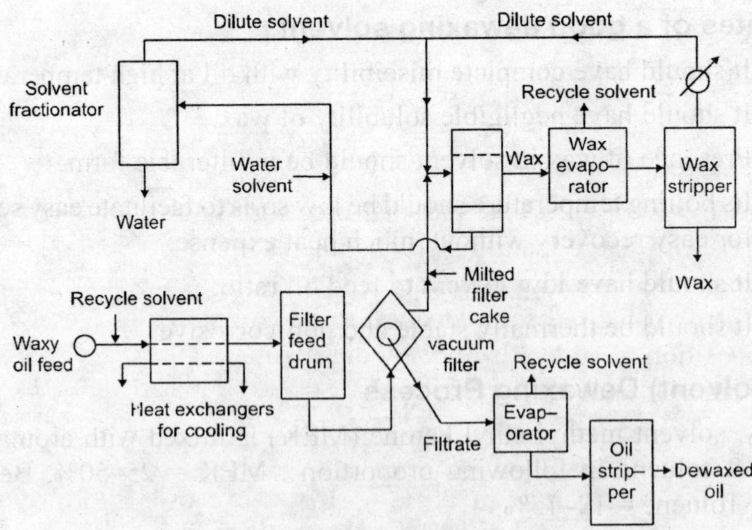

Fig. Flow sheet of MEK dewaxing process

Propane Dewaxing Process

Propane dewaxing unit is almost similar to MEK dewaxing plant. However, compared to MEK dewaxing process, it is advantageous in the sense that a part of solvent propane itself vaporises and provides refrigeration effect. The comparative disadvantages are that a high pressure is maintained in the propane evaporation system.

A flowsheet of the propane Dewaxing process is given below in Figure.

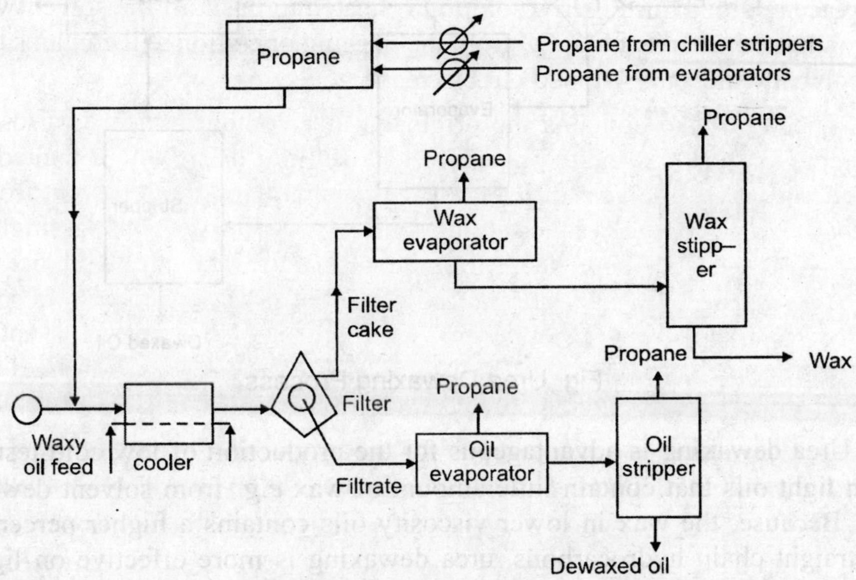

Fig. Propane Dewaxing Process

Solvent to feed ratio is generally higher (between 2 to 3). If it is less than 1.6, then gel formation occurs. Propane dewaxing requires a slightly lower temperature than MEK dewaxing.

Urea Dewaxing Process

This process is based on the fact that urea forms solid, filterable complexes (adducts) with large paraffins (constituents of wax) having little or no branching. Urea dewaxing is very effect on higher stocks. Activator like methyl alcohol is used for adduction to be completed in a reasonable time. Decomposition of adduct is done by dissolving the urea in a suitable solvent or by heating it to about 90°C.

A flowsheet of the urea dewaxing process is given below in Figure.

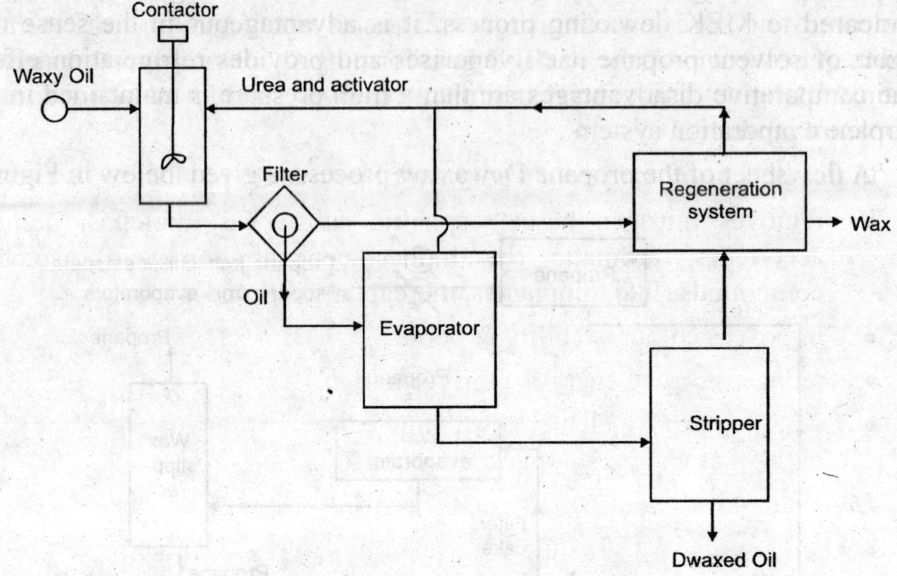

Fig. Urea Dewaxing Process

Urea dewaxing is advantageous for the production of low cold test oils from light oils that contain little amount of wax *e.g.* from solvent dewaxed oils. Because, the wax in lower viscosity oils contains a higher percentage of straight chain hydrocarbons, urea dewaxing is more effective on lighter stocks.

Deoiling of Wax

Deoiling of wax is done to remove the oil left in the wax produced in dewaxing process. The oil content in wax may be upto 50%, if chilling is done without solvent. Finished wax should not contain more than 0.5% oil. Sweating, resettling and solvent extraction are the three deoiling processes.

In sweating process, the crude wax (slack wax) is melted and cast into thick sheets which are then subjected to slow heating on pans with perforated false bottoms. As the temperature rises, the mixture of oil and low melting wax drains out. The heating is continued until the remaining wax has the desired melting point. The sweating process is repeated for the production of good quality wax.

In settling process, the slack wax is diluted with cold naphtha and recrystallised by chilling .Final separation of wax crystal is by centrifuging.

Solvent deoiling involves similar steps as in solvent dewaxing. MEK or propane is used for deoiling wax.

Acid, Alkali and Clay Treatment of Petroleum Products

Acid Treatment: Petroleum products of all boiling ranges (naphtha to lubricating oils) are given finishing treatment with sulphuric acid. A mild acid treatment removes the unstable gum forming & colour forming materials (mainly diolefins) and severer treatment removes sulphur as well from the petroleum products.

In nutshell, *sulphuric acid treatment* of petroleum products :

- removes nitrogen bases, asphaltic substances, diolefins, olefins, acetylenes, aromatic (by higher concentration acid), sulphur compounds (like thiophenes, mercaptans, sulphides).
- improves colour, stability & odour.
- reduces corrosiveness & gum formation.
- does not remove elemental sulphur, carbon disulphide and some high molecular weight sulphur compounds.

Drawback of sulphuric acid treatment of petroleum products are :

- large refining losses due to formation of acid sludge.
- problem of disposal of acid sludge.

Acid treatment comprises of bringing together acid & raw stock in desired proportion at a suitable temperature and then separating the phases by settling. The oily phase is neutralised and the spent acid is discarded. The product is then fractionated to remove high boiling material produced during acid treatment.

Generally, 93% strength (66°Be) H_2S0_4 is used. If the aim is to only improve the colour of light distillates, then a more dilute acid is used. Lube oil refining needs upto 98% strength acid. Lower contacting temperature yields better colour products and favours the removal of sulphur compounds. Gasolines are treated at 20-30°C. Higher temperatures (30-50°C) are used for treating kerosenes because removal of aromatics, unsaturated hydrocarbons & asphaltic material is favoured at those temperatures. High viscosity lubricating oil need still higher temperature (40-80°C) to ensure intimate contact with the acid. Higher temperature enhances refining losses. Heavier products need a longer time of contact.

Use of acetic anhydride and hydrofluoric acid to improve colour stability & odor of distillates have found limited use.

Alkali (caustic soda) Treatment: After processing with acid, the refined petroleum products are given a caustic soda wash to neutralise the acidity. Alkali wash treatment also removes naturally occurring acidic substances *viz.* hydrogen sulphide, carbonyl sulphide,lower aliphatic mercaptans, phenol, aryl mercaptans, fatty acids and naphthenic acids from petroleum products.

Clay Treatment: Clay is used as an adsorbant of undesirable substances from petroleum products. Clay treatment is done to :

- decolorise and stabilise cracked gasoline.
- desulphurise straight run gasolines and kerosenes.
- decolorise heavy oil products.
- neutralise acid treated heavy oils that cannot be caustic washed because of emulsification difficulty.
- remove (adsorb) arsenic from feedstock to catalytic reforming.
- remove asphaltic resinous and acidic materials from heavy oils & waxes.
- improve oxidation resistance of products.

In the treatment of gasolines & kerosene, the clay acts as a catalyst , whereas for reformer feedstocks & heavy oil , it is used as an adsorbant.

High asphalt oils are deasphalted by acid treatment or solvent extraction and then given a clay treatment. Conventional acid treatment of petroleum products are followed by clay treatment for neutralisation of the charge by the absorption of sludge and acidic materials.

Clays *(e.g.*, fuller's earth and bauxide) catalyse the polymerisation of color & gum forming materials in cracked gasolines. It also catalyses conversion of organic sulphides & mercaptans to hydrogen sulphide & hydrocarbons at 300–350°C.

The liquid phase clay treatment is carried out by either percolation or by contact filtration. The former consists in percolating the liquid charge through a fixed or moving bed of clay. In filtration, liquid is mixed with the absorbant and filtered.

In vapour phase clay treatment, vapors are passed over a fixed bed of clays. Under desulphurisation conditions, coke deposits deactivate the catalyst which can be regenerated by burning off the coke.

Spent clay disposal is a big problem, as it has limited demand in the manufacture of cinder bricks. Spent clays from the processes other than desulphurisation are usually not regenerated.

Deasphalting of Petroleum Products

Asphalts are high molecular weight compounds present in petroleum products in concentration from 0 to 65% and have high boiling points in the range of lube oil & heavy gas oils (HGO). Asphalt readily oxidise & form carbonaceous sludges and hence must be removed from lubricating oils. Asphalts are undesirable in catalytic cracking feedstocks (i.e. HGO) because they form coke. The processes by which asphalts are removed from heavy stocks is called 'deasphalting'. Deasphalting is done by distillation (atmospheric & vacuum), acid treating and solvent (propane) extraction.

Acid treatment for deasphalting is obsolete due to heavy losses of hydrocarbons. Bulk of deasphalting is done by atmospheric and vacuum distillation. However, lot of desirable materials for lubricating oils is still left in the residue from vacuum distillation. Solvent deasphalting is the best process from this point of view.

Propane (solvent) Deasphalting Process.

Propane dissolves hydrocarbons and rejects asphaltic materials near its critical temperature. It also dissolves smaller molecules in preference to larger ones. In the range of conditions used, raising the temperature of propane reduces its dissolving capacity and improves its selectivity. Increasing the solvent (propane) to oil ratio increases the sharpness of separation.

Reduced crude (feed) is fed near the top and liquid propane near the bottom of a liquid-liquid contacting column (Deasphalting Tower). The propane passes up the column, dissolving the oil, and carrying it overhead. Asphaltic materials pass out to the bottom. The top of bottom of the column is steam heated to reduce the oil carrying capacity of propane. As a result, some oil refluxes down the column to improve the sharpness of separation. Some propane is dissolved in the reject asphalt which is removed by evaporation and steam stripping. Similarly propane from deasphalted oil is also removed. These are condensed and recycled.

Typical propane deasphalting process parameters are given below.

Deasphalting tower details

Temperature, at the bottom	= 40-55°C
Temperature, at the top	= 65-85°C
Pressure	= 37.5 kg/cm^2
Height	=12 metres
No. of perforated trays	= 16

Propane to oil ratio is 4-12 (by volume) and the softening point of asphalt is about 60°C.

At a higher deasphalting tower top temperature (90°C), the heaviest lubricating stock is rejected with the asphalt at the bottom. The bottoms then are charged to a second column operating at the usual conditions to separate heaviest lube oil stock from asphalt. This is then called propane fractionation because such an operation separates the potential lubricating oil into two fractions.

A flowsheet of the propane deasphalting process is given below in Figure.

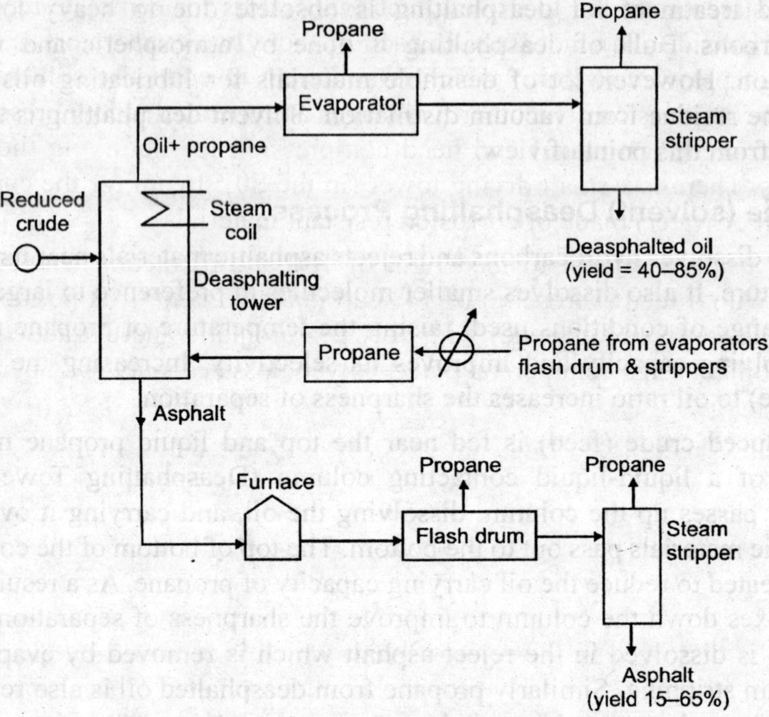

Fig. Propane Deasphalting Process

Asphalt thus recovered is further purified by air blowing in presence of catalyst like ferric chloride and phosphorous pentoxide thereby raising its softening point.

Dearomatisation(Purification) of Kerosene (Edeleanu Process)

Removal of aromatic compounds from kerosene is called dearomatisation of kerosene. Presence of aromatics in kerosene makes it smoky (*i.e* reduces its smoke point) hence, they are removed from it by liquid sulphur dioxide solvent extraction in a process called ***Edeleanu Process***. At its boiling point (–10°C), liquid sulphur dioxide completely dissolves aromatics, unsaturates, mercaptans & many other sulphur compounds but the paraffins &naphthenes remain insoluble in it. Demerit of this solvent (liquid SO_2) is that moisture cannot be tolerated in the system because the solvent then becomes highly corrosive.

In the **Edeleanu Process** of kerosene purification (ie, dearomatisation) to improve its smoke point, raw kerosene is dried, cooled and then charged to a packed extraction tower near the bottom. The recycled solvent (liquid SO_2) is refrigerated and fed at the top. Both the overhead raffinate & the bottom extract streams are fed to triple effect evaporators for complete recovery of sulphur dioxide. The first stage in each of these two evaporators is operated at a high pressure (6 atm.) which is sufficient to condense solvent vapour without compression. The other two stages are operated at low pressures (25 mm Hg in the last stage) and need compression for liquifying the solvent vapours. There is a small drying tower (in the $S0_2$ stream for the continuous removal of water) made of corrosion resistant material.

By using a solvent to feed ratio of 100 and extraction temperature of –18°C with a feed (kerosene) of 14% aromatic content, it is possible to get a product (purified kerosene) with only 2% aromatic and extract containing 82% aromatics in a commercial plant. Solvent loss is about 0.4 kg per kilolitre of the charge.

A flowsheet of the 'Edeleanu Process' is given below in Figure.

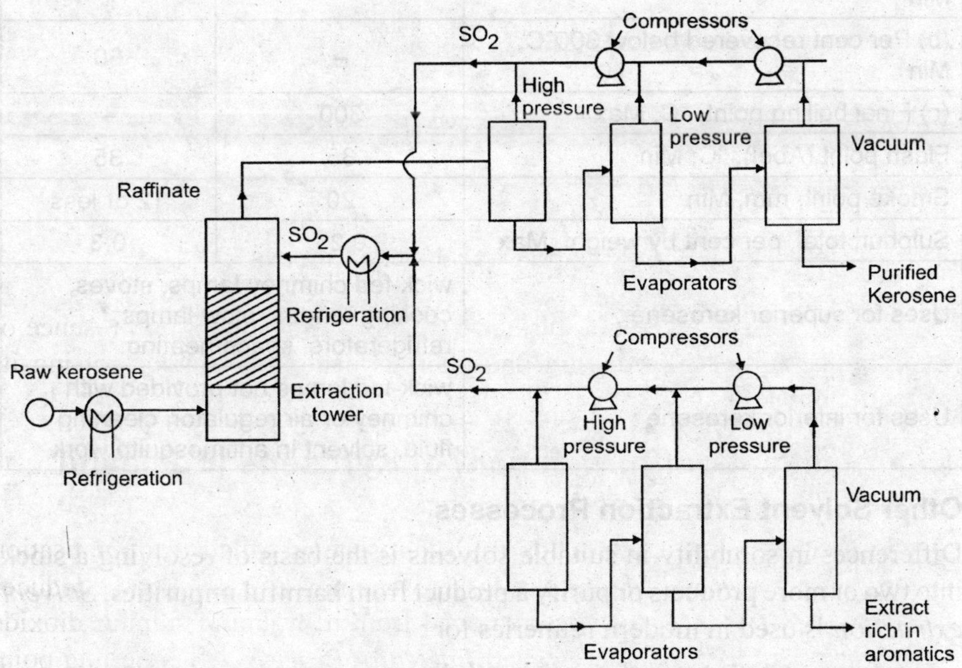

Fig. Flow sheet of Edeleanu process of dearomatisation of kerosene

Sulphur dioxide extraction can be used for recovering aromatics from gas oil, diesel oil and naphtha also.

Specification for Kerosene is tabulated below.

Table: Specification for Kerosene (IS: 1459-1968, Ammendment July, 1972)

Characteristic	Superior Grade	Inferior Grade
Acidity, inorganic	Nil	Nil
Burning Quality		
(a) Char value, mg/kg of oil consumed, Maximum	20	–
(b) Bloom on glass chimney	Not darker than grey	–
Colour (ASTM), not deeper than	–	5.0
Viscosity (Saybolt), Minimum	21	–
Copper-strip corrosion for 3 hours at 50°C	Not worse than No. 1	Not worse than No. 2
Distillation :		
(a) Per cent recovered below 200°C, Min	20	–
(b) Per cent recovered below 300°C, Min	–	60
(c) Final boiling point, °C, Max	300	–
Flash point (Abel), °C, Min	35	35
Smoke point, mm, Min	20	12 or less
Sulphur, total, per cent by weight, Max	0.25	0.3
Uses for superior kerosene :	wick-fed chimney lamps, stoves, cooking ranges, blow lamps, refrigerators, space heating.	
Uses for inferior kerosene :	wick-fed lamps not provided with chimney or air regulator, cleaning fluid, solvent in antimosquito work.	

Other Solvent Extraction Processes

Differences in solubility in suitable solvents is the basis of resolving a stock into two or more products or purify a product from harmful impurities. *Solvent extraction* is used in modern refineries for :

(a) ***upgradation** of lubricating oils* by removing aromatic materials and sulphur compound

(b) ***separation** of reduced crude and residuums* into lubricating oils and asphalts (deasphalting),

(c) ***upgradation** of kerosene, heating oils and diesel fuels* by removing aromatics, olefins and sulphur compounds,

(d) ***improving cracker feedstocks*** by removing asphaltic and other coke-producing materials (decarbonization),

(e) ***separation of wax bearing oils*** into wax and wax-free oil (dewaxing),

(f) ***recovery of benzene, toluene and xylenes*** from suitable naphthas,

(g) ***separation of i-butane*** from butane-butene stream, etc.

Particularly for improving quality of lubricating oils, other solvent extraction processes like furfural extraction, phenol extraction,Duo–sol extraction processes are used. Polycyclic materials are undesirable in lubricating oils because of their low viscosity index. An oil rich in polycyclic naphthenes is unsuitable for the production of motor oils. Polycyclic aromatics can be removed by furfural extraction, phenol extraction, & Duo-sol extraction processes.

Furfural Extraction Process

Furfural extraction is being increasingly used in the *upgrading of lube oils.* Aromatics are highly soluble in it. Low moisture in the system does not pose a problem and in fact a small amount of it is purposefully left in the solvent for increasing its selectivity. Moreover, it is readily available, relatively cheap and non-toxic.

In the furfural extraction process, feed stock (lube oil) is charged to the middle of an extraction column and passes up counter current to the solvent (furfural) which is introduced at the top. Solvent extracts polycyclic aromatics together with some desirable stock. The bottom to furfural-extract stream (about 90% furfural) is heated in a furnace & flashed to drive off furfural vapors and final traces are stripped by steam. The raffinate stream also contains 10% furfural which is recovered by flashing off. They are recycled after condensation.

Furfural is stable, nontoxic, relatively cheap, readily available and is selective at higher temperatures than are other solvents. Consequently, it has two advantages. Liquids are less viscous at higher temperatures and contacting of the feed & solvent is facilitated. Also, waxy stocks may be processed without difficulty. Because the selectivity of furfural is destroyed by water, the circulating solvent must be dehydrated continuously.

Wet furfural from steam stripping is condensed, and the two liquid phases that result - one, principally furfural, the other, principally water - are charged to a settling drum . The furfural layer that separates is stripped of water by countercurrent contact in a tower with dry furfural vapors. Wet furfural taken overhead is condensed and returned to the settling drum. The water layer from the drum is stripped of its furfural content by countercurrent contact with steam. The wet furfural taken overhead from this tower is also condensed and returned to the settling drum. Thus, the furfural-water condensates are separated into dry furfural and furfural-free water. Dry furfural is recycled to extraction tower and the water is discarded.

A flowsheet of furfural extraction process is given below in Figure.

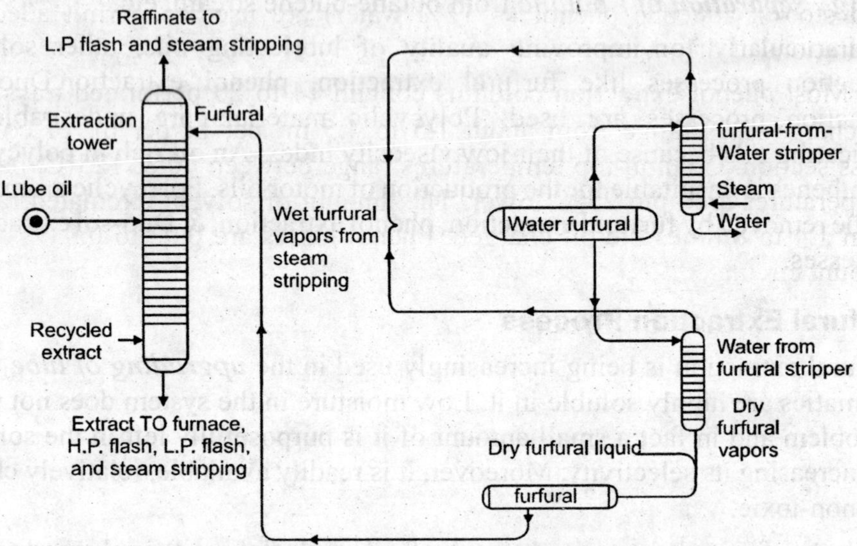

Fig. Furfural Extraction Process

Furfural extraction columns contain about 10-15 metres of raschig rings with redistribution equipment at 1.5 metres intervals. Charge rates range from 1.5 to 7.5 m^3 of oil per hr per m^2 of column cross section. The temperature of the entering hydrocarbon ranges from 45 to 105°C and the temperature of the entering solvent from 95 to 145°C. Typical temperatures at the top of the extractor range between 110 to 145°C and bottom temperatures are 5.0 to 25°C lower. Two to five volumes of solvent are circulated per volume of oil charged. Furfural loss is about 0.3% of the amount circulated.

Phenol Extraction Process

But for the fact that phenol is toxic and solidifies at atmospheric temperature, its solvency properties are similar to that of furfural. However, solvent recovery is easier and cheaper. It is extensively used in the upgrading of lubricating oils.

In phenol extraction process, the feed (lube oil) is first contacted with waste vapor (mainly water vapor) to recover phenol and is then charged to the lower part of a refluxed extraction column through which it rises counter-current to the solvent phenol. Reflux is introduced by the introduction of water at the bottom of the column. The rich extract is fractionated to drive off its water content, the water-phenol vapor taken overhead being sent to the phenol absorber. Phenol is recovered from extract & raffinate phases by heating & flashing off and is recycled to the extraction tower. Residual phenol is stripped from both streams by steam in vacuum towers. The phenol extraction system operates at slightly lower temperature than furfural extraction. With certain feed stocks, aqueous phenol (< 12% water) are used to permit adequate separations of the phases.

Most phenol extraction columns contain 14 to 20 perforated trays, and oil charge rates range from about 1.5 to 4.5 m^3 per hr per m^2 of column cross section. Column-top temperatures range between 70 & 125°C, bottom temperatures are upto 25°C lower. The amount of solvent circulated ranges from 1.4 to 8 times the oil charged. Phenol losses are 0.02 to 0.03% of the amount circulated.

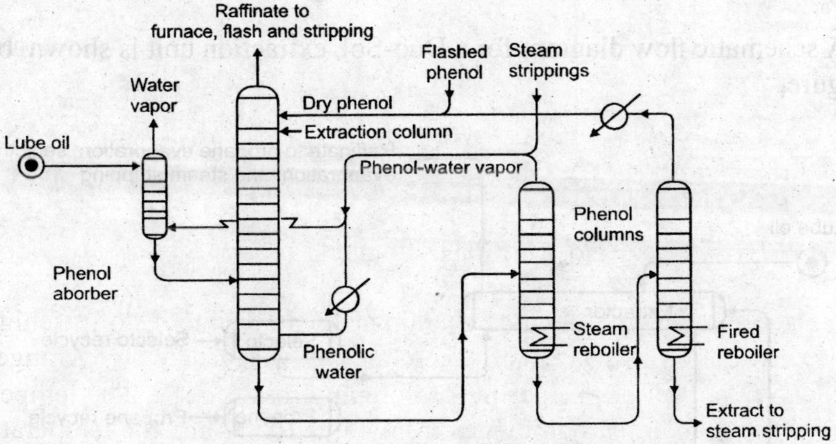

Fig. Flow sheet of phenol extraction plant

A schematic flow diagram for a phenol-extraction plant is shown above in Figure.

Duo-Sol Extraction Process

A mixture of propane and phenol-cresol mixture (called selecto) is used as a solvent in the double solvent (Duo-Sol) extraction process for upgrading lube oils. Selecto (phenol -cresol mixture) dissolves asphalt, polycyclic aromatics, naphthenic hydrocarbons, & colour bodies whereas other solvent (liquid propane) dissolves paraffinic hydrocarbons and precipitates asphalts. Thus the Duo-sol process has a dual function of removing asphalt as well as undesirable hydrocarbons removed in other extraction processes.

The extraction equipment consists of 7-9 mixer-settler stages operating in series. Propane is charged to one end stage, and a mixture of phenol & cresol to the other. The two solvents pass through the extractor counter currently. Feed (lube oil) is charged to an intermediate stage where it distributes itself between the two solvents. The raffinate rich propane stream passes counter-current to the entering phenol-cresol stream, which purifies the raffinate by extracting residual asphalt and polycyclics. The extract rich phenol cresol streams passes counter-current to the entering propane stream which strips raffinate type material from the extract and thus improves recovery. The raffinate & extract are stripped of their propane & phenol-cresol contents by several stages of evaporation and stream stripping.

Duo-Sol process operates at 15-65°C. The upper temperature is limited by the vapor extraction pressure of propane and the lower temperature by crystallisation of wax or phenol. Solvent to oil ratio is about 4. Propane & phenol-cresol mixture solvent losses are 0.2 & 0.03% respectively of the circulation. The water content of the phenol-cresol mixture is held below 0.15%.

A schematic flow diagram for a Duo-Sol extraction unit is shown below in Figure.

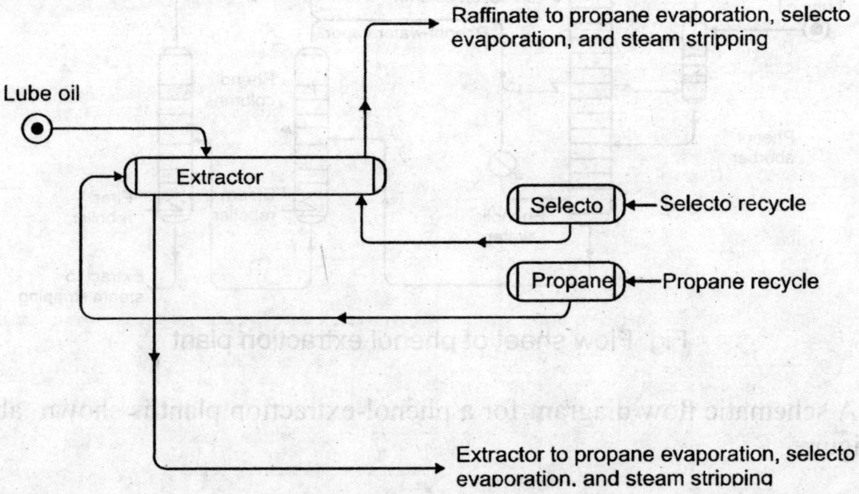

Fig. Duo-Sol Extraction Process

Udex (Glycol) Extraction Process

Aqueous solution (10% water) of diethylene glycol are used for the extraction of light aromatics *(e.g.* benzene, toluene & xylene) from catalyst reformate or cracked naphtha by Udex process.

The product is used as a high quality gasoline or further processed for recovering pure aromatics.

In Udex process, the feed is fed to a refluxed extractor. The rich solvent passes to a solvent stripper from which extract is taken overhead with the aid of steam and a portion of extract is refluxed to extractor. Raffinate from the extractor is water washed to recover dissolved glycol.

The washings are combined with the solvent from the bottom of the solvent stripper and the mixed stream is recycled to the extractor.

Since, most of the aromatics products from Udex process serves as a feed to petrochemical industry, hence they must be very pure. For that reason, refluxed system is used.

The feed stock is usually a catalytic reformate, although cracked naphthas may also be charged. Rich solvent from the extractor passes to a solvent stripper, from which extract is taken overhead with the aid of steam.

A schematic flow diagram of Udex extraction process is shown below in Figure.

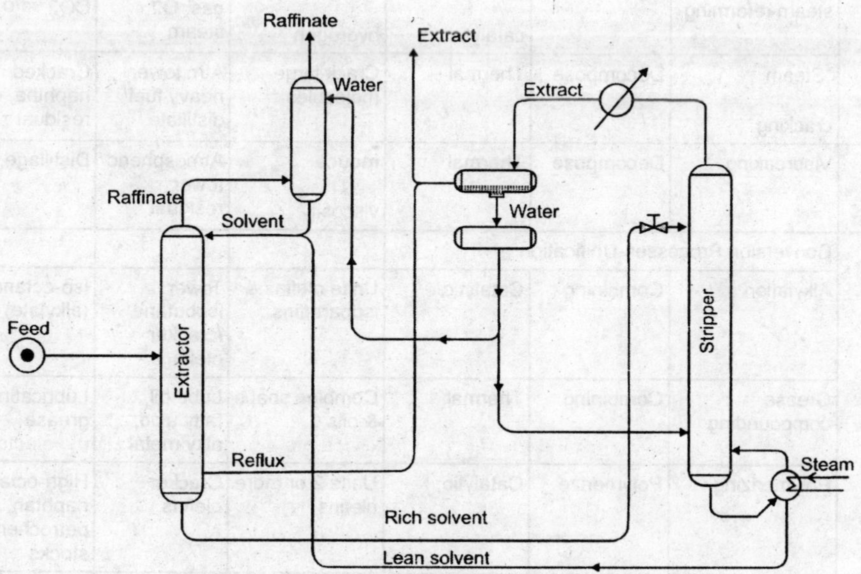

Fig. Udex (Glycol) Extraction Process

Overview of various petroleum refining operations and processes have been summarised below in table.

Table: Summary of Petroleum Refining Processes

S.N.	Process name	Action	Method	Purpose	Feedstock (s)	Product (s)
1.	Fractionation Processes					
	Atmospheric distillation	Separation	Thermal	Separate fractions	Desalted crude oil	Gas, gas oil, dis-. tillate, residual
	Vacuum distillation	Separation	Thermal	Separate w/o cracking	Atmospheric tower residual	Gas oil, lube stock, residual
2.	Conversion Processes-Decomposition					
	Catalytic cracking	Alteration	Catalytic	Upgrade gasoline	Gas oil, coke distillate	Gasoline, Petrochemical feedstock
	Coking	Polymerize	Thermal	Convert vacuum residuals	Gas oil, coke distillate	Gasoline, Petrochemical feedstock
	Hydro cracking	Hydrogenate	Catalytic	Convert lighter HC's	Gas oil, cracked oil, residual	Lighter, higher-quality products
	*Hydrogen steam reforming	Decompose	Thermal/ catalytic	Produce hydrogen	Desulfurized gas, O2, steam	Hydrogen, CO, CO2
	*Steam cracking	Decompose	Thermal	Crack large molecules	Atm tower heavy fuel/ distillate	Cracked naphtha, coke, residual
	Visbreaking	Decompose	Thermal	reduce viscosity	Atmospheric tower residual	Distillage, tar
3.	Conversion Processes-Unification					
	Alkylation	Combining	Catalytic	Unite olefins & isoparaffins	Tower isobutane/ /cracker olefin	Iso-octance (alkylate)
	Grease compounding	Combining	Thermal	Combine soaps & oils	Lube oil, fatty acid, alky metal	Lubricating grease
	Polymerizing	Polymerize	Catalytic	Unite 2 or more olefins	Cracker olefins	High-octane naphtah, petrochemical stocks
4.	Conversion Processes-Alteration or Rearrangement					
	Catalytic reforming	Alteration/ dehydration	Catalytic	Upgrade low-octane naptha	Coker/ hydrocracker naphtha	High octance Reformate/ aromatic
	Isomerization	Rearrange	Catalytic	Convert straight chain branch	Butane, pentane, hexane	Isobutane/ pentane/hexane

S.N.	Process name	Action	Method	Purpose	Feedstock (s)	Product (s)
5.	Treatment Processes					
	*Amine treating	Treatment	Absorption	Remove acidic contaminants	Sour gas, HCs w/C02& H2S	Acid free gases & liquid Hydrocarbons (HCs)
	Desalting	Dehydration	Absorption	Remove contaminants	Crude oil	Desalted crude oil
	Drying & sweetening	Treatment	Absortion/ thermal	Remove H20 & sulfur cmpds	Liq Hcs, LPG, alky feedstk	Sweet & dry hydrocarbons
	*Furfural extraction	Solvent extraction	Absorption	Upgrade mid distillate & lubes	Cycle oils & lube feed-stocks	High quality diesel & lube oil
	Hydrodesulfu-rization	Treatment	Catalytic	Remove sulfur, contaminants	High-sulfur residual/ gas oil	Desulfurized olefins 1
	Hydrotreating	Hydrogen ation	Catalytic	Remove impurities, saturate HC's	Residuals, cracked HC's	Cracker feed, distillate, lube
	*Phenol extraction	Solvent extraction	Absorpiton/ thermal	Improve viscosity index, color	Lube oil base stocks	High quality lube oils
	Solvent deas-phalting	Treatment	Absorption	Remove asphalt	Vacuum tower residual, propane	Heavy lube oil, asphalt
	Solvent dewaxing	Treatment	Cool/ filter	Remove wax lube stocks	Vacuum tower lube Oils	Dewaxed lube basestock
	Solvent extraction	Solvent extraction	Absorption/ precipitation	Separate unsaturated oils	Gas oil, reformate, distillate	High-octance gasoline
	Sweetening	Treatment	Catalytic	Remove H_2S, convert mercaptan	Unteated distillate/ gasoline	High-quality distillate/ gasoline

History of oil refinning processes have been given below in Table.

Table: History of Oil Refining Processes

Year	Process name	Purpose of refinning processes	By-products, produced
1862 1870	Atmospheric distillation Vacuum distillation	Produce kerosene Lubricants (original) Cracking feedstocks (1930's)	Naphtha, tar, etc. Asphalt, residual coker feedstocks
1913 1916	Thermal cracking Sweetening	Increase gasoline yield reduce sulfur & odor	Residual, bunker fuel Sulfur
1930 1932 1932 1933 1935 1935 1937 1939 1940 1940	Thermal reforming Hydrogenation Coking Solvent extraction Solvent dewaxing Catalytic polymerization Catalytic cracking Visbreaking Alkylation Isomerization	Improve octane number Remove sulfur Produce gasoline basestocks Improve lubricant viscosity index Improve pour point Improve gasoline yield & octance number Higher octance gasoline reduce viscosity Increase gasoline octance & yield Produce alkylation feedstock	Residual Sulfur Coke Aromatics Waxes Petrochemical feedstocks Petrochemical feedstocks Increased distillate, tar High-octane aviation gasoline Naphtha
1942 1950	Fluid catalytic cracking Deasphalting	Increase gasoline yield & octane number Increase cracking feedstock	Petrochemical feedstocks Asphalt
1952 1954 1956 1957	Catalytifc reforming Hydrodesulfurization Inhibitor sweetening Catalytic isomerization	Convert low-quality naphtha Remove sulfur Remove mercaptan Convert to molecules with high octane number	Aromatics Sulfur Disulfides Alkylation feedstocks
1960	Hydrocraking	Improve quality and reduce sulfur	Alkylation feedstocks
1974 1975	Catalytic dewaxing Residual hydrocracking	Improve pour point Increase gasoline yield from residual	Wax Heavy residuals

Appendix 7.1: Features of solvent extraction & dewaxing processes for purifaction of petroleum products

Solvent treating is a widely used method of refining lubricating oils as well as a host of other refinery stocks. Since distillation (fractionation) separates petroleum products into groups only by their boiling-point ranges, impurities may remain. These include organic compounds containing sulfur, nitrogen,& oxygen; inorganic salts & dissolved metals; and soluble salts that were present in the crude feedstock. In addition, kerosene & distillates may have trace amounts of aromatics & naphthenes, and lubricating oil base-stocks may contain wax. Solvent refining processes including solvent extraction & solvent dewaxing usually remove these undesirables at intermediate refining stages or just before sending the product to storage.

Solvent Extraction

The purpose of solvent extraction is to prevent corrosion, protect catalyst in the subsequent processes, and improve finished products by removing unsaturated, aromatic hydrocarbons from lubricant grease stocks. The solvent extraction process separates aromatics, naphthenes, and impurities from the product stream by dissolving or precipitation. The feedstock is first dried and then treated using a continuous countercurrent solvent treatment operation. In one type of the process, the feedstock is washed with a liquid in which the substances to be removed are more soluble than in the desired resultant product. In another process, selected solvents are added to cause impurities to precipitate out of the product. In the adsorption process, highly porous solid materials collect liquid molecules on their surfaces.

The solvent is separated from the product stream by heating, evaporation, or fractionation, and residual trace amounts are subsequently removed from the raffinate by steam stripping or vacuum flashing. Electric precipitation may be used for separation of inorganic compounds. The solvent is then regenerated to be used again in the process.

The most widely used extraction solvents are phenol, furfural, and cresylic acid. Other solvents less frequently used are liquid sulfur dioxide, nitrobenzene, and 2,2-dichloroethyl ether. The selection of specific processes & chemical agents depends on the nature of the feedstock being treated, the contaminants present, and the finished product requirements.

Feedstock	From	Process	Typical products	To(uses)
Naphthas, distillates, kerosene	Atm. tower	Treating/ blending	High octane gasoline . Refined fuels .... blending Spent agents .... blending	Storage Treating Treatment

Flowsheet of Aromatics Extraction by Furfural

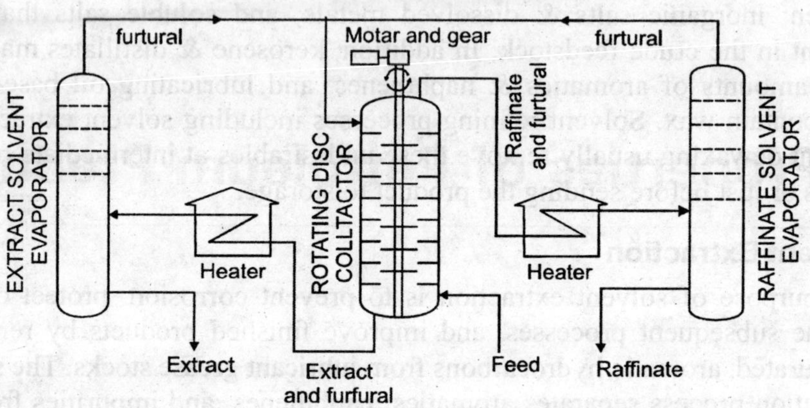

Solvent Dewaxing

Solvent dewaxing is used to remove wax from either distillate or residual base stocks at any stage in the refining process. There are several processes in use for solvent dewaxing, but all have the same general steps, which are: (1) mixing the feedstock with a solvent, (2) precipitating the wax from the mixture by chilling, and (3) recovering the solvent from the wax & dewaxed oil for recycling by distillation and steam stripping. Usually two solvents are used: toluene, which dissolves the oil & maintains fluidity at low temperatures, and methyl ethyl ketone (MEK), which dissolves little wax at low temperatures & acts as a wax precipitating agent. Other solvents that are sometimes used include benzene, methyl isobutyl ketone, propane, petroleum naphtha, ethylene dichloride, methylene chloride, and sulfur dioxide. In addition, there is a catalytic process used as an alternate to solvent dewaxing.

8

Properties of Petroleum Products

The important properties of liquid fuels/petroleum/petroleum products that make them suitable for specific uses differ from case to case. No single test or property can evaluate them. Some of the important properties of petroleum products are discussed below.

Specific Gravity

It determines the maximum power per unit of weight/volume. Hydrocarbons of low specific gravity (paraffins) possess the maximum thermal energy per unit volume. Hydrocarbons of high specific gravity (aromatics) possess the maximum thermal energy per unit weight. However, high calorific value of normal paraffins can not be fully utilised because of their low anti-knock ratings. Aromatics produce more carbon deposits than paraffins. These considerations together favour the use of isoparaffins for aviation gasolines. The lighter the crude oil (the higher API), the larger is the content of the lighter constituents like gasoline and kerosene. Aromatics have higher specific gravity than paraffins.

Hydrometer is used for determination of specific gravity upto 0.001. For higher values, specific gravity bottle is used. For a semi-solid mass like tar, an indirect method is used. Tar is mixed with equal volume of kerosene to make a thick solution. Specific gravity of tar is given as :

$$\rho_T = 2\rho_M - \rho_K$$

where, ρ_T = specific gravity of Tar

ρ_M = specific gravity of mixture of tar and kerosene

ρ_K = specific gravity of kerosene.

°API gravity and gravity on Baume scale (°Be)is given as (for liquids lighter than water) :

$$^{o}Be = \frac{140}{\rho} = -130 \qquad \text{and, } ^{o}API = \frac{141.5}{\rho} - 131.5$$

Specific gravity can be used to find the gross calorific value (GCV) of petroleum products using following formula of U.S Bureau of Mines.

G.C.V. = $12400 - 2100\rho^2$

where, ρ = specific gravity of oil at 15.5°C.

Typical specific gravity of crude and its products are given below :

crude	0.78-0.96
Naphtha	0.70-0.72
gasoline	0.7-0.76
diesel	0.82-0.86
lube oil	<0.9
reduced crude	0.92-0.96

Higher specific gravity means higher carbon to hydrogen ratio. Hence, heavier oils have lower gross calorific value on weight basis but higher gross C.V. on volume basis.

Aromatics have higher specific gravity than paraffins. Hence, a knowledge of specific gravity can predict the quality of a product. Increase in specific gravity means decrease in paraffin content. An increase in specific gravity increases the amount of heat per unit volume.

Density of Petroleum Products Versus Temperature

Following figure shows the approximate relationship of density with respect to temperature. It can be used to convert density of fuel oil or any other Petroleum product at 15°C to density at any other desired test temperature. t°C.

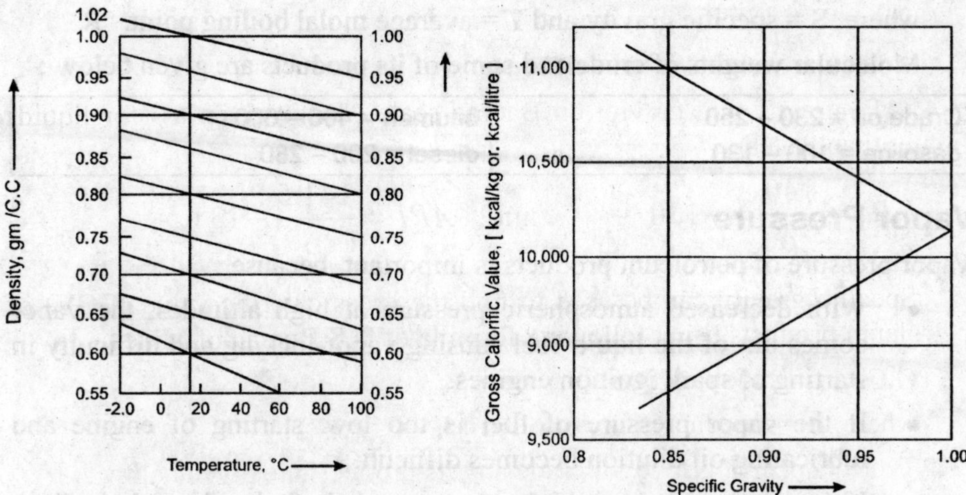

Fig. Tmemperature-Density(D) Relationship for Petroleum Products.

Fig. Specific Gravity Vs Calorific Value of Fuel Oils

To determine the density of fuel oil (having density 0.95 at 15°C) at 40°C, draw a vertical line at 40°C to the point where it meets 0.95 Density-Temperature line- From the point of intersection, draw a horizontal line. This line, meets the Density t°C line at 0.935. So the density of the fuel oil at 40°C is 0.935. The inverse approximation can be made similarly.

Another Figure above shows gross calorific value in kcal/kg & kcal/litre of fuel oils of different specific gravities at 15°C.

Molecular Weight

It is given by (for $\rho < 0.92$),

$$\text{Molecular weight} = \frac{44.29 \times \rho}{1.03 - \rho} \text{ where, } \rho = \text{sp. gr. at } 15.5\ ^\circ\text{C}$$

It is also given by *Eigenson's equation* as :

Molecular weight, $M = a + bT + CT^2$, where, T = average molal boiling point, °R

a, b, and c are constant depending upon the characterisation factor, *K*, which is given by :

$$K = \frac{T^{1/3}}{S}$$

where, S = specific gravity and T = average molal boiling point, °R

Molecular weights of crude and some of its products are given below :

Crude oil = 230 – 250	bitumen = 400 – 600
gasoline = 100 – 130	diesel = 200 – 280

Vapor Pressure

Vapor pressure of petroleum products is important, because :

- With decreased atmospheric pressure at high altitudes, the vapor comes out of the liquid fuel causing vapor-locking and difficulty in starting of spark ignition engines.
- If the vapor pressure of fuel is too low, starting of engine and lubricating oil dilution becomes difficult.
- If the vapor pressure is high, vapor occurs in fuel tanks and pipelines causing pumping & metering difficulties.
- If the vapor pressure is high that means flash point is low resulting in danger of fire and explosion. Besides, breathing loss is high, *i.e.,* evaporation losses during storage of petrofuels is high.
- Lubricants having high vapor pressure will be lost quickly particularly in case of thin film boundary lubrication.

Vapor pressure of a liquid fuel is reported as ***Reid vapor pressure*** (RVP) in psi or mm.Hg at 100°F (37.8°C). This apparatus consists of a 'Reid bomb' i.e. a metallic cylinder consisting of an upper expansion air chamber & a lower liquid fuel chamber and fitted with a dial pressure gauge or a mercury monometer. The volume of air expansion chamber is four times that of fuel chamber. It is shown below in Figure.

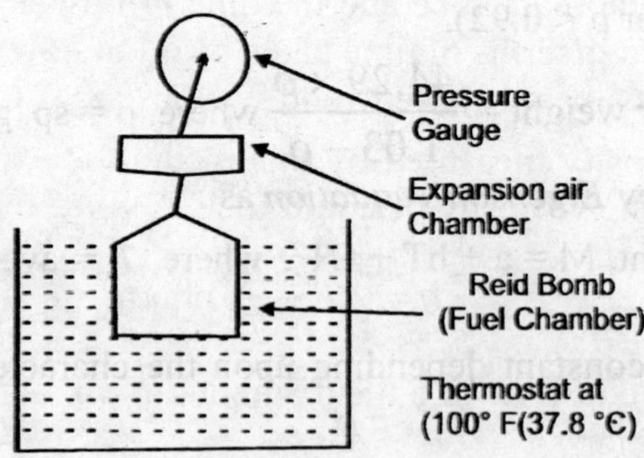

Fig. Reid Bomb.

The liquid oil is cooled and filled up in the lower chamber. The temperature of air in the upper chamber is taken and the two chambers are connected together in a gas tight manner. The bomb is immersed upright in a water bath at 100°F and shaken repeatedly until a constant pressure reading is obtained. This is corrected for initial pressure and temperature of air to get the Reid vapor pressure. True vapor pressure is about 5-9% higher than Reid vapor pressure.

Viscosity

Viscosity of oil/petrofuels is an important property, because :

- high viscosity fuel oil can not be properly atomised resulting in loss of fuel by way of incomplete combustion.
- highly viscous oil has to be preheated to reduce the viscosity to minimise the pressure drop in pumping to different refineries from the oil field.
- high viscosity lubricating oil reduces its fluidity besides causing undue friction.
- in case of lubricants, it determines the bearing friction, heat generation and rate of flow under particular conditions of load, speed & design.
- it affects the amount of fuel that can be drawn through a wick to a flame by capillary action.
- it affects the flow through pipelines, injector nozzles & orifices.
- it decides the pump clearance; which if out of range, results in pump seizer.

Viscosity is determined by three makes of commercial viscometers. The *Redwood viscometer* (used in commonwealth countries),*Saybolt viscometer* (used in U.S.A.) and *Engler viscometer* (used in Europe). In case of viscometers, a fixed volume of liquid at a fixed temperature is allowed to flow through a standard capillary tube and the time of flow is noted. The result is sometimes expressed in terms of time taken by oil to flow through a particular instrument *e.g.* viscosity = 100 Redwood seconds at 20°C.

Kinematic viscosity measured by this method is expressed in stokes or centistokes.Kinematic viscosity of liquid fuels is given by :

$$\eta = AT - \frac{B}{T} \text{ n to } \eta$$

where, T = Time of flow of a fixed amount of oil (at a fixed temperature) through the viscometer.

A and B are viscometer constants and depend upon the dimensions of the viscometer capillary through which oil flows.

For Redwood Viscometer No. I, Tis the time of flow of 50 c.c. of oil and values of constants B & A for various values of "T" are given below in Table.

Table - Values of Redwood Viscometer Constants 'B' & 'A'

Value of T	Value of B	Value of A
34-100 seconds	1.78	0.0026
> 100 seconds	0.50	0.00247

Approximately, $\eta = AT$, for fluids having $\eta > 10$ cst. Change η to meu

Example: *The following observations were made with a sample of oil using Redwood viscometer No. I :*

Volume of oil sample collected = *50 c.c.*

Temperature of oil = *20°C*

Time of flow of 50 c.c. oil = *600 seconds*

Calculate the Kinematic viscosity of the oil.

Solution. From the above table, the values of viscometer constants corresponding to the time of flow 600 seconds *(i.e.* > 100 seconds) are B = 0.50 & A = 0.00247.

Hence, Kinematic viscosity of oil, $\eta = AT - (B/T)$ = 0.00247 x 600 – 0.05/600 stokes or, η= (1.4820 - 0.0008) = 1.474 stokes = 147.4 cst(centistokes)

Redwood viscometer No. I is used for low viscosity oils whereas Redwood viscometer No. II is used for highly viscous oil (its oil flow port area is larger). The value of constants A & *B* are different for the two viscometers as their flow port diameters are different. These values are supplied with the viscometers. Number of seconds required for flow of fixed quantity of oil (usually 50 c.c. at a fixed temperature) is a measure of viscosity. High viscosity oil takes longer time to flow out.

The commonly used methods measure the viscosity in *Arbitrary Units.* The time required for a standard volume of oil to flow through a standard orifice is measured and the viscosity is stated as a number of seconds at a certain temperature as measured by a named apparatus.

In British practice, the standard instrument for measuring viscosity is Redwood Viscometer which is shown below in figure. It consists of a metal cup with an axially placed orifice in the base. The hole can be closed by a metal ball or a rod. The metal cup can be heated and the oil stirred to ensure uniform temperature conditions. When the ball is removed, a thin stream of oil runs into a small graduated glass flask and the time to fill the flask is recorded which represents the viscosity of oil. Two types of Redwood Viscometer are used. No. 1 is used for less viscous oils and No. 2 is used for more viscous oils. The time required for a standard volume of oil to flow from Redwood No. **2** viscometer is 1/10 of that required for an equal volume of same oil at the same temperature to flow from the Redwood No. 1 viscometer. Viscosities measured by the Redwood viscometer are reported as *"n* seconds Redwood No. (1 or **2**) at a given temperature.

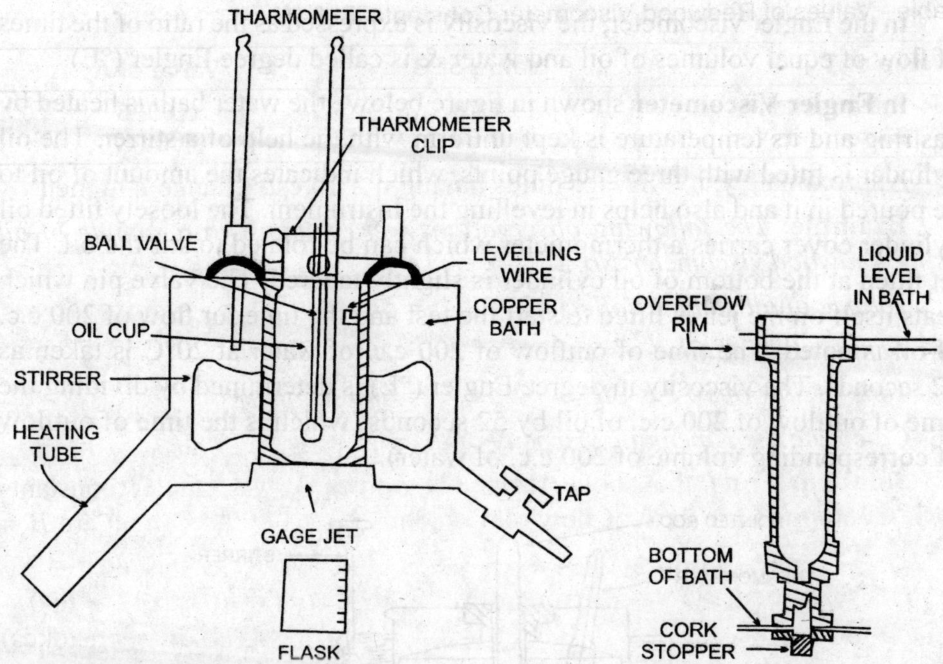

Fig. Redwood Viscometer. Fig. Saybolt Viscometer.

Kinematic viscosity (expressed in stokes/centistokes) is related to *absolute /dynamic viscosity* (expressed as poise or centipoise) as :

$$\eta = \frac{\mu}{\rho}$$

Where ρ, = density of oil, gm/cm^3

μ = absolute viscosity of oil, poise = gm/cm sec.

η = Kinematic viscosity of oil, stoke =cm^2/sec

[1 poise = 100 centipoise ; 1 stoke = 100 centistokes]

Saybolt universal and Saybolt Furol are similarly different viscometers used for different viscosity oils.

In American practice,*Saybolt viscometer* (figure given above) is used. Similar to Redwood viscometer No. 1 & No. **2,** Saybolt also uses two standards, *i.e. Saybolt Universal Viscosity* (time taken in seconds for **60** ml of sample flowing through a calibrated universal orifice under specified conditions) and *Saybolt Furol Viscosity* (time taken in seconds for 60 ml of sample through a calibrated Furol orifice under specified conditions). The *Furol* viscosity is approximately one tenth the universal viscosity and is recommended for those liquids having viscosities more than 1000 seconds (Saybolt Universal), such as fuel oils and other residual materials. The word ‘Furol’ is a contraction of fuel and road oils.

In the Engler viscometer, the viscosity is expressed as the ratio of the times of flow of equal volumes of oil and water & is called degree Engler (°E).

In **Engler Viscometer** shown in figure below , the water bath is heated by gas ring and its temperature is kept uniform with the help of a stirrer. The oil cylinder is fitted with three gauge points, which indicates the amount of oil to be poured in it and also helps in levelling the instrument. The loosely fitted oil cylinder cover carries a thermometer which can be rotated to stir the oil. The jet fitted at the bottom of oil cylinder is slightly tapered. The valve pin which seats itself on the jet, is lifted to start the test and the time for flow of 200 c.c. of oil is noted. The time of outflow of 200 c.c. of water at 20°C is taken as 52 seconds. The viscosity in degree Engler (°E) is determined by dividing the time of outflow of 200 c.c. of oil by 52 seconds (which is the time of outflow of corresponding volume of 200 c.c. of water).

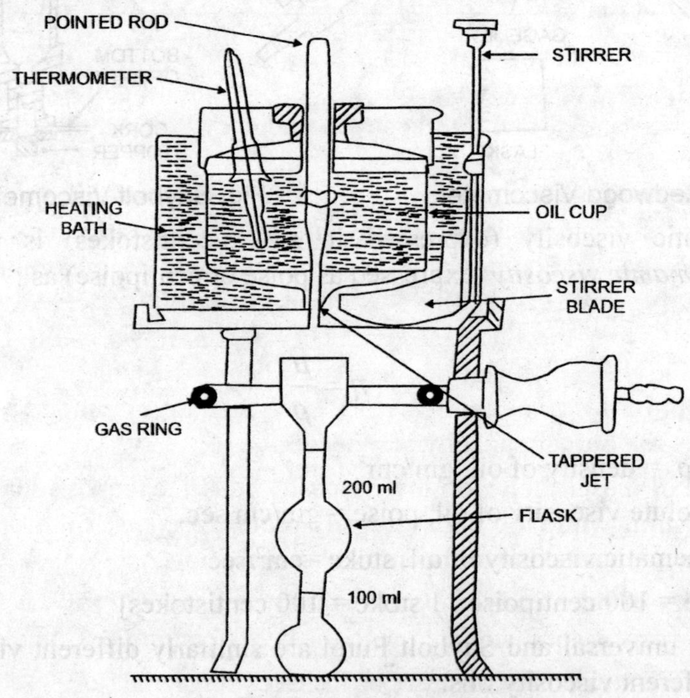

Fig. Engler's Viscometer.

Viscosity data of one standard is converted to another by following formulae.

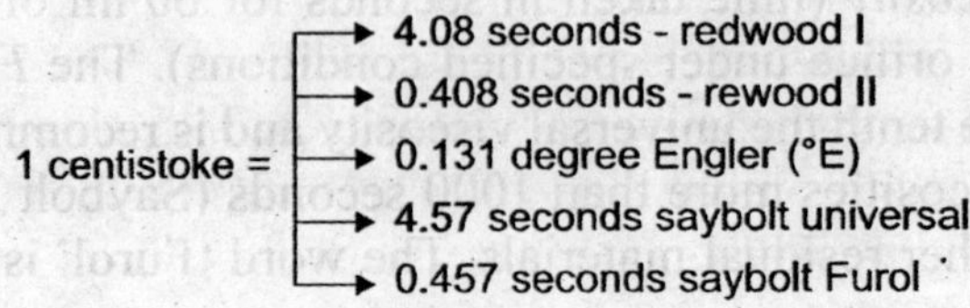

Viscosity scale	Range of flow time, T, seconds	Kinematic viscosity, stokes
Saybolt Universal	T = 32 to 100	0.00226 T – 1.95/T
	T => 100	0.00220 T – 1.35/T
SayboltFurol	T = 25 to 40	0.0224 T – 1.84/T
	T = >40	0.0216 T – 0.60/T
Redwood number I	T = 34 to 100	0.00260 T– 1.79/T
	T => 100	0.00247 T – 0.50/T
Redwood number (II)	T = 32 to 90	0.02458 T – 1/T
	T => 90	0.0244 T – 0/T
Engler	°E = 1.35 to 3.2	0.08 T – 0.0864/T
	° E => 3.2	0.076 T – 0.04/T

Table -Kinematic viscosity (η) conversion formulae. Viscosity of fuel oils influences the ease of handling, transport and nature of storage. Oil viscosity affects its atomisation and combustion efficiency. Preheating of fuel oils is done to reduce its viscosity (above 120°C viscosity is almost constant) to achieve better atomisation (finer oil droplets for better combustion). Maximum viscosity for easy atomisation in conventional burner is 25 centistokes (100 seconds Redwood I). For easy and economic pumping the maximum viscosity should be 1200 centistokes (486 Seconds Redwood II). At higher preheating temperature of oil, the decreased viscosity facilitates easy flow and better atomisation in the burner but the decreased density results in the delivery of lesser weight of oil to the burner. Hence, beyond an optimum oil preheating temperature, the burner capacity is reduced.

In case of diesel fuels, too low viscosity causes excessive leakage at the injection pistons while too high viscosity produces coarse oil droplets which forms engine deposits due to incomplete combustion.

Viscosity of oil blends is experimentally determined (as viscosity is not an additive property or calculated with the help of emperical blending charts).

At high pressures, the viscosity of oil increases very rapidly. At 70 atm. pressure, viscosity increases by 15%. Most oils begin to solidify at 3500 kg/cm^2 pressure.

Viscosity Index (V.I.)

The viscosity of a liquid decreases with increase in temperature. The change in viscosity with change in temperature is expressed by viscosity index. It is an emperical number indicating the rate of change of viscosity of an oil from 100°F to 210°F. Low viscosity index means a large change in viscosity with change in temperature. A high viscosity index shows relatively small change in viscosity with temperature.

Paraffinic oil has very high viscosity index =100 ,while naphthenic oil has very low viscosity index = 0 (These are arbitrarily assigned values for standardisation).

Viscosity index of an oil is determined by measuring its viscosity at two temperatures and comparing the results with those for a standard oil of V.I.= 100, & for a standard oil of V.I.= 0.

$$\text{Viscosity index} = \frac{L-U}{L-U} \times 100$$

where, U = Viscosity of oil sample at 100°F

L = Viscosity of a standard oil of V.I. = 0, at 210° F

H = Viscosity of a standard oil of V.I. = 100 at 210°F.

Values of L & H are obtained from tables after the viscosity of sample has been determined at 100°F.

Viscosity index measures the paraffinity of an oil. Lubricating oil should in general have high viscosity index. Viscosity index improvers *(e.g.* polybutene) are used for improving it. High V.I. oils are used where there is a wide temperature variation. High V.I. lubricating oils are called multigrade oils.

Viscosity Gravity Constant (V.G.C.)

It is given by the formula :

$$VGC = \frac{10G - 1.0752 \log (V-38)}{10 - \log (V-38)}$$

where, G = specific gravity at 15.5°C, V = Saybolt viscosity at 37.8°C.

This correlation of specific gravity and viscosity gives the measure of paraffinity in lubricating oil as the paraffins are less dense than naphthenic hydrocarbons. V.G.C. is related to other physical properties of the oil also.

Viscosity Index Extension(VIE)

For those oils which have a viscosity index more than 100, another parameter called viscosity index extension *(VI_E)* is used which is defined as,

$$VI_E = 100 + \frac{(anti \log N) - 1}{0.0075} \quad \text{where, } N = \frac{\log H - \log U}{KVI_{200}} (H > U)$$

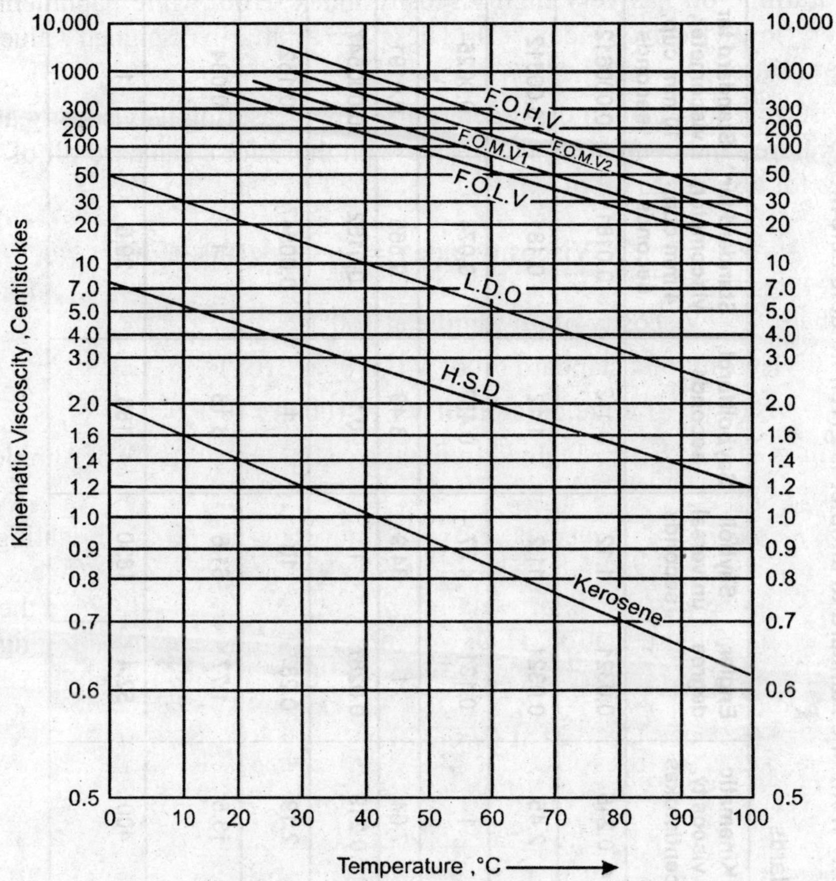

Fig. Viscosity-temperature relationship of petrofuels.

Legends:

H.S.D — high speed diesel fuel

F.O.M.V1 — Furnace oil, Medium viscosity grade I

L.D.O. — Light Diesel oil

F.O.M.V2 — Furnace oil, Medium viscosity grade II

F.O.L.V. —Furnace oil, Low viscosity grade

F.O.H.V —Furnace oil, High viscosity grade

where, H = kinematic viscosity of an oil having a VI of 100 at 100°F using the standard method and having the same viscosity as the unknown oil at 210°F.

U = kinematic viscosity at 100°F of unknown oil.

KVI_{200} = kinematic viscosity of oil sample at 210°F.

Viscosity temperature relationship of various petrofuels is shown above in the figure.

A comprehensive table for conversion of viscosity from one standard to another is given in table below.

Table- Conversion factors for viscosity of various standards

Viscosity of various standards	Redwood I, seconds	Redwood II, seconds	Kinematic viscosity, centistokes	Engler, degree	Saybolt universal, seconds	Sayboltfurol, seconds	Standard tar viscometer, 4 mm cup, seconds	Standard tar viscometer, 10 mm cup, seconds
Redwood I, seconds (for > 90 seconds)	1	0.1	0.245	0.0321	1.12	0.112	0.0181	0.000612
Redwood II, seconds (for > 30 seconds)	10	1	2.45	0.0321	11.2	1.12	0.181	0.00612
Kinematic viscosity, centistokes	4.08	0.408	1	0.131	4.57	0.457	0.074	0.0025
Engler, degrees (for > 6°)	31.2	3.12	7.64	1	34.9	3.49	0.565	0.0191
Saybolt universal, seconds (for > 120 seconds)	0.893	0.089	0.219	0.0287	1	0.1	0.0162	0.000547
Saybolt furol, seconds (for > 40 seconds)	8.93	0.893	2.19	0.287	10	1	0.00547	0.162
Standard tar viscometer, 4 mm cup (for > 100 seconds), seconds	55.1	5.51	13.5	1.77	61.6	6.16	1	0.034
Standard tar viscometer, 10 mm cup (for > 10 seconds), seconds	1630	163	400	52.4	1830	183	29.6	1

Flash Point

It is the minimum temperature at which an oil/petrofuel gives out sufficient vapour to form an inflammable mixture with air and catches fire momentarily *i.e.* flashes, when flame is applied. Flash point of an oil gives an idea about the :

- nature of boiling point diagram of the system.
- amount of low boiling fraction present in the liquid fuel.
- explosion hazards during storage and handling.
- volatility of the liquid fuels.

Determination of flash point. It consists of heating a given volume of petroleum product at a standard rate of temperature rise until the vapor produced gives a flammable mixture with air in an enclosed space *(i.e.,* closed flash point temperature) or with air in an open cup *(i.e.,* the higher open flash point temperature) ignition as a result of application of flame. At fire point, not only the vapor air mixture flashes but the liquid will continue to burn for 5 seconds.

Flash points of different liquid fuels are as follows :

Crude oil = -10 to 0°C	Gasoline = - 40 to 30°C
Kerosene = 28-55°C	Diesel = > 55°C
Lube oil = 130-300°C	Fuel oil = > 66°C

Flash point less than 23°C is dangerous and highly inflammable. The safe value is > 60-66°C. **Pensky-Marten's** apparatus is used for the determination of flash point above 50°C *e.g.,* for fuel oils, lubricating oils, bitumens etc. **Abel closed cup apparatus** is used for more volatile oils with flash point below 50°C. It is determined by **Cleveland *open cup*** apparatus also.

In ***Cleveland apparatus*** used for the determination of flash point & fire point of petroleum products (except fuel oils) having open cup flash point below 79°C, an open brass cup is filled with oil and heated at a certain constant rate. Naked flame is repeatedly applied to the oil vapor at definite interval of temperature. When the vapor catches fire momentarily, the temperature of oil is noted as flash point.

In ***Pensky-Marten closed cup apparatus,*** dried oil is filled up in the cup with lid and heated by a burner. Oil is exposed to flame at every 2°F rise in temperature of the oil. The temperature at which the distinct flame first appears inside the cup by applying test flame through the opening meant for the purpose is the flash point.

Pensky-Marten's apparatus shown below in figure.

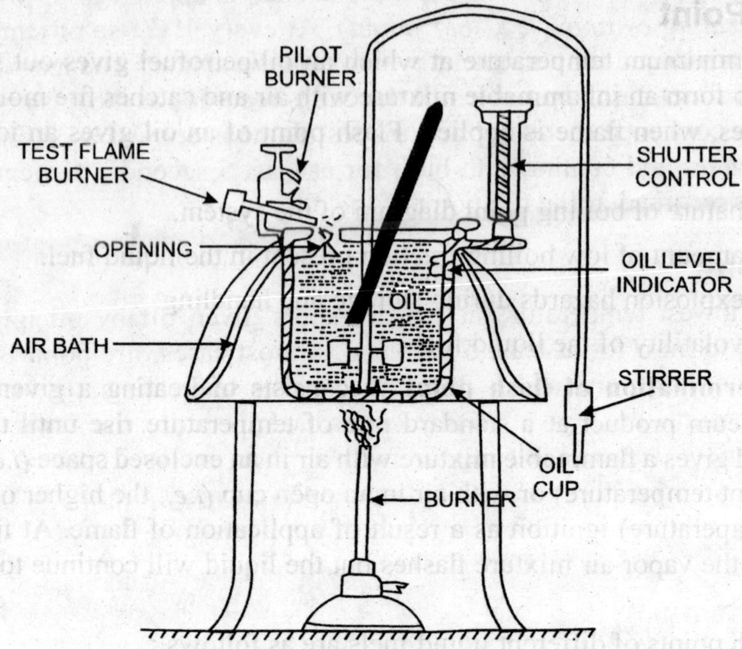

Fig. Pensky-Marten's flash-point apparatus.

Pensky Marten's apparatus comprises of :

- An ***oil cup*** of 5 cms diameter and 5.5 cms deep in which oil is filled up to the level marked inside. Four openings of standard sizes are provided in the lid of the cup. Through one of these passes a thermometer and the second opening is used for introducing test flame. Through third opening passes the stirrer carrying two brass blades. Fourth opening is meant for admission of air.
- ***Shutter*** is a lever mechanism, provided at the top of the cup. By moving the shutter, opening in the lid opens and flame (carried by a flame exposure device) is dipped into this opening & brings the flame over the oil surface.
- ***Flame exposure device*** is a small flame and this is connected to the shutter by a lever mechanism.
- ***Air bath***. Oil cup is supported by its flange over an air bath, which is heated by a gas burner.
- ***Pilot burner***. As the test flame is introduced in the opening, it gets extinguished, but when the test flame is returned to its original position , it is automatically lighted by the pilot burner.

Working. Oil under test is filled up to the mark in the oil cup and it is heated by heating the air bath by a burner. Stirrer is worked between tests at a rate of about 1 to 2 revolutions per second. *Heat is applied so as to raise the oil temperature by about 5°C per minute.* At every 1°C rise of temperature, flame is introduced for a moment by working the shutter. The temperature at which a distinct flash appears inside the cup is recorded as the flash point. The heating is continued thereafter and the test flame is applied as before. When the oil ignites and continues to burn for at least 5 seconds, the temperature reading is recorded as the fire point of the oil.

Fire Point

It is the lowest temperature at which vapors given off by oil ignites and continues to burn for at least 5 seconds. In most cases, fire point is 5-40°C higher than flash point and is determined in the same apparatus as for flash point determination. It gives an idea of fire hazards during the storage and use of oil.

It is customary to determine both the temperatures, although the lower temperature (flash point) is itself a measure of the fire risk of an oil.

The permissible conditions of storage and use are different for products with flash points below 23°C (*e.g.* gasoline), between 23° & 66°C (*e.g.* kerosene) and above 66°C (*e.g.* fuel oils and gas oils). The oils of the first category are regarded as dangerous, highly inflammable for transport & storage, and are subject to strict government regulations. Fuel oils with flash points above 66°C are regarded as 'safe'.

Besides oil's volatility & the inflammability limits of the vapour-air mixture, the flash point also depends upon the design of the apparatus, the test procedure and the barometric pressure. *Kerosene is tested in Abel apparatus* , and *fuel oil, diesel fuels & gas oils are tested in Pensky-Martens apparatus (closed-cup type)*. The *American 'Tag' apparatus* is similar to Abel.

Flash points of lubricating oils, some crude oils & residues are often determined by the open-cup tests. Either Pensky Martens cup or Clevelend open cup is used. The open cup test is liable to error and gives only approximate values. The open cup flash point is higher than the closed-cup value for the same sample.

The flash point and the fire point give indications of the range and nature of the boiling point curve. The flash point of lubricants helps to indicate the relative amount of low-boiling material present in it. Flash point or fire point should not be confused with completely unrelated spontaneous ignition temperature.

Cloud Point, Pour Point and Freezing Point

These properties are related to the flow conditions of crude oil and its products at low temperature.

Cloud Point: When an oil is cooled at a specified rate, the temperature at which it becomes cloudy or hazy is called the *cloud point of oil.* This haziness is due to the separation of crystals of wax or increase of viscosity at low temperature. It is determined by cold filter plugging point test to find out the extent to which diesel fuel or gas oil will flow even though the temperature is below which the wax crystal will appear *i.e.,* cloud point. It is normally determined for those petroleum products that are transparent in a 40 mm thick layer and have cloud point below 49°C.

Cloud point is important for fuel oils which have to pass through unheated filters of fine mesh *e.g.* a jet plane may be exposed to -60°C and if solid wax separates from fuel oil, the carburettor may be blocked up.

Pour Point: The temperature at which the oil just ceases to flow (or pour) is called the pour point. It determines the temperature below which, an oil can not be used as a lubricant as well as the waxy nature of oil.

To determine the pour point, the oil is cooled in the test jar and with every 3°C fall in temperature tilted to observe the flow of oil. The temperature at which it does not flow even when kept horizontal for 5 seconds is taken as the pour point. Naphthenes and aromatics have lower pour point than paraffins but they are undesirable in fuel oils & diesel fuels. Asphalts act as pour point depressants as they inhibit the growth of wax crystals. Increasing lighter hydrocarbons reduces pour point.

Determination of pour point is carried out with the help of pour-point apparatus shown below in figure . It consists essentially of flat-bottomed tube (about 3 cm in diameter and 12 cm high) enclosed in an air-jacket. The jacket is surrounded by freezing mixture (ice+$CaCl_2$) contained in a jar. The tube is half-filled with oil. A thermometer is introduced in the oil. As the cooling proceeds slowly *via* air jacket, the temperature falls continuously. With every degree fall of temperature in the oil, the tube is withdrawn from the air-jacket for a moment (about 2-3 seconds) and examined. It is then replaced immediately. The temperature at which cloudiness is noticed is recorded as the cloud-point. After this, cooling is continued and the test-tube is withdrawn after every 3°C fall of temperature and tilted to observe the flow or pour of oil. The temperature at which oil does not flow in the test-tube, even when kept horizontal for 5 seconds, is recorded as the pour-point.

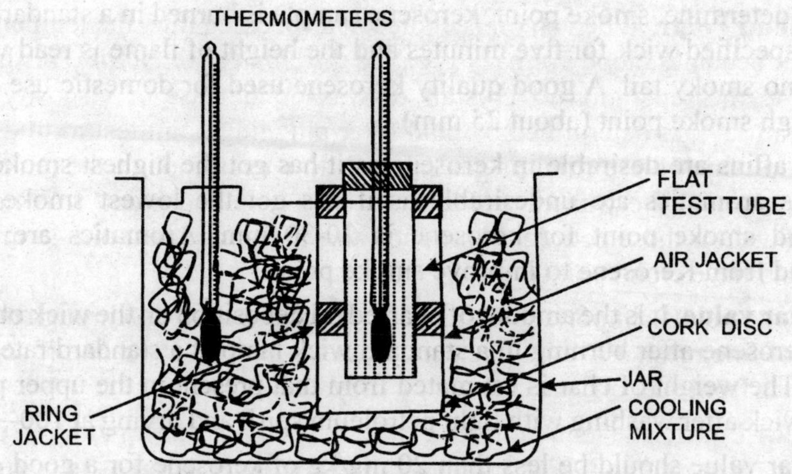

Fig. Pour Point Apparatus.

Cloud point and pour-point indicate the suitability of lubricants in cold conditions . Lubricants used in a machine working at low temperatures should possess low pour-points; otherwise solidification of lubricant will cause jamming of the machine. Presence of waxes in the lubricating oil raise the pour-point.

The difference of pour-point and cloud point is 4-6° F usually.

Freezing point: It is the temperature at which the fuel oil freezes completely and can not flow at all. This is important in case of aviation gasoline because at high altitudes where LOW temperatures are encountered, the fuel supply from fuel tank to engine may be impeded due to choking of pipeline if the freezing point of the fuel oil is not sufficiently low.

Paraffins possess higher freezing points than naphthenes and aromatics. Aviation gasoline should have freezing point below 60°C to avoid trouble due to crystal formation in feed lines and filters.

Smoke Point and Char Value

These are the properties of kerosene which determine the suitability of kerosene as a fuel and as an il- illuminant.

Smoke point. This is the maximum height of flame in mm without smoke formation when the kerosene is burned in a standard lamp under closely controlled conditions. Smoking of kerosene is due mainly to the presence of aromatic hydrocarbons.

To determine, smoke point, kerosene sample is burned in a standard lamp with a specified wick for five minutes and the height of flame is read when it leaves no smoky tail. A good quality kerosene used for domestic use should have high smoke point (about 25 mm).

Paraffins are desirable in kerosene as it has got the highest smoke point whereas aromatics are undesirable as it has got the lowest smoke point. Standard smoke point for kerosene is 20-30 mm. Aromatics are, hence removed from Kerosene to improve smoke point.

Char value. It is the amount of charred oil deposition on the wick obtained from kerosene after burning in a standard wick lamp at a standard rate for 24 hours. The weight of char is estimated from that present in the upper portion of the wick after washing with light petroleum spirit and drying at 100–110°C.

Char value should be less than 20 mg/kg of kerosene for a good quality kerosene.

Carbon Residue

It is the percentage of carbonaceous residue left after distillation of crude oil or its products in absence of air. This determination is important for fuel oil, lubricants, cracking feed stocks etc. Carbon deposit fouls the surface, results in wear & choking of cylinder and affects the regeneration of cracking catalyst as it gives an indication of relative coke forming tendency on evaporation & pyrolysis of an oil. It is also a measure of carbon depositing tendency of fuel oil and engine deposits in diesel fuel fired engine.

It is determined by Conradson apparatus and Ramsbottom apparatus.

In ***Conradson method***, a weighed quantity of oil is taken in a crucible which is placed in an iron crucible having a cover with a small opening. These are, then placed in a third crucible and covered with a hood & strongly heated at a specified rate and the residue is cooled & weighed. The result (Conradson carbon) is expressed as percentage of the residue on the original weight taken.

In ***Ramsbottom method***, a weighed quantity of oil is heated in a hemispherical glass bulb of standard dimension for 20 minutes at 550°C. The residual carbon is weighed and recorded as percent carbon residue (Ramsbottom carbon).

Conradson carbon residue in high speed diesel fuels should be < 0.1% (on a 10% distillation residue). Atomising burners are insensitive to carbon residue of fuels but gas oil used in vaporising burners should have < 0.05% carbon residue. Conradson method is normally used for fuel oil and Ramsbottom method for lubricating oil.

Conradson carbon and Ramsbottom carbon are related by *ASTM* relation as:

$$3.8\ C = R^2 + 4.25\ R - 0.35$$

where, C = Carbon residue by Conradson method

R = Carbon residue by Ramsbottom method.

Aniline Point

It is the lowest temperature at which an oil is completely miscible with an equal volume of aniline. It is the measure of the aromatic content of the oil, hence is a characteristic property of diesel. Since aromatics dissolve aniline (itself an aromatic substance) more readily than paraffins or *iso*-paraffins, the higher the aniline point, lower the aromatics and higher the paraffin content with very high cetane number making the oil suitable for use in diesel engine.

Aniline point of an oil also gives an indication of the possible deterioration of rubber sealing, packing etc. in contact with the oil. The aromatics have a tendency to dissolve natural rubber and certain types of synthetic rubber. Therefore in such cases, a low aromatic content in the lubricant (*i.e.* with high aniline point) is desirable. Aniline point of cetane is 95°C and for hexyl benzene is –12°C. High aniline point indicates highly paraffinic diesel fuel with high diesel index and good ignition quality. For aromatics, aniline point is low and the ignition quality is poor.

Hence, higher the aniline point, better the diesel fuel and the lubricant.

Diesel Index

An alternative method of expressing the quality of diesel oil is by use of the "Diesel index". It is given by:

$$\text{Diesel index} = \frac{\text{Aniline point, }^{\circ}\text{F} \times \text{API gravity}}{100}$$

Diesel index is roughly related with cetane number as Diesel index = Cetane number + 3

The use of diesel index does not necessitate the use of a test engine for determining the cetane number and hence the quality of diesel.

Octane Number

This is a property of gasoline which is used in a spark ignition engine and expresses its knocking characteristics. When a gasoline engine is made to operate at high load & low speed, a sort of rattling noise may develop from the combustion chamber resulting from unsteady & uncontrolled combustion. This noise is called knocking or detonation. Knocking is harmful to engine as its parts are set to vibration thereby reducing the life of engine. Knocking depends upon the quality of fuel also. Octane number requirement of petrol engines depends on their compression ratio.

It has been found that *n*-heptane knocks very badly so its anti-knock property has been arbitrarily taken as zero and *iso* octane (2, 2, 4 trimethyl pentane) gives very little knocking so its anti-knock value has been taken as 100. Octane number of the fuel is defined as "the percentage by volume of *iso*-octane in a mixture of *n*-heptane and *iso*-octane with the same knocking tendency as fuel". Thus if a sample of gasoline gives as much knocking as a mixture of 80 parts of *iso*-octane and 20 parts of *n*-heptane, its octane number is 80.

Higher the octane number, better is the fuel as maximum permissible power increases with the octane number besides causing minimum knocking.

Octane number of hydrocarbon fuels increases in the order *n-paraffins — olefins — naphthenes — iso-paraffins — aromatics.*

To determine the octane number of a fuel, it is burnt in a standard engine for the purpose and the knock is measured by a knockmeter. This is compared with the standard knockmeter readings of fuels of different octane numbers to determine the octane number.

In India, motor gasoline of octane number 87 is sold whereas the same of aviation gasoline is more than 100. In U.S.A. octane number of standard and premier gasoline is 94 and 100 respectively. Octane number of premier gasoline sold in India is 93.

Octane number of gasoline is improved by the addition of *tetra ethyl lead (TEL),* the quantity being 3.6 c.c. TEL/gallon of motor gasoline and 5.5 c.c. TEL/gallon of aviation gasoline. TEL is a very good anti-knocking agent.

Constituents of gasoline respond differently to the addition of TEL. This is called lead response and decreases in the order *paraffins - naphthenes - olefins - aromatics.* Alcohols respond negatively as do sulphur compounds.

Performance Number

The addition of small quantity of TEL to gasoline increases its octane number very rapidly, the maximum quantity used being 6 c.c./gallon. It is possible to increase the octane number to a value greater than 100 *i.e.* in excess of that of iso-octane. Values above 100 octane number are known as "Performance number". Aviation gasoline invariably have octane number greater than 100 hence their anti-knock characteristics are expressed by performance numbers. On performance number scale, 100 octane number equals 100 performance number.

The best aviation gasoline has performance number of 160.

For octane number above 100, the following relationship exists between performance number (P.N) and octane number (O.N) :

$$O.N + 100\frac{P.N - 100}{3}$$

Cetane Number

It is a characteristic property of diesel and is used to indicate its quality and performance in compression ignition engine.

In case of diesel engine, the fuel should ignite as soon as it is injected into the cylinder. If it does not ignite instantaneously, (the time lag between fuel injection and fuel ignition is called *ignition delay period*) abnormal combustion takes place resulting in shock waves, due to fluctuating pressure rise in the cylinder. With diesel of long ignition delay period, much of the charge is injected into the cylinder before ignition is initiated, causing violent combustion, sudden increase in pressure and rough and bumpy running. This is called *diesel knock.*

Cetane ($C_{16}H_{34}$) has a very small ignition delay period hence it is given a cetane number rating of 100. α-methyl naphthalene has a very high ignition delay period, hence it is given a rating of zero. Cetane number of diesel oil is the percentage by volume of cetane in a cetane/α- methyl naphthalene mixture that has the same ignition delay period and performance in a standard compression ignition engine as that of the fuel.

Diesel oils for high speed (> 1500 rpm), medium speed and low speed engine should have cetane number not less than 50, 35 and 20 respectively.

Cetane number increases in the order aromatics - *iso*-paraffins - naphthene-olefins - *n*- paraffins, whereas the octane number decreases in the same order. The result is that a oil of high cetane number has a low octane number and *vice-versa.* A petroleum crude which gives diesel oil of high cetane number, gives gasoline of low octane number. A good spark ignition fuel makes a poor compression ignition fuel and *vice-versa.*

11.5% of additives called "dopes" like ethyl nitrate, acetone, iso-amyl nitrate etc. are added to improve the cetane number. Cetane number is determined by a standard test compression ignition engine by measuring the ignition delay.

High cetane number diesel fuel will facilitate easy starting of C.I. engines particularly in cold weather, faster engine warm up, increased engine efficiency & power output and reduced exhaust smoke, odour& combustion noise.

Emulsification

Some oils mix intimately with water and the mixture is called emulsion which has the tendency to collect dirt, grit, particles and other foreign matters.

The emulsification tendency is undesirable in most applications of the oil. Formation of emulsion in a circulating oiling system is undesirable, because sticky, emulsion and sludges interfere with oil circulation. A good lubricating oil should not form emulsion under lubrication condition and even if it forms an emulsion, it should break off very rapidly. Additives are added to check tendency of emulsification. In case of steam turbines, the oil should separate quickly from the emulsion so that it can be recirculated for use. Formation of stable emulsion may be desirable when the oil is to be used in presence of water as in the case of certain steam cylinder oils and water soluble cutting lubricants.

In emulsion test, steam is passed into the oil under specified conditions and the time taken for the separation of oil from the emulsion in minutes is recorded as steam emulsion number which depends upon the refinement and viscosity of oil. Good refinement and lower viscosity are conducive to quicker separation of the oil from the emulsion.

Oxidation Stability

Oils tend to oxidise in contact with air rapidly at higher temperature. Oxidation products are insoluble, gummy and sticky materials which may clog oil holes, filters, pipelines and other parts of the oil circulation system. Soluble oxidation products are corrosive. Hence resistance to oxidation is a desirable property of oil. Oxidation inhibitors are added to improve their resistance to oxidation. The inhibitors get oxidised first and the oil is protected.

A common test for it is the *Sligh oxidation test* in which 10 gm sample of oil is heated in a flask containing pure oxygen at 200°C for 2.5 hours. The flask is cooled, the contents are diluted with naphtha and allowed to stand for an hour. The precipitate is filtered, washed, dried and weighed. The weight is expressed as percentage of the weight of the sample and multiplied by 100 to give *Sligh oxidation number.*

Acid Value/Number

Acid value or number determines the acidity of an oil. It is defined as the number of milligrams of the KOH required to neutralise the free acid present in 1gm of oil.

New or used petroleum products contain acidic constituents present as additive or as degradation products formed during service. Total acidity is a measure of the combined organic and inorganic acidity.

Generally, mineral acids are not present in the lubricants unless refined in a faulty manner. Organic acids are not so corrosive. Acid value of petroleum products should be less than 0.1. Increase in acid number indicates oxidation of the oil which will lead to gum & sludge formation besides corrosion.

Colour and Fluorescence

Colour of petroleum products indicates the degree of refinement. Distillates are lighter in colour than the residual oils. A pale colour does not indicate lower viscosity.

The mineral oils show fluorescence in reflected light whereas fixed oils do not show fluorescence. Hence fluorescence helps to detect adulteration.

Distillation Range

This enters into the specification of distillates like gasoline, kerosene etc. In the distillation test, a measured volume of the oil is distilled at the specified rate which is suitably increased as the distillation proceeds. The *initial boiling point* (I.B.P.) is the temperature when the first drop of condensed liquid drops from the condenser and the *final boiling point* (F.B.P.) is the maximum temperature recorded at the end of distillation. *Mid-boiling point* is the temperature at which 50% of the oil distils off. Distillation is stopped at 360°C as further heating causes cracking of the residue. If necessary, distillation is done under vacuum also. Distillation range helps in evaluation of petroleum products. Mid-boiling point is often used for correlating with other parameters.

Distillation data are used for making important curves like *true boiling point* (T.B.P.) curve, equilibrium/flash vaporisation curve. T.B.P. curve gives the variation of boiling point with the yield of distillates. Each point in the flash vaporisation curves gives the total amount of vapor in equilibrium with the liquid at that particular temperature. These curves are used in the design of continuous distillation equipments in refineries.

Distillation characteristics is a measure of volatility of petrofuels and determines the system of fuel metering required (wick feed, carburation or atomisation) and indicates the vapor losses during fuel storage & vapor locking tendency as well as uniformity of distribution to cylinders 10% distillation for motor spirit indicates the ease of starting the engine while too high a FBP will cause crankcase oil dilution.

Sulphur Content

Even after purification, petroleum products do contain some or other quantity of sulphur. Sulphur content increases with the increase in the boiling range of the products. Some typical sulphur content values are, Kerosene = 0.05 to 0.2%, gas oil/diesel oil = 0.3 to 1.5% and fuel oil = 2 to 4% (LSHS = 1%).

Sulphur content in oil is determined accurately by *bomb calorimeter* and *roughly by lamp method.* In lamp method, a known weight of oil is burned in an air current and the combustion products are passed through an alkaline solution, the sulphur being finally estimated as H_2SO_4 in the solution. Sulphur is largely present in oils in the form of corrosive organic compounds. On combustion, it gives off foul gases corrosive in nature.

Moisture

Water solubility in oil is very low (0.005 to 0.05%). However, it is present in fuel oil upto 1% in dispersed state. Undesirable water is separated from oil by preheating. *Dean and Stark method* is used for its determination in which oil sample is heated with xylene which forms an azeotrope with water. The distilled vapor on condensation and settling gives water as a separate layer which is measured.

The presence of water in fuel oil is undesirable because, among other harmful effects, arrival of water globules at the outlet of an oil burner may lead to '*flame failure*' (the flame is extinguished) either by interruption of the oil stream or by the '*quenching*' (extinguishing) effect of a suddenly evolved mass of steam. The amount of water present in oil is determined by dissolving the oil in Xylene (it acts as a carrier for water) and distilling off all the water and some xylene. In a common form of apparatus the distillation flask is fitted with a vertical, butoffset, reflux condenser which discharges into a vertical, graduated tube. Water collects at the bottom of the tube and is measured, while excess xylene is returned to the boiling flask or a metal still continuously. The standard apparatus of this type is that of ***Dean and Stark*** as shown in below Fig. This method is suitable for variety of bituminous materials, but it is especially applicable to crude petroleums and products derived from them, such as fuel oils, road oils, creosotes, road tars, asphalts and tars.

A suitable quantity of sample is transferred to the flask and 100 ml of solvent (xylene) is added to that. When large samples are analysed, a solvent volume in excess of 100 ml maybe necessary.

The components of the apparatus are assembled as shown in below Fig., making all connections vapour and liquid tight.

The condenser tube and trap must be chemically clean to ensure free drainage of water into the bottom of the trap. A loose cotton plug is inserted in the top of the condenser to prevent condensation of atmospheric moisture inside it. Cold water is circulated through the jacket of the condenser.

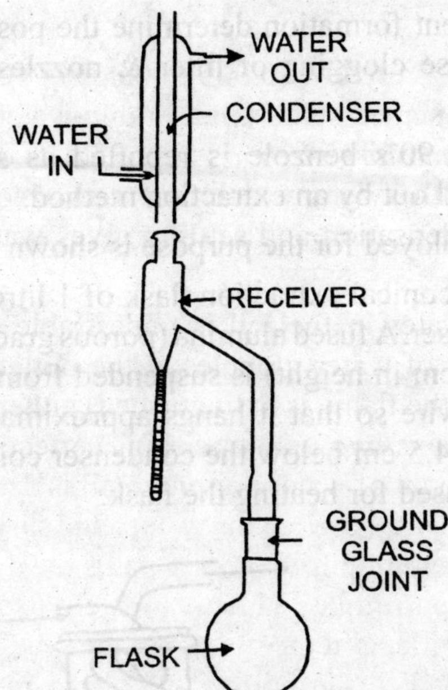

Fig. Dean and Stark Apparatus for Determination of Moisture in Liquid Fuels.

The heat is applied to the flask/still adjusting the rate of boiling so that condensed distillate discharges from the condenser at the rate of 2–5 drops per second. The distillation is continued until no water is visible in any part of the apparatus except in trap. If there is a persistent ring of water in the condenser tube, the rate of distillation is increased or condenser water is cut off for few minutes.

When the evolution of water is completed, the trap and contents are allowed to cool to room temperature. Any water drops adhering to the sides of the trap are dislodged with a glass or polytetrafluoroethylene or other suitable means and transferred them to water layer. The volume of the water in the trap is read to the nearest scale division and the water in the sample is reported as per cent by volume.

$$\% \text{ Water} = \frac{\text{Volume of water in trap}}{\text{Volume of sample}} \times 100$$

Volatile water soluble material persent may also be measured as water in this method.

Sediment in Liquid Fuels

Sediments comprising of pipe & storage tank scale and a few fibers is defined as material insoluble in toluene. It should be < 0.25% (by weight) in fuel oil. Sludge and sediment formation determine the possible life of oil during storage. Sediment cause clogging of filter & nozzles and corrodes storage tanks & pipelines.

The insolubles in 90's benzole is reported as sediment content. The determination is carried out by an extraction method.

The apparatus employed for the purpose is shown below in Figure .

This consists of a conical extraction flask of 1 litre capacity fitted with a block of tin coil condenser. A fused alumina (porous grade) thimble (cylindrical tube 2.5 cm dia and 7 cm in height) is suspended from the condenser coil by means of a platinum wire so that it hangs approximately 4.5 cm above the extracting solvent and 4.5 cm below the condenser coil. An electric hot plate or steam bath may be used for heating the flask.

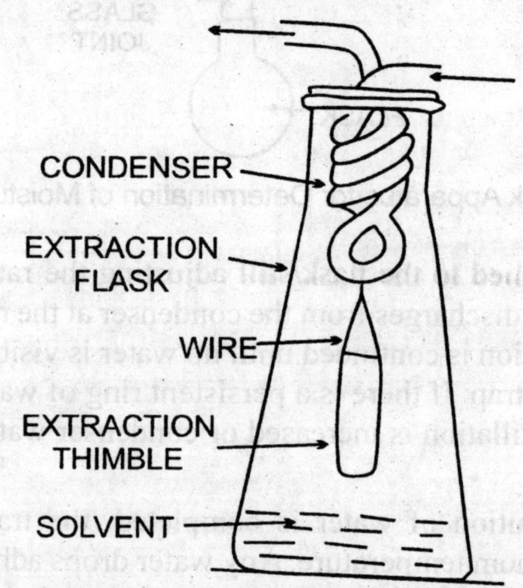

Fig. Sediment Determination Apparatus.

A preliminary extraction with 90's benzole is given to thimble before its use and the solvent is allowed to drip for 1 hour at a temperature of 100-110°C. The thimble is cooled to room temperature and weighed to the nearest 0.1 mg. The extraction is repeated until the weights of thimble after two successive extractions do not differ by more than 0.2 mg. Now 10 gm of the sample is placed in the extraction apparatus and the sample is extracted with 90's benzole until the solvent dripping from the thimble is colourless. The rate of extraction is so maintained that the surface of the mixture of oil and benzole in the thimble does not rise more than 2 cm from the top. After the completion of extraction, the thimble is dried for 1 hour at 100–110°C, cooled and weighed. The extraction is repeated for 60–75 minutes and again followed by drying, cooling & weighing. The extraction is repeated for 1 hour, until the weights of the dried thimble after two successive extractions do not differ by more than 0.2 mg. The sediment content is represented as per cent of original sample.

$$\% \text{ Sediment} = \frac{\text{Weight of sediment}}{\text{Weight of sample}} \times 100$$

Gross C.V. of petroleum fuels are in the range of 10000 to 11300 kcal/kg , the higher figure being for gasolines and lower for heavy fuel oils. Petroleum products are high in hydrogen content (11.8–14.5%) and hence their net calorific value is less than the gross by 620–760 kcal/kg. C.V. of oil is determined in a *bomb calorimeter.*

Following emperical formulae are also used for its determination.

(a) *Mendeleev formula,* $Q_G = 81C + 300H + 26\ (S–O)$

(b) Also, $Q_N = Q_G - 6\ (9H + W)$ where, $H = 2.6 - 1.5\ \rho$

(c) Also, $Q_G = 12400 - 2100 \times \rho^2$

where, Q_G = Gross calorific value, kcal/kg, Q_N = Net calorific value, kcal/kg

C, H, S, O and *W* are weight percentages of carbon, hydrogen, sulphur, oxygen and water in oil. ρ = specific gravity at 15.5°C

Gross calorific value Q_G of most hydrocarbon liquid fuels is also given by *Cragoe formula,* which is as follows :

(d) $Q_G = 51860 - 8790 \cdot S^2(1-x-y-z) + 9410\ z$ KJ/kg

$= 12390 - 2100S^2\ (1-x-y-z) + 2248\ z$ Kcal/kg

where*s, x, y, z* are the specific gravity, moisture, ash and sulphur respectively in fuel expressed as fractions.

ASH in Petroleum Products

Ash in petroleum products can result from oil, water soluble metallic compounds or extraneous solids such as dirt & rust. An idea of the ash forming material present in oil can give information about its suitability for use in a given application.

This method describes a procedure for determining the ash content in distillate, residual fuel oils, crude oils, lubricating oils, waxes, and other petroleum products. However, this process is undesirable where ash forming additives are used and the compounds of lead and zinc are present because these volatalise appreciably during incineration. The ash content of an oil is given as the percentage by weight of the inorganic residue left after ignition of the oil. The amount of ash in petroleum oil is normally very low. It increases with the viscosity of the fuel oil, since most of the original inorganic impurities allowed to remain in the crude are concentrated in the primary residues. Ash in oil is determined by gradually burning a weighed quantity of oil in a silica crucible, finally igniting the crucible and its charred contents at 800°C.

For estimation of ash, 50 gm sample (nearest to 0.1 gm) is taken in a weighed dish of platinum, porcelain or silica which is already heated at 700°C for 10 minutes and cooled. The dish and sample are heated until the contents can be ignited with a flame. This dish is maintained at such a temperature that the sample continues to burn at a uniform and moderate rate, leaving only ash and carbon when the burning ceases. Then the residue is heated in a furnace at 800°C until all carbonaceous material has disappeared. The dish is cooled to room temperature in a desicator and then weighed to the nearest 0.1 mg. The dish is reheated for 10–20 minutes at 800°C, cooled in desiccator and reweighed. Heating & weighing are repeated until consecutive weighings differ by not more than 0.5 mg. The ash content in oil is given by

$$\% \, Ash = \frac{w}{W} \times 100$$

where, w = weight of ash, grams and W = weight of sample, grams.

If the fuel oil contains very low ash *i.e.* 0.01–0.02%, more quantity of oil may be taken in a beaker and transferred to the dish in successive portions and igniting them until a suitable quantity of oil has been used.

If the sample contains sufficient moisture to cause foaming & loss of material, the sample is discarded, and a fresh sample is taken to which 1–2 ml of isopropyl alcohol (99%) is added before heating. If this is not satisfactory, 10 ml of an equivolume mixture of benzene & isopropyl alcohol is added and mixed thoroughly. Several strips of ashless filter paper are placed in the mixture and heated, when the paper begins to bum, most of the water would have been removed.

Mechanical Stability of Lube Oils

To judge the stability of a lubricant under conditions of very high pressure, different mechanical tests are carried out. One of such tests is *"four ball extreme-pressure lubricant test"* shown below in Figure. in which the lubricant is poured in a machine containing four balls in which lower three balls are stationary ; while the upper ball is rotated. Load is gradually increased and the balls withdrawn & examined at specific intervals for scars formation, etc. on them. If the lubricant is satisfactory under the given load, the ball bearings after the test comes out clean. However, when the load is progressively increased, liberated heat welds the balls, together and the lubricant is said to fail completely. Thus, this test enables to determine the load that can be carried safely by a lubricant.

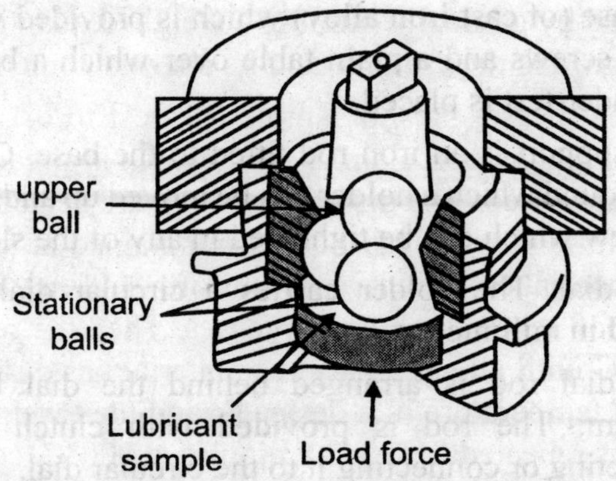

Fig. Four ball extreme-pressure tester.

While selecting greases for practical jobs the two properties of main consideration are consistency and drop point.

Consistency or yield value of grease. It is expressed in terms of penetration, which is defined as *"the distance in tenth of millimetre that a standard cone.penetrates vertically into the sample under the standard conditions of load, temperature and time"*. The value of load, temperature and time are taken respectively as 150 g, 25°C, and 5 seconds. Consistency of greases depends on the structure and interaction of the gelling elements in it and to some extent on the viscosity of oil used. The consistency is determined by using "Penetrometer" shown below in Fig. The apparatus consists of:

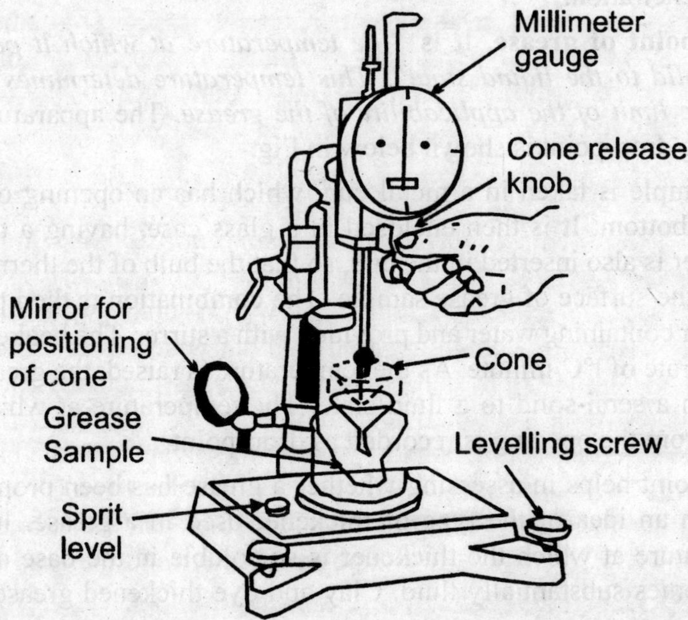

Fig. Penetrometer used for determining consistency of grease.

- Heavy base (of cast iron alloy) which is provided with a spirit level, levelling screws and a plain table over which a box containing the grease under-test is placed.
- Vertical support is an iron rod fitted to the base. On this are slotted marks, around which a holder can be moved up and down. The holder has a screw which can be tightened in any of the slots.
- Circular dial. The holder carries a circular dial gauge which is graduated in millimeters.
- Moving dial rod is arranged behind the dial by a mechanical mechanism. The rod is provided with clutch arrangement for disconnecting or connecting it to the circular dial.
- Standard cone. The moving dial rod is fitted at its lower end with a standard cone (weighing 150 g).
- A mirror. Vertical rod is provided with an adjustable mirror for removing parallax while positioning the cone in contact with grease sample surface.

The apparatus is levelled, the cone is cleaned and the sample of grease under-test in a box is placed below the cone. The height of the cone is so adjusted, with the help of mirror, that tip of the cone just touches the sample. Initial dial reading is noted. The cone is then released for exact 5 seconds, by pressing a button provided for this. After 5 seconds, the press button is released and final dial reading noted. The difference of the two dial readings gives the penetration.

Drop-point of grease. It is *"the temperature at which it passes from the semi-solid to the liquid state". This temperature determines the upper temperature limit of the applicability of the grease.* The apparatus used for determining drop-point is shown below in Fig.

The sample is taken in a metal cup, which has an opening of standard size in the bottom. It is then enclosed in a glass case, having a tight lid. A thermometer is also inserted in the cup, so that the bulb of the thermometer is just above the surface of grease sample. The combination is then placed in a glass beaker containing water and provided with a stirrer. The beaker is heated slowly at a rate of 1°C/minute. As the temperature is raised, the grease sample passes from a semi-solid to a fluid state. The temperature at which its first drop falls from the opening is recorded as drop-point.

Drop point helps in assessing whether a grease has been properly made or to obtain an idea of the type of thickener used in a grease. It indicates the temperature at which the thickener is so soluble in the base oil that the grease becomes substantially fluid. Clay and dye thickened greases have no measurable drop points.

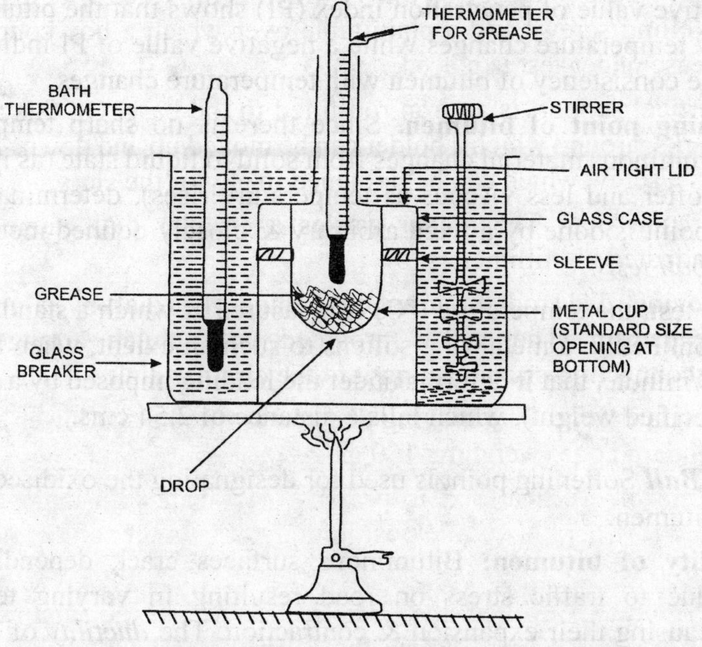

Fig. Drop-point apparatus.

Suitability of several grades of bitumens commercially available are assessed by penetration, penetration index, softening point and ductility for various applications.

Penetration of bitumen. It determines the type and grade of bitumens used for a particular application. The bitumen sample is maintained at 25°C in a container kept in water bath. The weighted needle is brought to the surface and the penetration of the needle into the bitumen at the end of 5 seconds internal in units of 0.1 mm is termed as *"penetration"* of the bitumen. The penetration of a bitumen at the softening point is about 800.

Penetration index of bitumen. Penetration at 25°C and the softening point or penetration at two different temperatures (say at 25°C & 10°C) defines the extent to which the consistency of bitumen changes with temperature defined by a factor called "Penetration Index". It is defined as follows :

$$\frac{\log(800) - \log(PEN, 25°C)}{T_{RB} - 25} = \left[\frac{20 - PI}{10 + PI}\right]\left[\frac{1}{50}\right]$$

where, *PEN,* 25°C = Penetration of bitumen at 25°C

PI = Penetration index of bitumen

T_{RB} = Ring and ball softening point of bitumen, °C

A positive value of penetration index (PI) shows that the bitumen is less effected by temperature changes while a negative value of PI indicates large effect in the consistency of bitumen with temperature changes.

Softening point of bitumen. Since there is no sharp temperature at which a bituminous material changes from solid to liquid state (as it gradually becomes softer and less viscous as temperature rises), determination of its softening point is done by a fixed arbitrary & closely defined method called "*ring and ball test*".

In this test, the temperature (°C) is measured at which a standard disc of bitumen confined in a metal ring softens to such an extent, when heated at a rate of 5°C/minute, that it deforms under the loading imposed by a small steel ball (of specified weight), which falls a distance of 2.54 cms.

Ring and Ball Softening point is used for designating the oxidised bitumen and hard bitumen.

Ductility of bitumen: Bituminous surfaces crack depending on its ductility due to traffic stress on road resulting in varying temperature exposure causing their expansion & contraction. The *ductility* of bitumen is expressed as the distance in cm to which a standard briquette of bitumen can be elongated before the thread formed breaks under the specified conditions. Bitumen having high ductility may be more temperature susceptible.

To determine the ductility of bitumen, a molten bitumen sample is poured into a standard would, allowed to cool to room temperature and then placed in a thermostatically controlled water bath to bring the briquette of bitumen to the test temperature before mounting in the ductility testing machine which consists of a moving carriage moving over a lead screw. Smooth constant speed and continuous operation is ensured by an electric motor driven reduction gear unit.

Setting Point and Melting Point of Wax

Setting point of wax is the constant temperature observed within a range of 0.1°C for one minute, when the molten wax is allowed to cool in a specified apparatus and the temperature is recorded at frequent interval. This method is not suitable for micro crystalline wax or intermediate wax or blends of paraffin waxes with these or any additives.

Congealing point of all petroleum waxes or petrolatum is determined by applying a drop of molten wax to the thermometer bulb and noting the temperature at which it congeals, when the thermometer is rotated under standardised cooling conditions.

Drop melting point of petroleum waxes or petrolatum is the temperature at which a drop of the waxes sample falls from the bulb of a thermometer, when heated under standard conditions.

Gum Content of Petrofuels

Gums present in petrofuels cause piston ring sticking & deposits on engines and indicate the amount of olefins present which have poor storage stability. There are two types of gums in petrofuels namely *existent gum* and *potential gum.*

Existent gum is that gum which is already formed and present in the fuel and can be deposited as the petrofuel evaporates. It is the amount of non volatile heptane insoluble residue left when the sample of oil is evaporated in a jet of hot air at 160°C. For jet fuels, evaporation of oil sample is done in a jet of superheated steam at 232°C.

Potential gum is the amount of gum that may be formed under long storage condition by polymerisation& oxidation of unstable hydrocarbons like olefins present in the oil. It is the amount of gum formed after the sample is aged in an oxidation stability bath and evaporated under specified conditions.

Copper Strip Corrosion Test of Petrofuels

Petrofuels contain varying amount of sulphur and sulphur compounds which are corrosive in nature. Copper strip corrosion test serves as a measure of difficulties encountered with copper, brass or bronze parts of the petrofuels handling & utilisation systems.

In this test, a smoothly polished & cleaned copper strip is immersed in the oil sample which is then maintained at the specified temperature for the specified length of time. The strip is then removed, washed with aromatic & sulphur free petroleum spirit and examined for etching, pitting or discoloration. It is then matched with ASTM (American Society for testing materials) copper strip corrosion standard color code number 1, 2, 3 & 4 designating slight tarnish, moderate tarnish, dark tarnish & corrosion respectively. Subscripts 'a' to 'e' describes a standard color reproduction in the standard chart. The classification code 'la' for example indicates slight tarnish with a light orange color. ·

Since some of the parts of fuel pump in aircrafts are made of silver, the corrosive tendency of the *aviation turbine fuel* (ATF) is determined by *silver strip corrosion test* in which a polished silver strip is immersed completely in ATF at 45 ± 1°C for 16 hours. At the end of 16 hours, the silvers strip is removed, washed and evaluated for corrosion against the set of standard.

Refractive Index of Petrofuels (Hydrocarbon Liquids)

Refractive index (defined as ratio of velocity of light in air to that in liquid under examination) gives an indication of aromatic content & molecular weight of petrofuels. The refractive index of liquids (also defined as the ratio of sine of angle of incidence to the sine of the angle of refraction) varies inversely with both wavelengths & temperature. Naphthene content in naphtha can be calculated by knowing its refractive index & density of the saturates fraction as determined by the refractive intercept method. Refractive intercept of liquid hydrocarbons is given by n - 0.5p, where n = refractive index and p = density (gm/cc) both at 20°C.

Water Separometer Index (Modified) (WSIM)

WSIM, which is a measure of fuel cleanliness (relativeto its freedom from surfactant materials) measures the water separation characteristics of fuels. In water separometer, a prepared emulsion of water &petrofuel is passed through a cell containing a coalescer. Entrained water is found in the effluent from cell by light transmission. The ease with which the fuel releases the emulsified water is rated on a numerical scale 1 – 100.

Thermal Stability of JET Fuels

Jet fuel thermal oxidation tester (JTFOT) measures the high temperature stability of jet fuel (gas turbine fuel) to assess the deposit level resulting from the fuel contacting a heated surface.

The jet fuel under test is pumped at a fixed rate (600 c.c in 2.5 hours) through an aluminium heater followed by a stainless steel filter (17 microns) meant for trapping the fuel degradation products (*i.e.,* deposits). Amount of deposit on heater tube and the rate of plugging of filter gives the indication of thermal stability of jet fuels.

Smoke Volatility Index (SVI) of ATF

SVI is a measurement parameter for combustion deposit by wide cut gasoline type of aviation turbine fuel (ATF). It is a function of its smoke point and volatility defined as :

SVI = smoke point + 0.42 (% of ATF boiling below 204.4°C)

Kauri-Butanol (KB) Number of Hydrocarbon Solvent i.e., Solvent

Power

KB number is a measure of solvent power (or solvency) of a hydrocarbon solvent *i.e.,* its ability to dissolve a given product. KB number is a point at which a solvent added as a thinner would cloud a solution of kauri resin dissolved in *n*-butanol. For most polymers & resins, solvent power decreases in the order aromatics >naphthene>*n*-paraffins>*i*-paraffins. Higher KB number implies higher solvent power. Standards are toluene (KB number = 105) and 25% toluene + 75% /*i*-heptane (KB number = 40). It measures the relative solvent power KB number ranges from 25 to 105 for most commercial solvents.

9

Requisites of Good Quality Petroleum Products

Knocking

When the fuel-air mixture undergoes very fast or intermittant combustion inside the combustion chamber of an internal combustion engine, then a sort of rattling noise develops from the combustion chamber (due to uneven, unsteady and uncontrollable combustion) which is called knocking or detonation.

Reasons for knocking. Knocking in petrol (gasoline) & diesel engines are due to different reasons. In case of petrol engines, the carburettor (which mixes petrol and combustion air) supplies the petrol air mixture to the combustion chamber which is ignited by the spark from the spark plug. If the first portion of the fuel burns in a normal manner and the last portion burns instantaneously, then a large momentary pressure imbalance is caused which sets up a pressure or shock wave in the combustion chamber resulting in knocking. This is a characteristic of petrol constituents.

Another reason is the preignition of petrol before the regular spark occurs. It occurs at high compression ratio (which is the ratio of cylinder volume at the end of suction stroke to the cylinder volume at the end of compression stroke) which increases the temperature of fuel air mixture in the combustion chamber above ignition point even before spark occurs. Hence, the fuel ignites before the spark occurs. This pre-ignition of petrol causes uneven and unsteady combustion resulting in knocking of engine.

In case of diesel engine (compression ignition engine), there is no spark plug and the combustion is initiated by increasing the temperature of air by compression inside the combustion chamber by piston till the ignition temperature is achieved.

In diesel engines, as soon as the fuel is injected into the combustion chamber, it should instantaneously ignite. If it does not ignite instantaneously *i.e.* if there is an ignition delay, then more and more quantity of unburnt diesel gets accumulated in the combustion chamber which burns at once causing knocking of the engine. This is called ***diesel knock***. Hence for a good diesel, its spontaneous combustion should take place as soon as it is injected into the cylinder.

Knocking depends upon the characteristics of fuel, design of engine and the operation of engine. If the engine is operated at high load and low speed, knocking results. If the petrol and diesel are adulterated then also knocking results.

Effects of knocking: Knocking in an engine results in wear and tear of cylinder, piston & other engine parts. Besides, it lowers the efficiency of the engine. Knocking reduces the life of the engine.

Octane Number

Octane number expresses the anti-knocking characteristics of petrol. Normal heptane (a constituent of petrol) when burnt in a petrol engine knocks it very badly *i.e.* the anti-knocking value of n-heptane is very poor and it is assigned an octane number equal to zero. On the other hand, *iso-octane* (also a constituent of petrol) has got a very good anti-knock characteristics, hence it is assigned on octane number of 100.

Definition of octane number: Octane number of petrol is the percentage by volume of iso-octane in a mixture of iso-octane and normal heptane with the same knocking tendency as the petrol in question.

If the octane number of a petrol is 80 then it means that this petrol gives as much knocking as a mixture of 80% iso-octane and 20% normal heptane.

Determination of octane number (O.N.): The petrol whose octane number is to be determined is burnt in a standard spark ignition internal combustion test engine and its knocking tendency is measured by an instrument called knockmeter. Now this reading of knockmeter is matched with the standard chart in which the value of octane number corresponding to the knockmeter reading is given. This gives the octane number of the petrol.

Three methods are used for O.N. determinations namely :

- *Motor method - for low O.N. motor fuels*
- *Research method - for high O.N. motor fuels*
- *Aviation method - for very high O.N. aviation fuels.*

Parameters of the above test methods are as follows :

Parameters in various methods	Motor method	Research method	Aviation method
rpm	900	600	1200
spark advance	variable	13° btdc	—
mixture heating	Yes	No	—

Octane number of Indian petrol: Petrol having octane number 87 is sold in India for use in motor vehicles. Premier gasoline (petrol) sold in India having octane number 93 is not much in demand due to its higher cost. In U.S.A. the octane number of premier gasoline is 94. The octane number of petrol used in air-crafts are more than 100 and they are designated by performance number.

Significance of octane number: Higher octane number of petrol means higher efficiency and higher power output by the engine. Hence, higher the octane number, better the petrol.

Limitation of octane number as a guide: Tendency of a petrol to knock depends not only on its octane number but it depends also upon the design and operation of the engine. So, the octane number of a gasoline is simply an emperical figure and not related to the actual knocking in an automobile engine.

Means to improve octane number. Octane number of petrol is increased by adding some additive compounds in it. The most important additive is Tetra Ethyl Lead (TEL). Quantity added is 3.6 cc/gallon of motor petrol and, 5.6 cc/gallon of aviation petrol.

TEL is extremely poisonous, hence petrol containing TEL is always coloured for identification and safety.

The use of petrol having higher octane number then necessary to avoid knocking is not at all useful as it is more susceptible to pre ignition and gum formation on storage.India has discontinued the use of leaded motor petrol in view of its environmental consequences .However TEL/TML is added in aviation gasoline to increase its octane number beyond 100 which is designated as performance number.

Advantages of high octane number of petrol: As per an estimate, as the octane number increases from 87 to 100, the cruising speed (which is very important for airplanes) increases by 25 km/hour. There is increase in speed of climb by 80% and the maximum attainable height increases from 8000 metres to 11000 metres. The required take-off distance decreases from 800 metres to 600 metres and pay load also increases by 2 tons. These all are very important for aeroplanes.

Anti-knocking value: Anti-knock value of the hydrocarbons present in petrol increases with increase in the compactness of the molecules, cyclic structure & double bond. Thus butane has an octane number of 90 whereas heptane has 0.

Anti-knock value *i.e.* octane number increases in the order paraffins → olefins → naphthene → iso-paraffins → aromatics. It means that aromatics have highest anti-knock value whereas paraffins have lowest anti-knock value (hence octane number). So, presence of maximum quantity of aromatic and minimum quantity of paraffins is desirable in petrol.

In case of paraffins and olefins, anti-knock value *(i.e.* octane number) decreases with increasing chain length. In case of naphthenes, the larger the size of the ring & the length of the side chain, the lower is the anti-knock value. In case of aromatics, side chain addition upto 3 carbon atoms length improves the octane number, but afterwards, the octane number decreases in proportion to the length of the longest chain.

Octane Number of Blends: When two paraffins are blended, then the octane number of the resultant blend is directly proportional to the percentage of each paraffin in the blend. With blends of paraffins and aromatics, the octane number of the resultant blend is lower than expected from a linear relationship. With blends of paraffins and olefins, the octane number of resultant blend is higher than expected from a linear relationship.

Lead Response: The response of gasoline to the addition of tetra ethyl lead $[(C_2H_5)_4Pb]$ for increasing its octane number is different for different categories of hydrocarbons. The order of decreasing effect of TEL addition (called lead response) is paraffins → naphthenes → olefins → aromatics. Alcohols, sulphur compounds and lubricating oils respond negatively.

Cetane Number

Cetane number is the measure of knocking in a diesel engine. Cetane ($C_{16}H_{34}$) has a very short ignition delay hence its cetane number is taken as 100 whereas alpha methyl naphthalene (C_8H10) has a very large ignition delay hence its cetane number is taken as 0.

The cetane number of a diesel is defined as the percentage of cetane (by volume) in a mixture of cetane and alpha methyl naphthalene which has the same ignition delay as the diesel under test.

Values of cetane number of diesel: The cetane number of low speed, medium speed and high speed diesels are 25, 35 and 45 respectively.

Engine speed, rprn	> 1500	800-1500	400-800	100-400	< 100
Suitable cetane number range	50–60	45–55	35–50	30–10	15–30

Cetane number of diesel can be increased by adding additives like ethyl nitrate and acetone peroxide.

Cetane number *increases* in the order—aromatics → iso-paraffins → naphthenes → olefins → paraffins (AINOP) whereas the octane number *decreases* in this order. Therefore, an oil of high octane number has a low cetane number and *vice versa.* Consequently, a petroleum crude which gives petrol of high octane number gives diesel of low cetane number.

Higher cetane number diesel fuel results in easier cold weather engine starting, faster engine warm up, reduced exhaust smoke & odor, increased engine efficiency & power output, less combustion noise and improved combustion control.

Two important motor fuels namely gasoline and diesel are produced from crude oil. *Petrol is the commercial name of gasoline.*

Gasoline (Petrol)

Production of petrol (gasoline) from crude oil

Petrol is produced from crude oil by the following processes :

- distillation of crude oil produces *straight run gasoline.*
- cracking of fuel oil produced from crude oil to produce *cracked gasoline.*
- alkylation and polymerisation of refinery gases produced during distillation of crude oil.
- reforming of heavy gasoline and naphtha to get *reformed gasoline.*-
- condensation of natural gas to get *natural gasoline* called *"casing head gasoline".*

The gasoline produced by the distillation of crude oil is called ***straight run gasoline*** which is about 20% of the crude and has a low octane number of 55-60. Gasoline comprises of C_5 to C_{10} hydrocarbons fraction boiling in the range of 30 to 180°C. It is also called motor spirit or motor fuel. On prolonged storage, it undergoes slow but progressive oxidation deterioration resulting in formation of non-volatile gummy residue & darkening of its color. Gummy residues cause deposit formation in carburettor parts, venturi, manifold of distribution system and in the combustion chamber.

Gasoline produced by catalytic cracking of heavy fuel oil at high temperature & pressure is called *cracked gasoline* and has a high octane number of 90. Refinery gases produced during distillation of crude oil are subjected to some auxiliary processes like polymerisation (gasoline produced in this case is called *polymer gasoline),* alkylation and isomerisation to produce high octane number (90-98) gasoline. Reforming process of heavy gasoline and straight run naphtha produces ***reformed gasoline*** of high octane number (90-98) by increasing the proportion of high octane iso-paraffins and aromatics. There are still other methods called hydrocracking and hydroreforming of heavier fractions of crude oil to produce superior gasoline with high octane number.

Requisites of good gasoline

- It should have high octane number to develop more power at higher compression ratio without knocking.
- It should have controlled volatility and vapor pressure. Very low vapor pressure of petrol causes starting trouble in the engine whereas very high vapor pressure of petrol causes vapor locking of fuel pipelines.
- Its calorific value should be high for higher efficiency. C.V. of petrol is around 10000- 10500 kcal/kg.
- Its freezing point should be low to avoid the choking of fuel lines of aircrafts flying at high altitude where the temperature is low and for engines operated during cold season.
- It should not contain abrasive, corrosive & gum forming substance otherwise it will cause abrasion, corrosion & choking respectively of fuel lines, carburettor jets etc.
- It should have optimum flash point for easy starting of engine and avoidance of explosion & fire hazards during storage.

Indian specification of motor gasoline is given in table below.

Table - Specification for Motor Gasoline

(IS : 1585—1960, Amendments in 1967 & 1972 IS: 2796—1971, First Revision)

Characteristics	**87 Octane no. Gasoline**	**83 Octane Gasoline**	**93 Octane Gasoline**
Colour, visual	Orange	Orange	Red
Octane number (Research Method). Minimum	87	83	93
Lead Content (a) as lead, g/1. Maximum	0.42	0.56	0.80
Distillation characteristics : (a) Initial Boiling Point (IBP)	—	Not limited, but to be reported	
(b) Recovery upto70°C, % by volume, Minimum	10	10	10
(c) Recovery upto125°C, % by volume, Minimum	50	50	50
(d) Recovery upto180°C, % by volume, Minimum	90	90	90
(e) Final boiling point (FBP), °C, Maximum	215	215	215
(f) Residue, percent by volume, Maximum	2	2	2
Reid vapour pressure at 38°C, kg/cm2(g), Maximum	0.70	0.70	0.70
Sulphur, total, percent by weight, Maximum	0.25	0.25	0.20
Residue on evaporation, mg/100 ml, Maximum	4	4	4
Oxidation stability, minutes, Minimum	360	360	360
Copper-strip corrosion for 3 hours at 50°C	Not worse than No. 1		Not worse than No. 1
Density at 15°C	Not limited, but to be reported		

Refining of petrol: The crude petrol is refined by treating with sulphuric acid to remove sulphur compounds present in it (the process is called *'sweetening'* of petrol). It is then passed through a bed of fuller's earth to remove its colour, which is due to the presence of gum in it. Finally, the dissolved gases in it are driven off by a process called *stabilisation.* Tetraethyl lead or other additives are then added to increase the octane number (if the need be) or to inhibit the gum formation before the petrol is ready for use.

Additives in petrol (gasoline)

Commercial grade motor gasoline contains the following additives :

i) 'Anti-knocking agents', e.g. tetra ethyl lead (TEL). Currently unleaded petrol (without TEL) is being sold/used in India, as toxic lead in petrol causes atmospheric air pollution.

ii) 'Antioxidants' to prevent gum formation and improve storage stability e.g., 2, 6-ditertiary butyl-4-methyl phenol.

iii) 'Metal deactivators' to prevent gum formation due to trace metals like iron and copper.

iv) 'Combustion chamber deposit modifier' e.g., tricresyl phosphate (TCP) to minimise misfire due to lead fouling of spark plugs and to reduce ignition & preignition.

v) 'Anti-icing agents' to minimise engine stalling & power loss due to ice formation in carburettor & fuel lines e.g., ethanol and hexylene glycol.

vi) 'Anti rust agents'.

vii) 'Cleansing agents' to prevent deposits in the fuel supply system.

viii) 'Dying agents' to give distinctive colors to different brands and names.

AVIATION GASOLINE

Petrol used in the aeroplanes has comparatively higher volatility and higher octane number (> 100) than motor gasoline. Though higher proportion of TEL (4.6 c.c/U.S. gallon) is added in aviation gasoline, other additives are lesser in number and quantity. The main components of aviation gasoline are isoparaffin alkylates and aromatic reformates. A typical specification of aviation gasoline is tabulated below.

Table: Typical Specification of Aviation Gasoline

S. No.	Parameters	Unit	Value
1.	Reid vapor pressure at 37.8°C (100°F)	kg/cm^2	0.385-0.490
2.	Freezing point, maximum	°C	60

S. No.	Parameters	Unit	Value
3.	Performance number	—	130-145
4.	Aromatic content, minimum	% by volume	5
5.	Diesel index, minimum	—	75-90
6.	Net calorific value, minimum	kcal/kg	10390-10500
7.	Total sulphur, maximum	% by weight	0.05
8.	Tetra ethyl lead	mg/litre	nil
9.	Oxidation stability (16 hours ageing)		
	(a) Gum, maximum	mg/100 ml	6
	(b) Precipitate, maximum	mg/100 ml	2
10.	I.B.P/F.B.P.	°C	<75/170

Diesel

It is a straight run product of crude petroleum oil distillation obtained in between kerosene and lubricating oil. Diesel obtained from paraffinic crude are most suitable as higher amount of paraffin present in it increases its cetane number. Diesel is also produced by cracking of heavy fuel oil, vacuum distillation of atmospheric distillation residue and coking process to supplement the limited yield of straight run diesel to meet its demand (six times of petrol in India). *High speed diesel* (cetane no.> 42) are used in diesel engines for automative purpose *e.g.*.bus, truck, locomotive, etc. Light Diesel Oil (LDO) are used in engines in marine ships and for electricity generation. Middle distillates of crude oil boiling in the range of 150 - 400°C are suitable as diesel fuels. Iso-octyl nitrate is added to improve the cetane number by 3 to 6 units. In cetane number increase requires 100 pounds per thousand barrels (Ptb) cetane improver. Other additives used in diesel fuel include ***detergents*** (act as cleaners to remove carbonaceous & gummy deposits) upto 50 Ptb,***Stabiliser*** (to retard oxidation during storage), ***anti-corrosion & rust inhibitor*** (3-5 Ptb), ***cold weather flow improvers*** (65-125 Ptb), ***Anti-wax settling additive*** (40-60 Ptb), ***Lubricity additive*** (3-8 Ptb), ***metal deactivator*** (2-3 Ptb) etc.

Though the most desired components of diesel fuels are the straight run distillates from paraffinic and mixed base crude oils, these are often used as feedstock for gasoline production via cracking. Since, there is some overlap in the boiling ranges of kerosene and diesel fuel hence diesel is prepared from heavy distillates obtained from catalytic cracking unit. Since, the distillates are rich in aromatics and isoparaffins (having low cetane number), additives are added to improve the cetane number of diesel fuel. The suitable volatility is obtained by blending with light fractions. Residual oils are also used in significant proportions in meeting large demand of diesel fuel.

Requisites of good diesel fuel

- It should have a high cetane number.
- It should contain very low amount of sulphur (<0.5%), ash & carbon residue so that fouling corrosion and wear of engine parts would be minimum.
- Its calorific value should be high.
- Its flash point should not be less than 65°C to avoid fire and explosion hazard.
- It should have a very low freezing point so that it can be successfully used as a jet fuel after blending with kerosene or naphtha.

Indian standard specification for diesel oils is tabulated below.

Table - Specification for Diesel Oils (IS : 1460—1968/1974)

Characteristic	**Unit**	**Grade Special**	**Grade A**	**Grade B (LDO) Light Diesel Oil**
Flash point, (Pensky-Martens, closed)	°C, Minimum	38	38	66
Kinematic viscosity	centistokes at 37.8°C	2.0 to 7.5	2.0 to 7.5	2.5 to 15.7
Carbon residue (Ramsbottom)	percent by weight, Maximum	0.20	0.20	1.50
Cetane number, Minimum	—	42	40	—
Diesel Index, Minimum	—	45	43	—
Distillation	per cent recovery, at 366°C, Minimum	90	90	—
Copper strip corrosion for 3 hours at 100°C	—	Not worse than No. 1	Not worse than No. 1	Not worse than No. 2
Water content	per cent by volume, Maximum	0.05	0.05	0.25
Sediment	per cent by weight, Maximum	0.05	0.05	0.10
Ash	per cent by weight, Maximum	0.01	0.01	0.02
Acidity, inorganic	—	Nil	Nil	Nil
Acidity	total mg of KOH/g, Maximum	0.50	0.50	—
Sulphur	total per cent by weight, Maximum	1.0	1.0	1.8

Characteristic	Unit	Grade Special	Grade A	Grade B (LDO) Light Diesel Oil
Pour point	°C, Maximum	6	6	12 for winter, 18 for summer
Winter — November to February, Summer — March to October . Special grade and grade A cover distillates of low volatility. Grade B covers the class of more viscous distillates and blends of these distillates with residuum oil. Diesel fuel for naval application shall have a flash point of 66°C, Minimum and cetane number of 45, Minimum.				

Fuel Oil (Furnace Oil)

Fuel oils are the fractions distilled between lubricating oil and bitumen. The heavy fuel oil is thermally cracked/decomposed at high pressure and temperature (process is called Visbreaking) to reduce its viscosity before it is used as fuel oil. At times, heavy fuel oil is blended with naphtha, gas oil and light fuel oil to improve its atomising capability through burners. Fuel oil is composed of higher molecular weight compound known as asphaltic bitumens.

On one extreme is the light fuel oils produced from cycle gas oil of cracking units and on the other extreme is heavy fuel oils produced from residues of crude distillation units, thermal catalytic cracking units, and coker (coking unit). Intermediate grade/viscosity of fuel oils are made by blending of distillates and residual stocks. 'Visbreaking' of residual stocks is a very important process in the production of fuel oils. Cycle gas is filtered through clay and hydrofined. Tricresyl phosphate and sludge dispersing agent are added to improve combustion in burners. Magnesium bearing additives are used in case of residual fuel oils to eliminate fouling and corrosion.

Properties of fuel oils

Desirable properties of fuel oil are tabulated below.

Table: Desirable properties of fuel oil

Properties	Remarks
Viscosity	It should be low (25 cst) for its proper atomisation through burner and low pressure drop during pumping. But viscosity of fuel oil below 18 CST causes leak back through the pump and wear on the moving parts which are lubricated by the fuel oil will increase.
Pour point and freezing point	It should be low to permit easy flow in cold weather. Desirable pour point is 18°C.

Properties	Remarks
Sulphur content	Should be low (< 4%) to avoid corrosion, pungent odor and increase in dew point of flue gases.
Sludge, gummy material and ash	Should be low to avoid chocking of pipelines, erosion of furnaces/ engines and contamination of products. Ash comprising mainly of vanadium and sodium should be < 0.05%.
Calorific value	Should be high. It varies between 9000-9400 kcal/kg.
Specific gravity	Should be low as the pressure drop on pumping it will be low.
Flash point	Should be high for avoiding evaporation loss and fire hazards.
Water content	Should be very low as it decreases the calorific value of oil.

Analysis of fuel oil. A typical analysis of fuel oil is given below :

Carbon = 86%, Hydrogen = 12%, Sulphur = 1%, Water = 0.6%, Sediments = 0.4%.

Typical specification of fuel oil

Calorific value = 9200 kcal/kg

Specific gravity at 25°C = 0.9

Flash point = 75°C

Viscosity at room temperature= 85 centistokes

Theoretical flame temperature=1900°C

Maximum freezing point = 15°C

Storage temperature = minimum 20°C

Atomisation temperature = 125°C

Emissivity factor = 0.5–0.7

Furnace oil produced in India has a maximum viscosity of 1500 seconds (Redwood no. I) at 50°C. However, *Low sulphur heavy stock (LSHS)* furnace oil (S<1%) is semi-solid at ambient temperature and requires heating to 60°C for pumping. For proper atomisation through burners, fuel oil needs to be heated to 100°C for achieving a viscosity of about 25 Cst.

Indian standard specification of fuel oil is given in table below.

Table: Indian Standard Specification of Fuel Oils (furnace oils)

(IS : 1593-1960)

Characteristic	Unit	Grades of furnace oil		
		Low Viscosity	Medium Viscosity	High Viscosity
Flash point (Pensky-Martens) (closed)	°C, Minimum,	66	66	66
Kinematic viscosity	centistokes at 50°C Maximum	80	125	370
Water content	percent by volume, Maximum	1.0	1.0	1.0
Ash.	percent by wt., Maximum	0.1	0.1	0.1
Sulphur	total percent by wt., Maximum	3.5	4.0	4.5
Sediment	percent by wt., Maximum	0.25	0.25	0.25
Acidity, inorganic	—	nil	nil	nil

A typical specification of low sulphur heavy stock (LSHS) fuel oil produced at *Barauni oil refinery* is given below in table. LSHS is a class-C petroleum product.

Table 26.6: Typical specifications of LSHS produced at Barauni Oil Refinery

Characteristics	Unit	Value
Kinematic viscosity (at 100°C)	centistokes	26
Inorganic acidity	—	NIL
Specific gravity	—	0.93
Kinematic viscosity, maximum	Cst at 50°C	500
Gross calorific value	kcal/kg	10000-10500
Sediment, maximum	weight %	0.25
Pour point	°C, maximum	60
Flash point (Pensky-Marten, closed cup)	°C, minimum,	66-72
Sulphur content	weight %, maximum	1
Ash content	weight %, maximum	0.1
Moisture	weight %, maximum	1

Combustion characteristics of fuel oil

- Combustion of fuel oil takes place initially in vapor state, hence atomisation must be sufficient to initiate combustion.
- In case of heavy fuel oil, cracking of hydrocarbons takes place at the surface of droplets of oil resulting in liberation of carbon particles which burn with a luminous flame having higher emissivity factor, hence higher radiative heat transfer capacity.
- Higher the aromatic content of fuel oil *i.e.* higher the carbon/hydrogen ratio, the more luminous is the flame & higher is the emissivity factor but C.V. decreases and specific gravity increases with increase in C/H ratio.
- About 15 kg of air per kg of fuel oil is required for complete combustion with low excess air to the tune of 10% only.
- The shape and size of flame affects the thermal efficiency. Optimum flame can be produced with fuel oil by suitable burner and combustion air control.
- With 10% excess air, the flame temperature obtained from fuel oil is around 1700°C which will radiate energy at the rate of 1000000 kcal/ m^3 of combustion volume (e = 0.5).
- High amount of sulphur (upto 4%) present in fuel oil leads to high dew point (upto160°C) of flue gas. Hence the flue gas going out of chimney cannot be cooled below this temperature otherwise SO_2 & SO_3 present in it will condense and corrode the boiler parts, induced draught fan blades & chimney checker bricks etc.
- Sodium and vanadium in the ash of fuel oil are corrosive and get deposited on the hot pressure parts of boilers.
- Pre heating of oil in storage tank is needed to make it less viscous and easily pumpable. Heating an entire oil tank or outflow heater may be needed to heat oil which is to be pumped away. Outflow heaters with steam or electricity are also used to ensure smooth pumping.
- For pre-heating of oil, in line heaters are to be used to raise oil temperature from pumping to combustion temperature. In line heaters are of either electrical or steam tracing type. The pre-heat temperatures of oil are determined by the viscosity of oil supplied at room temperature and the recommended viscosity at the burner tip.
- Fuel oil viscosity and pumping temperature are tabulated below :

Viscosity (Centistokes) of oil at room temperature	Pumping Temperature, °C
50	60
230	104
900	121

Parameters for obtaining good efficiency in fuel oil combustion is given below :

- Maintain viscosity of oil of 100 Redwood sec I at burner tip by proper setting of oil preheat temperature.
- Provide atomizing air at 1-3 kg/cm^2 (about 2% of total combustion air requirement)
- To control excess air, flue gases should be continuously analyzed for $C0_2$ or 0_2.
- Acid formation occurs at around 160°C, and cold end corrosin should be avoided by maintaining stack temperature above 160°C.
- Slightest damage to the burner tip may increase fuel consumption by 10-15%, and hence any worn out burner tips should be replaced immediately.
- Oil pressure at the burner tip for pressure jet burners should be 17-20 kg/cm^2.

Correct flame is normally short. Flame impingement on walls and tubes is to be avoided since it causes carbon/soot deposits.

Naphtha

Naphtha is a light hydrocarbon boiling in the gasoline range having boiling point 60 to 200°C. When produced by atmospheric distillation of crude oil, it is called *straight run naphtha* (octane no. = 50 - 70), while when produced by conversion processes like visbreaking, fluidised bed catalytic cracking, hydrocracking & coking, it is termed as *cracked naphtha* (octane no. = 92 – 94). Depending upon its boiling point (bp), naphtha is also classified as *light naphtha (bp*= <100°C), *intermediate naphtha (bp* = 100 – 150°C) and *heavy naphtha (bp* > 150°C). It contains C_5 to C_{13} hydrocarbons comprising typically of 55 – 65% paraffins, 20 – 30% naphthenes and 10 – 15% aromatics depending upon type of crude oil processed.

Typical specifications of naphtha which is a class-A product as per petroleum act 1934 is tabulated below.

Table: Typical specifications of naphtha

S.No.	Characteristics	Unit	Value
1.	Specific gravity	dimensionless	0.70-0.75 (0.7165 at 15°C)
2.	Flash point (Pensky Martens closed cup)	°C	< 23
3.	Gross calorific value	Kcal/kg	10,200 (7200 Kcal/litre)
4.	Kinematic Viscosity at 37.8 °C	Centistokes	<0.5
5.	Carbon/hydrogen (C/H) ratio	dimensionless	5.5
6.	Reid vapor pressure (RVP) at 37.8°C	Kg/cm^2	0.42
7.	Residue on evaporation (maximum)	mg/ml	5
8.	Total sulfur content	ppm	100
9.	Lead content	ppm	3
10.	Initial boiling point (IBP)	°C	45
11.	Final boiling point	°C	155
12.	Aromatics content	% by volume	11.5
13.	olefins content	% by volume	NIL

Main uses of naphtha is as follows:

- Straight run naphtha is subjected to catalytic reforming to produce gasoline of high octane number in the range of 93 – 94, which can be blended with straight run naphtha/gasoline (0ctane no. = 50 – 70) to produce 87 octane number motor gasoline presently used in India.
- Catalytic steam reforming of naphtha is done to produce synthesis gas for ammonia production in nitrogenous fertiliser plants.
- Naphtha (C_5 – 140°C) is used as feedstock for steam crackers producing hydrogen, methane, ethylene, propylene, butene, butadiene, benzene and isoprene in petrochemical plant. Dearomatised naphtha (4 – 6% aromatics) is used in both fertiliser plants and petrochemical plants.
- Intermediate naphtha is used as motor & aviation gasoline whereas heavy naphtha is used as kerosene & jet fuels. Heavy naphtha is also used as solvent (for dry cleaning of clothes) as well as thinner for paints, varnishes & lacquers.

- It is vaporised and mixed with blast furnace gas to improve its calorific value thereby replacing high C.V. coke oven gas used for mixing in B.F. gas in steel plant. Mixed gas of desired C.V. is prepared during shortage of coke oven gas by naphtha vapor injection in steel plant.

Kerosene

Straight run kerosene produced by distillation of crude oil contains C_{10}–C_{14} hydrocarbons boiling in the range of 150 –250°C. It burns with smoky flame because of high aromatics content in it. *Smoking tendency decreases in the order : aromatics→naphthenes→iso-paraffins→paraffins.* That means, aromatics having same carbon atoms have the highest smoking tendency, while paraffins are the least smoking constituent of kerosene.

- Tendency of aromatics to smoke decreases with increase in molecular weight.
- Tendency of naphthenes to smoke decreases with increasing molecular weight and increases on addition of double bond.
- Tendency of paraffins to smoke increases with increasing molecular weight. Branching and olefin linkage in paraffins also increases, its smoking tendency.

A good quality kerosene should have high proportion of paraffinic hydrocarbons and less of aromatics. Aromatics are removed from kerosene by liquid sulphur dioxide extraction *(Edeleanu Process)* thereby improving its smoke point *i.e.* reducing its smoking tendency. Dearomatisation of kerosene is also done by *hydrotreating,* in which aromatics are converted into naphthenes. Kerosene is also subjected to a process called *hydrofining* which improves its sulphur content, color, odour & burning quality (as measured by wick char value).

Kerosene is used for the following purposes

- It is a domestic fuel for cooking as well as an illummant for lighting.
- It is used in the manufacture of insecticides; fungicides & herbicides used for controlling pests, fungi & weeds respectively.
- It is used in the manufacture of detergents (keryl benzene sulphonate)
- Extracted aromatics from kerosene are used for manufacture of aromatic solvents & oil.

Aviation Turbine Fuel (ATF)

Aviation turbine fuel (ATF) is the fraction of crude oil boiling in the range of 150 –270°C, comprising mainly of paraffins and naphthenes with <20% aromatics & <5% olefins. Kerosene type ATF and wide cut gasoline type ATF are in general use, the former being a modified development of the illuminant kerosene originally used in gas turbine engines while, the later is a wider boiling range fuel including some gasoline fraction.

Since, straight run ATF alone can not meet its large demand, it is also produced by hydrocracking of residual stocks (Isomax process) and hydrotreating of petroleum distillates boiling in the range of 60 - 350°C (Arofing process) thereby catalytically converting aromatics into naphthenes.

Important desirable properties of good ATF are:

- High smoke point (i.e., low aromatics content), n-paraffins are the best from smoke point & luminousity point of view.
- High heat content per unit volume in high speed aircraft and weight is more important in low speed aircraft. C.V. on weight basis *(i.e.* Kcal/kg) is the highest for paraffins, lowest for aromatics and intermediate for naphthenes. However, for C.V. on volume basis *(i.e.,* Kcal/m^3), the order is reversed.
- Denser molecular structure as it permits more energy to be packed in a given tank space. A high density fuel is normally desirable as C.V. on volume basis (Kcal/m^3) increases with increasing density and it decreases on weight basis *(i.e.,* Kcal/kg).
- High luminometer number *(i.e.*, high *n*-paraffin content) to minimise the temperature rise in combustion zone from flame radiation.
- Low olefins content (<5%) as they are unstable, chemically reactive and have gum forming tendency.
- High thermal stability, which is achieved by the use of an anti-oxidant and a metal deactivator.
- Low freezing point (–50°C maximum for kerosene type ATF and –60°C maximum for gasoline type ATF) as air crafts fly at high altitude with very low outside air temperature. Freezing point parameter affects the yield of ATF from a particular crude oil, because it controls the final boiling point which decreases with decrease in freezing point of ATF.
- Controlled low volatility to minimise fuel loss from unpressurised air craft tanks due to boiling off at high altitudes. Desirable value for kerosene type ATF is 20 volume percent recovery upto200°C, which controls front end volatility and ensures that ATF will readily vaporise for cold starting of engine as well as limit evaporation losses at high altitudes.

Additives added in ATF include:

- Antioxidants which prevent gum formation and retard peroxide formation which attacks nitrile rubber hoses used in fuel supply system in the air-crafts.
- Metal deactivators for removing any dissolved copper (by chelating) that catalyses the gum formation in ATF which may block burner jets & foul up heat exchangers.
- Anti-icing additives (0.5 - 2 volume % of ethanol/isopropanol or 0.05 - 0.2 volume % of hexylene glycol/dipropylene glycol) to avoid freezing of dissolved water in ATF, which may clog the fuel filter.
- Fuel soluble corrosion inhibitor (like alkyl aminophosphate) to protect the fuel distribution pipeline. It also increases fuel lubricity thereby helping in the operation of high pressure fuel pump, which is entirely fuel lubricated.
- Anti-smoke additive which helps in reducing the amount of smoke emitted from jet engines while taking off.
- Fuel soluble biocides, which inhibit fungicidal and microbial growth in fuel supply system of aircraft.
- Static charge dissipator additive which increases the electrical conductivity of ATF & thus reduces the possibility of an electrostatic discharge causing explosion during high speed mechanized refuelling through a micronic filter or in refinery ATF storage tanks.

Some important requirements/specification of kerosene type ATF (K-50 grade) as per IS : 1571-1976 is tabulated below.

Table: Important Specifications of kerosene type ATF

S.No.	Parameters	Unit	Specifications
1.	Aromatics & olefins content	Volume % (maximum)	20 & 5 respectively
2.	Net calorific value	Kcal/kg (minimum)	10225
3.	Copper content	µg/kg ATF (maximum)	150
4.	Freezing point	°C (maximum)	50
5.	Smoke point	mm (minimum)	20
6.	Total sulphur	weight % (maximum,'	0.20
7.	Kinematic viscosity	Centistokes (maximum)	6 at –17.8°C ,15 at –34.4°C
8.	Existent gum	mg/100 cc (maximum)	7
9.	Flash point (Abel)	°C	38

S.No.	Parameters	Unit	Specifications
10.	Specific gravity	—	0.775–0.830
11.	Antioxidants	mg/100 c.c,	8.6-24.0
12.	Water separometer index modified (WSIM)	—	70–85
13.	Total acidity	mg KOH/gm (maximum)	0.012
14.	Copper strip corrosion for 2 hr at 100°C	—	Not worse than no. 1
15.	Silver strip corrosion for 16 hrs at 45 + 10 °C	—	Not exceeding (0)
16.	Initial boiling point	°C	Not limited but to be reported
17.	Final boiling point	°C (maximum)	288

Table - Additives used in the formulation of lubricants

S.No.	Additives used for various purposes	Specific examples of additives
1.	Anti-foaming agents	Polyorganosiloxanes
2.	Anti wear agents	Tricresylphosphate
3.	Antioxidants	2, 6 ditertiary butyl paracresol, phenyl naphthylamine
4.	Bacteriocides	Phenols
5.	Corrosion inhibitors	Lead napthenate, sodium sulfonate
6.	Detergents	Metal sulfonates, metal phenates
7.	Emulsifiers	Fatty oils, Fatty acids
8.	Extreme pressure agents	Sulphurised fatty oils
9.	Viscosity index improver	Tricresylphosphate

Appendix to Chapter 9

Appendix 9.1: Gasoline Fuel Quality Parameters

Appendix 9.2: Diesel Fuel Quality Parameters

Appendix 9.3: Indian Gasoline Specification required for meeting Bharat Stage II,III, & IV. Emission Norms

Appendix 9.4: Phases of reduction of lead content in gasoline (petrol) in India

Appendix 9.5:	Indian Diesel Specification required for meeting Bharat Stage II, III, & IV Emission Norms
Appendix 9.6:	Phases of reduction of % sulphur (S) content in diesel in India.
Appendix 9.7:	Indian automotive CNG (compressed natural gas) Specifications.
Appendix 9.8:	Bureau of Indian Standards (BIS) Specifications for automotive LPG (IS 14861-2000)
Appendix 9.9:	Effect of sulphur content in diesel on particulate emissions from diesel engine
Appendix 9.10:	Characteristic of hydrocarbons present in diesel fuel
Appendix 9.11:	Auto Fuel (Diesel & Gasoline) Quality for Environmental Emission Standard in India During Various Periods
Appendix 9.12:	Gasoline Blending Components & their Role in Automobile Pollution Control
Appendix 9.13:	Factors affecting the octane requirement of petrol
Appendix 9.14:	Use of Oxygenates in Motor Gasoline to improve its performance level
Appendix 9.15:	Influence of Crude Oil Type on Diesel Fuel Characteristics
Appendix 9.16:	Influence of Diesel Fuel Properties on Exhaust Emissions
Appendix 9.17:	Requirements of Automotive fuels
Appendix 9.18:	Important characteristics of Liquid Petroleum Fuels – Petrol and diesel
Appendix 9.19:	Petrol & diesel Fuel Quality Interactions with emissions
Appendix 9.20:	Effects of petrol Properties on Emissions from catalytic cars
Appendix 9.21:	Comparison of effects of reformulated gasoline on emissions for catalytic cars with those with regular grade unleaded petrol
Appendix 9.22:	Indian Specification for Euro IV/Bharat Stage IV Emission Norms
Appendix 9.23:	Effect of diesel cetane number (CN) increase from 50 CN to 58 CN on NOX emissions
Appendix 9.24:	Overall Effects of Diesel Fuel Property Changes on vehicular emissions
Appendix 9.25:	Comparison of Diesel Fuel Specifications in Europe and India

Appendix 9.1: Gasoline Fuel Quality Parameters

S.No.	Parameters	Characteristic Features
1.	Lead	Lead alkyl additives have historically been used to enhance the octane rating of gasoline, but in course it was eliminated from gasoline for health effects associated with it. Also, even small lead contamination in the gasoline can destroy a modern vehicle emission control technologies, like catalytic converters and oxygen sensors.
2.	Benzene	Benzene, the smallest aromatic compound with no alkyl group, is a carcinogen. It is emitted from gasoline as a result of evaporation and as unconverted benzene from the exhaust pipe. Benzene has an extremely high octane rating, and is hence a good gasoline blending component from point of view of combustion. An effective way to reduce human exposure to benzene is to control benzene in gasoline.
3.	Sulphur	Sulphur affects efficiency of the catalyst used in after treatment devices and also adversely affects heated exhaust gas oxygen sensors. Vehicle manufacturers recommend for conventional catalytic converters that the level of sulphur in gasoline be kept below 5000 ppm, and preferably below 100 ppm. Sulphur not trapped in catalytic converters is emitted as SOx, some of which undergoes chemical transformation to become secondary particulate matter or acid rain.
4.	Reid Vapor Pressure (RVP)	The Reid vapor pressure is a measure of gasoline volatility. During operation of spark ignition engines both performance and emissions are affected by the volatility of gasoline. Lowering RVP is a cost effective way of controlling VOC emissions, including light olefins. This requires lowering level of butanes, and possibly C5 hydrocarbons. Butanes are the cheapest source of octane and their removal typically adversely affects refinery economies. Lowering RVP, however, can make starting carbureted vehicles difficult, particularly in cold weather, and may cause misfires & higher tailpipe HC emissions. RVP should be within a certain range depending on vehicle technology and ambient conditions.

S.No.	Parameters	Characteristic Features
5.	Aromatics	Aromatics with two or more alkyl groups are photo-chemically reactive and contribute to ozone formation. The photochemical reactivity of aromatics and their decomposition to benzene are the two primary concerns with aromatics. Aromatics have extremely high octane ratings, and hence are good gasoline blending components from point of view of combustion. High aromatic content can increase engine deposits, these deposits formation in combustion chamber increase tailpipe emissions, including HC and NOx.
6.	Olefins	Olefins in gasoline are formed primarily during cracking processes. Olefins are photo chemically reactive and are ozone precursors. In addition at elevated levels, loefins increase the emission of NOx, which is a precursor for ozone, particulate matter and acid rain. At the same time olefins have fairly high octane rating and higher olefins in gasoline may lead to deposit formation and increased emissions of reactive hydrocarbons and undesirable compounds.
7.	Alkylates	Alkylates are high octane paraffin hydrocarbons with essentially no adverse effects on air quality. However, alkylation is an expensive process and requires presence of a catalytic cracking unit at the refinery.
8.	Volatile Organic Compounds (VOCs)	Volatile organic compounds contain photo chemically reactive hydrocarbons. Reduction in VOCs will therefore reduce the amount of ozone precursors in the atmosphere. VOCs could also absorb onto particulate and increase particulate mass. Evaporation emissions consist entirely of VOCs; VOCs are also found in exhaust gas.
9.	Oxygenates	Oxygenates organic compounds such as ethers and alcohols may be added to increase octane number. The presence of oxygen in oxygenates also cause clean combustion in vehicles. However the use of oxygenates can increase the emissions of undesirable compounds such as aldehydes. Oxygenates also dilutes gasoline, thereby decreasing the amount of such undesirable gasoline components i.e. benzene, sulphur, total aromatics, and olefins. Oxygenates are however more miscible with water than gasoline.

S.No.	Parameters	Characteristic Features
10.	Additives	Gasoline in industrial countries contains additives to prevent the accumulation of deposits in engine and fuel supply system. Deposits can increase tail pipe emissions. Ash forming additives can adversely affect the operation of catalyst and other components (e.g. Oxygen sensors) in a way that increase emissions.
11.	Octane Number	Octane rating denotes the ability to resist auto ignition which can cause engine knock. Petrol octane number is measured as Research Octane Number (RON) and Motor Octane Number (MON). RON correlates best with low speed, mild knocking conditions and MON correlates with high speed knocking conditions and with part-throttle operation. RON values are higher than MON and the value difference between these is the sensitivity which is mostly less than 10. The average of RON and MON is called the Anti Knock Index (AKI). Vehicles are designed for a particular octane number, petrol with an octane number lower than that required may result in knocking which could lead to severe engine damage. Octane number has no direct correlation with emissions.

Appendix 9.2: Diesel Fuel Quality Parameters

S.No.	Parameters	Characteristic Features
1.	Density	Diesel fuel is metered volumetrically, so that the higher the density, the more mass is injected. The use of a fuel with greater density than that used in the pump calibration could even result in over fueling at maximum load, resulting in substantially higher smoke emissions. The reduced density will reduce particulate emissions from all diesel vehicles and NOx emissions from heavy duty vehicles. The reduction in fuel density decreases engine power output and increases volumetric fuel consumption.
2.	Viscosity	Problems like inadequate fueling and pump distortion may be faced with high viscosity fuel. Low viscosity will increase leakage from the pumping elements, and in worse cases (low viscosity, high temperature) can result in total leakage. To allow optimization of engine performance, it is important to minimize the range between minimum and maximum viscosity limits as it is impacted by ambient temperature.

S.No.	Parameters	Characteristic Features
3.	Sulphur	Sulphur in diesel contributes to fine particulate emissions, through the formation of sulfates both in the exhaust stream and later in the atmosphere. It also decreases the efficiency of oxidation catalysts and degrades advanced treatment devices. The percentage reduction of particulate matter on account of sulphur reduction is dependent on the level of engine technology.
4.	Aromatics	Poly aromatic hydrocarbons (PAHs) have been linked to higher particulate emissions.
5.	Distillation characteristics T90, T95	T90 & T95 are the temperature at which 90 & 95 percent respectively, of diesel evaporates. Decreasing T90 or T95 could have a favorable impact on emissions, especially if heavy PAHs are removed as a result.
6.	Cetane number	Cetane number is aeasure of the compression ignition quality of a diesel fuel; it influences cold startability, exhaust emissions and combustion noise. A high cetane number indicates a shorter lag between fuel injection and fuel ignition. A cetane index is an approximation of the cetane number calculated from the density and distillation temperatures. Cetane improvement additives will increase the cetane number but not the cetane index and hence many countries set cetane specifications as minimal cetane index or cetane number.
7.	Cleanliness	With respect to satisfactory operation of diesel vehicles, cleanliness refers to the absence of water and particulate contamination. Dirt and water can plug fuel filters and cause seious damage to the fuel injector system because of close tolerance of fuel pumps and injectors. Diesel engine is equipped with fuel filters to protect the fuel delivery system.
8.	Stability	Stability is the ability of a fuel to resist the formation of gums and insoluble oxidation products. Fuels with poor oxidation stability contain insoluble particles that can plug fuel filters, potentially leading to decreased engine performance or engine stalling from fuel starvation.
9.	Flash point	The flash point is the temperature to which a fuel must be heated to produce an ignitable mixture of fuel and air above the surface of the liquid. This property has no impact on engine performance because the auto ignition temperature of the fuel - air mixture is not affected by variations in flash point.

Appendix 9.3: Indian Gasoline Specification required for meeting Bharat Stage II,III, & IV. Emission Norms

S.No.	Parameters	Unit	Bharat stageII		Bharat Stage III		Bharat Stage IV	
			Unleaded Regular	Unleaded Premium	Unleaded regular	Unleaded Premium	Unleaded regular	Unleaded Premium
1.	Colour, visual		Orange	Red	Orange	Red	Orange	Red
2.	Density 15oC	Kg/m³	710-770	710-770	720-775	720-775	720-775	720-775
3.	Distillation:							
	a) recovery up to 70⁰C (E70)	% volume	10-45	10-45	10-45	10-45	10-45	10-45
	b) recovery up to 100⁰C (E100)	% volume	40-70	40-70	40-70	40-70	40-70	40-70
	c) recoveryupto180⁰C(E180)	% volume	90	90	-	-	-	-
	d) recovery up to 150⁰C (E150)	% volume	-	-	75min	75min	75min	75min
	e) final Boiling Point (FBP) Max,	°C	215	215	210	210	210	210
	f) residue Max. % volume	% volume	2	2	2	2	2	2
4.	Research Octane Number (RON),	min	88	93	91	95	91	95
5.	Anti knock Index (AKI) or (MON)	Min	84(AKI)	88(AKI)	81(MON)	85(MON)	81(MON)	85(MON)
6..	Gum content (solvent washed)max	Mg/100ml	Existent gum, max 40g/m3 potential gum, max 50 g/m3			5	5	5
7.	Oxidation stability	Min	-	-	360	360	360	360
8.	Sulphur, total, max	% mass	0.05	0.05	150 mg/kg	150 mg/kg	150 mg/kg	150 mg/kg
9.	Lead content (as Pb), max	g/l	0.013	0.013	0.005	0.005	0.005	0.005
10.	Reid vapourpres, (RVP) max	kpa	35-60	35-60	60	60	60	60
11.	Benzene, content, max	% volume	For the metros 3.0 & for the rest 5.0		1	1	1	1

S.No.	Parameters	Unit	Bharat stageII		Bharat Stage III		Bharat Stage IV	
			Unleaded Regular	Unleaded Premium	Unleaded regular	Unleaded Premium	Unleaded regular	Unleaded Premium
12.	Copper strip corrosion for 3hrs 50 oC	Rating	Not more than no.1		Class I	Class I	Class I	Class I
13.	Olefin content, max	% volume	-	-	21	18	21	18
14.	Aromatic content, max	% volume	-	-	42	42	35	35
15.	Oxygen content, max oxygenates content:	% mass	-	-	2.7	2.7	2.7	2.7
	a) Methanol, max	% volume	-	-	3	3	3	3
	b) ethanol, max	% volume	-	-	5	5	5	5
	c) iso-propyl alcohol, max	% volume	-	-	10	10	10	10
	d) Iso- butyl alcohol, max	% volume	-	-	10	10	10	10
	e) tertiary-butyl alcohol, max	% volume	-	-	7	7	7	7
	f) Ether containing 5 or more carbon atoms max	% volume	-	-	15	15	15	15
	g) Other Oxygenates, max	% volume	-	-	8	8	8	8

Appendix 9.4: Phases of reduction of lead content in gasoline (petrol) in India

Phase	Date of Introduction	Lead Content	Areas Covered
I	June 1994	Low lead (0.15 g/l)	Delhi, Mumbai, Kolkata, Chennai
II	01.04.1995	Unleaded (0.013 g/l) + low leaded	Delhi, Mumbai, Kolkata, Chennai
III	01.01.1997	Low leaded	Entire country
IV	01.09.1998	Only unleaded	National Capital Territory (NCT)
V	31.12.1998	Unleaded + Low leaded	Capitals of states and Union Territories
VI	01.01.1998	Unleaded	National Capital Region (NCR)
VII	01.02.2000	Unleaded	Entire Country

Appendix 9.5: Indian Diesel Specification required for meeting Bharat Stage II, III, & IV Emission Norms

S.No.	Parameters	Unit	Bharat Stage II	Bharat Stage III	Bharat Stage IV
1.	Ash, max	% mass	0.01	0.01	0.01
2.	Carbon Residue (Ramsbottom) on 10% residue, max	% mass	0.3	0.03 without additives	0.03 without additives
3.	Cetane Number (CN), min		48*	51	51
4.	Cetane Index (CI), min		46*	46	46
5.	Distillation 95% vol. Recovery at oC, max	°C		360	360
6.	Flash point Abel, min	°C	35	35	35
7.	Kinematic Viscosity @ 40 oC	Cst	2-5	2-4.5	2-4.5
8.	Density @ 15oC	Kg/m³	820-860 (820-870)*	820-845	820-845
9.	Total Sulphur, max	Mg/kg	0.05% mass	50	50
10.	Water content, max	Mg/kg	0.05% vol	200	200

S.No.	Parameters	Unit	Bharat Stage II	Bharat Stage III	Bharat Stage IV
11.	Cold filter plugging point (CFPP) a) Summer, max b) Winter, max	°C °C	18 6	18 6	18 6
12.	Total contamination, 25max	Mg/kg	-	24	24
13.	Ox11idation stability, max	g/m³	-	25	25
14.	Polycyclic Aromatic Hydrocarbon (PAH), max	% mass	-	11	11
15.	Lubricity, corrected wear scar diameter (wsd 1,4) @ 60° C, max	μ m (microns)	460	460	460
16.	Copper Strip corrosion for 3 hrs @ 50°C	rating	Not worse than No. 1.	Class I	Class I

Note: * For diesel process from Assam crude, either CN of 45 minimum or CI of 43 minimum and density of 820-870 kg/m ^{3} shall be applicable.

Appendix 9.6: Phases of reduction of % sulphur (S) content in diesel in India.

Phase	Date of Introduction	Sulphur content in diesel, %	Areas covered
I	April 1996	0.5	Four metros and Taj
II	April 1997	0.25	Delhi and Taj Trapezium
III	April 1998	0.25	Metro cities
IV	January 2000	0.25	Entire country
V	April 2000 March 2001 July 2001	0.05 0.05 0.05	NCR Private vehicles NCR all vehicles Chennai and Kolkata
VI	October 2001	0.05	All retail outlets of four metros
VII	2003	0.05	Ahmadabad, Surat, Agra Pune and Kanpur
VII	2005	0.05	Entire country
IX	2005	0.035	10 metro cities and Agra

Phase	Date of Introduction	Sulphur content in diesel, %	Areas covered
X	2010	0.035	Entire country
XI	2010	0.035	10 metro cities

Appendix 9.7: Indian automotive CNG (compressed natural gas) Specifications

Constituents	Value	Tolerance
Wobbe number (Btu/ft^3)	1340	Variation shall be limited to ± 30 units during normal operation
Free water, mg/m^3	8	Maximum
Total sulphur including H_2S, mg/m^3	20	Maximum
N2+ CO_2, vol%	3.5	Maximum*
Oil mist content, ppm	Insignificant	Insignificant
Oxygen, vol. %	0.5	Maximum
Methane (C_1)	87%	Maximum**
Ethane (C_2)	6%	Maximum
C_3 and higher HC	3%	Maximum
C_6 and higher HC	1%	Maximum
Total unsaturated HC	1%	Maximum
Hydrogen (mole%)	0.1	Maximum
Carbon monoxide (mole%)	0.1	Maximum

Appendix 9.8: Bureau of Indian Standards (BIS) Specifications for automotive LPG (IS 14861-2000)

S.No.	Characteristics	Requirement
1.	Vapour Pressure (gauge) @ 400 °C, kpa minimum	520* 1050
2.	C_5 Hydrocarbons and heavier, mol% max.	2.0
3.	Dienes (as 1,3 Butadiene), mol% max	0.5
4.	Total volatile sulphur (after stanching) ppm, max	150
5.	Copper strip corrosion @ 400 °C for 1 hour, max	Class I
6.	Hydrogen Sulphide	Pass the test

S.No.	Characteristics	Requirement
7.	Evaporation residue, mg/kg, max	100
8.	Free water content	Nil**
9.	Motor octane number (MON), min	88
10.	Odour	Unpleasant and distinctive down to20% LEL

LPG - Liquified Petroleum Gas

In winter, the gauge vapour pressure requirement shall be minimum 700kpa at 40°C: winter period shall be from 1st November to 15th February.

** the water content shall be determined at the refinery / First dispatching location.

*** Product shall contain 20 ppm, minimum. Ethyl mercaptan at the first dispatching location to ensure the detection of leakage by odour.

Note: Liquefied petroleum gas (LPG) is a mixture of commercial propane and butane. It is obtained either from the natural gas processing or during petroleum refining. Composition of LPG varies very widely depending on the use and demand of butane. LPG is a gas at normal ambient temperature and pressures (the boiling point of propane & butane at atmospheric pressure is about -45°C & -2° C, respectively). When subjected to pressure of 4 to 20 bars mixture of propane and butane becomes liquid. The pressure at which it becomes liquid at room temperature depends upon propane to butane ratio. The LPG is usually fed to the engine in gas phase.

- Most LPG vehicles operate on bi-fuel systems for operation either on gasoline or LPG. It is important as the number of LPG filling stations is usually small.
- One drawback with a bi-fuel system is that the engine is neither optimized on LPG nor on gasoline.
- Better cold start and warm-up characteristics due to its gaseous state compared to gasoline hence lower HC emissions.
- HC emissions from LPG vehicles have significantly lower potential of smog formation compared to gasoline and diesel fuels.
- Negligible PM emissions compared to diesel.
- Small reductions in CO compared to gasoline as no enrichment of mixtures during warm up or acceleration phase is required.
- No significant difference in NOx emissions.
- Variation in propane/butane ratio in LPG poses problem as the octane number of the two main constituents; propane (RON is 112) and butane (RON is 94) is quite different.
- (RON- Research Octane No.)

Appendix 9.9: Effect of sulphur content in diesel on particulate emissions from diesel engine

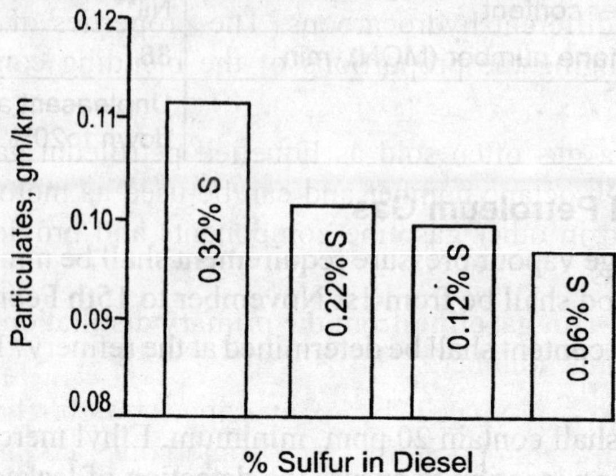

Appendix 9.10: Characteristic of hydrocarbons present in diesel fuel

Hydrocarbons Characteristics	Ignitability	Properties at cold	Volumetric heating value	Density	Smoking tendency
n-paraffins	Good	poor	Low	low	Low
i-paraffins	Low	Good	Low	Low	Low
Olefins	Low	Good	Low	Low	Average
Naphthenes	Average	Good	Average	Average	Average
Aromatic compunds	poor	Average	high	High	high

Appendix 9.11: Auto Fuel (Diesel & Gasoline) Quality for Environmental Emission Standard in India During Various Periods

Diesel Specification:- Year	1996	2000	2005	2010
Cetane No. , Minimum	45	48	48	51
Sulphur, % W/w, Maximum	0.50	0.25 0.25	0.05	0.035
Distillation T95, °C		(metro)	370	360
Polyaromatic	-	370	-	11
Gasoline Specification:-	35-70	-	35-60	60
RVP at 38°C, Kpa	5.0	-	3.0(all) 1.0	1.0
BENZENE, % by volume maximum	0.15 % (low pb) 0.013%	5.0 3.0	(Metro)	0.005
Lead, gm/m³, maximum	(unleaded)	(Metro)	0.013	0.015
Sulphur, % by mass, maximum	0.10 (unleaded) 0.20	0.013	0.05	42
Aromatics, % v/v., Maximum	(leaded)	0.10	-	2.7
Oxygen, % by vol., Maximum	-	-	-	
	-	-		

Appendix 9.12: Gasoline Blending Components & their Role in Automobile Pollution Control

Gasoline is produced by blending the output from several refinery processes, each containing different hydrocarbons. The properties of gasoline can be modified by changing the proportions of the blending components, which include:

- **Butane-a** gas often sold as liquefied petroleum gas (LPG), which has a high octane number and can be used as motor fuel. Butane is dissolved in other gasoline components and provides volatility for cold starts.
- **Straight-run gasoline-**from the primary distillation of crude oil; has a low octane number.
- **Isomerate-** produced by converting straight-chain paraffins to branched-chain paraffins by isomerizing straight-run gasoline fractions. Provides clean burning, higher octane quality.
- **Reformate-** from the catalytic reforming of n-alkanes, and cyclo-alkanes, into aromatics. The process, known as platforming when the catalyst contains platinum, produces high-octane gasoline, but the yields fall as the octane number of the product increases. Catalytic reforming is the most important process for producing gasoline; a typical summer-grade premium gasoline contains about 70 percent reformate.
- **Catcracked gasoline-** from catalytic cracking of residual oil containing large molecules with boiling points above 370^{0}C to form unsaturated olefins and branched molecules with high octane rating.
- **Hydrocracked gasoline-** from the hydrocracker which breaks up large molecules but also adds hydrogen. Contains mainly low octane alkanes and is unsuitable for gasoline but is a good feedstock for the catalytic reformer.
- **Polymerate-** produced by reacting light alkenes (olefins such as propene and butene) to give heavier olefins, which have a good octane quality and do not increase the vapor pressure of gasoline.
- **Alkylate-** from processes where light iso-alkanes from primary distillation, such as isobutane, are coupled with light alkenes (olefins such as butane) from the catcracker to make high-octane components.
- **Oxygenates-** certain alcohols and ethers containing carbon, hydrogen, and generally one oxygen atom. Can be used as octane boosters or gasoline extenders.

Appendix 9.13: Factors affecting the octane requirement of petrol

- An increase in humidity from 20 to 90 percent (at 20^0C) decreases octane demand by about 5 units (numbers).
- An increase in altitude from sea level to 1,000 m reduces octane demand by about 10 units.
- An increase in engine spark advance by 4 degrees can increase octane requirement by about 5 units.
- An increase in engine cooling jacket temperature from 60 to 90^0C can increase octane requirement by 6-7 units.
- Engine deposit accumulation at 16,000 kms can increase the octane requirement by as much as 10 units

Note: Engines designed for use with high-octane fuels can produce more power and have lower fuel consumption than engines designed for lower-octane fuels. For a given engine design, however, there is no advantage in using a higher octane fuel than the engine requires. Higher octane "premium" gasoline-unless it is required to avoid knocking – is not necessarily of higher quality and provides no benefit in most cases.

Appendix 9.14: Use of Oxygenates in Motor Gasoline to improve its performance level

Oxygenated supplements, which cover a range of lower alcohols and ethers, can substitute for lead additives in gasoline. Oxygenates are produced from a variety of feedstocks; ethanol is derived mostly from renewable agricultural feedstocks, and methanol is derived primarily from natural gas. Ether production facilities are integrated within a refinery. Large-scale units use butane isomerization/dehydrogenation technology, where both the butane and methanol feedstocks are derived from lower-value gas sources. The behavior of oxygenates in terms of blending and vehicle performance is different from hydrocarbons-only gasolines. The commonly used oxygenates in gasoline blends are methanol, ethanol, isopropanol, tertiary butyl alcohol (TBA), methyl tertiary butyl ether (MTBE), ethyl tertiary butyl ether (ETBE), di-isoprophyl ether (DIPE), and iso-butyl alcohol. Tertiary amyl methyl ether (TAME) also has been used in small amounts. The volume of oxygenates in gasoline blends is generally about 3 to 22%, although commonly used blends have 10-15% oxygenates. The oxygen content of the blend is generally about 1 to 2% by weight. Oxygenates serve three basic objectives: extending the gasoline stock or serving as a fuel substitute (for example, ethanol in Brazil), boosting octane value, and providing an effective means of reducing harmful emissions (for example, MTBE in unleaded gasoline in Mexico City). Some of the important vehicle performance characteristics of oxygenated fuels include:

Antiknock performance: Oxygenated fuels perform better than hydrocarbon-only fuels at low olefins levels. They give better antiknock performance in unleaded gasoline; this is particularly the case with increasing MTBE content. MTBE has proven to be an effective substitute for lead in gasoline because it has a high octane rating (RON 119, MON 101) and is less water-sensitive than alcohol. It is not clear, however, whether oxygenates give better antiknock performance under lean burn conditions.

Driveability: Vehicles fitted with fuel injection systems have better tolerance in cold weather conditions for low volatility fuels containing oxygenates. At high altitudes and in hot weather, oxygenated fuels (except for certain methanol blends) give similar or better handling performance compared with wholly hydrocarbon fuels.

Fuel economy: In the case of commercial oxygenated gasoline (which must comply with existing gasoline specifications), fuel economy is essentially unchanged with increasing amounts of oxygen. There could be marginal benefits in terms of reduced energy consumption with increased fuel oxygenate content.

Exhaust emissions: CO & HC are progressively reduced as oxygen content is increased, while NOx emissions may increase, particularly when oxygenates are blended with fuels with low aromatic content. Ethanol blends increase aldehyde emissions with increasing concentration; with other oxygenates there is only a marginal increase, which may be corrected with exhaust oxidation or three-way catalytic converter. Oxygenates are particularly useful in lowering emissions from older vehicles, especially toxic emissions such as benzene. For the same fuel oxygenate content, MTBE and ETBE exhibit similar exhaust emission characteristics.

Appendix 9.15: Influence of Crude Oil Type on Diesel Fuel Characteristics

S.No.	Crude oil source	Crude oil type	Cetane number	Sulfur content (percent)	Calorific value
1.	North sea	Paraffinic	High	0.10-3.0	Low
2.	Middle east	Paraffinic	High	0.86-1.40	Low
3.	Nigeria	Napthenic	Low	0.12-0.34	Medium
4.	Venezuela/Mexico	Napthenic/aromatic	Very low	0.30-1.3	High
5.	Indonesia/Australia	Paraffinic	high	0.10-0.30	low

Appendix 9.16: Influence of Diesel Fuel Properties on Exhaust Emissions

S.No.	Diesel Fuel Property	Smoke		Exhaust Gases			Particulate matter	Noise
		White	Black	HC	CO	NO_x		
1.	Density:-	Increase	Increase	-	-	-	Increase	-
	Up	Decrease	Decrease	-	-	-	Decrease	-
	Down							
2.	Distillation:-	Increase	Decrease	-	-	Decrease	-	-
	IBP up	Increase	Increase	Decrease	-	Increase	-	-
	10% (T10); UP	Increase	Increase	Decrease	-	Decrease	Increase	-
	50% (T50); Up	-	Increase	-	-	Increase	Increase	-
	90% (T90); Up	-	Increase	-	-	Increase	Increase	-
	FBP; Up							
3.	Viscosity:-		Increase		Increase		Increase	Increase
	Up		Decrease		Decrease		Decrease	Decrease
	Down							
4.	Cetane number:-	Decrease	-	Decrease	Decrease	Decrease	Decrease	Decrease
	Up	Increase	-	Increase	Increase	Increase	Increase	Increase
	Down							

Note: IBP = initial boiling point ; FBP= final boiling points.

A. Cold-engine.

B. Aromatic content is inversely correlated with the cetane number & density.

Appendix 9.17: Requirements of Automotive fuels

Fuel Quality parameters	Relationship with Engine & Vehicle Performance
Combustion Characteristics	Better ignition and combustion properties result in higher fuel efficiency and lower emission of pollutants. High octane number for SI engine and a high cetane number for CI engines is necessary for good combustion.
High heat of combustion	For the same mass of fuel on board a higher operation range of vehicle is obtained.
High volumetric energy content	A smaller fuel tank improving vehicle space utilization and packaging. Liquid fuels being sold on volume basis, it results in better economics for the operators.
Low temperature performance	A significant fraction of fuel is required to evaporate at low engine temperatures for better engine cold start &warm-up, good low-temperature drivability and better fuel economy & emissions.
High temperature drivability	For ease of hot starting, good hot weather drivability and low evaporative emissions, volatility of fuels is appropriately controlled to meet the needs of seasonal and geographical variations in ambient temperature.
Oxidation stability	Good low temperature oxidation stability reduces fuel deterioration during storage and deposit formation in the fuel system & engine combustion chamber.
Deposit formation control	Deposit control additives are now widely used for minimizing deposit formation &maintenance to get better fuel economy and low emissions during the vehicle service life.
Material Compatibility	It is essential to prevent corrosion of metallic and deterioration of fuel system rubber & elastomeric components
Flow characteristics	Fuel has to be in fluid condition at low ambient temperature. It is particularly critical for diesel fuels. Also, appropriate diesel fuel viscosity is essential for flow as well as for good injection characteristics.

Appendix 9.18: Important characteristics of Liquid Petroleum Fuels – Petrol and diesel

Petrol	Diesel
Distillation	Ignition quality (Cetane index, Cetane number)
Reid vapour pressure	Distillation (volatility), 90% boiling point

Petrol	Diesel
Heat of combustion*	Heat of combustion*
Density	Density
Antiknock quality (Research & Motor octane number)	Viscosity
Oxidation stability	Aromatic content
Gum content, mg/100ml	Sulphur content
Lead content	Storage stability
Sulphur	Injection system cleanliness
Benzene	
Sediments	Sediments

*Heat of combustion is not specified in the diesel fuel quality standards.

Appendix 9.19: Petrol & diesel Fuel Quality Interactions with emissions

Type of Interactions	Petrol characteristics	Diesel characteristics
Direct emission of pollutants	Lead content, Front end volatility, vapour pressure, benzene content	Sulphur, Poly aromatic hydrocarbon content
Effect on combustion	Octane number, volatility, oxygen content, final boiling point	Cetane number, Viscosity, Density, Volatility (mid and tail end boiling points), aromatic content
Effect on engine cleanliness and maintenance	Lead content, Final boiling point, Oxidation stability, Storage stability, Corrosion resistance	Tail end volatility and final boiling point, Carbon residue, Sulphur content, Oxidation and storage stability, Sediments
Durability of emission control devices	Lead content, Sulphur, Sediments	Sulphur

Note: Fuel properties can affect emissions directly as well as indirectly. Some fuel components by themselves may pose health hazards or adversely influence the environment. Some other fuel properties influence engine combustion and engine durability, while some other properties accelerate deterioration in the performance of emission control devices & systems.

Appendix 9.20: Effects of petrol Properties on Emissions from catalytic cars

Property	Change	Lead	CO	HC-Exh	HC-Evap	NOx	Benzene	Butadiene	Aldehydes
Reduce Lead	0.013- 0.005 g/l	0			-0		-0	-0	
Add Oxygenate	0 2.7% O_2	0	0-			+0	0	0	
Reduce Aromatics	50- 20% v/v		0		0		+0	+0	
Reduce Benzene	3 -2% v/v	0	0	0		0	0	0	0
Reduce Olefins	10- 5% v/v	0	0	+0		-0	-0	0	0
Reduce Sulphur	380- 20 ppm		0		0		0	0 -	
Reduce RVP	70- 60 kPa	0	0	-0		0	0	0	0
Increase E 100	35- 65%	0	0		-	0-	-0?	0 -	
Increase E 150	85- 90%	0	0-	0		?	0	?	?

*Contradictory results obtained in different investigation.

Key: 0 = No effect; -0 = -2 to 2% effect; 0- =2 to 10% effect; +0 = 10 to 20 % effect; 0+ = > 20% effect; ? = Insufficient data

Use of oxygenates in gasoline have the following features;

- Addition of oxygenates results in lower CO and HC emissions primarily due to mixture leaning effect more so in the carburetor vehicles.
- When Oxygenates are added, oxygen is attached with some of the fuel molecules and hence an enhanced oxidation of fuel may be expected resulting in somewhat higher reduction in CO & HC emissions than the leaning effect alone.
- The magnitude of CO emission reduction for the closed loop feedback controlled vehicles using 3-way catalysts is low as the mixture leaning effect occurs only during engine warm up and acceleration phases. Under normal operating conditions, the feedback system controls air fuel ratio close to stoichiometric.
- Excessive leaning caused by oxygenates can result in poor vehicle drivability and higher unburned hydrocarbons. Leaning effect may increase NOx.
- Use of oxygenates particularly alcohols also increases emissions of formaldehyde, which is a known carcinogen and has high photochemical reactivity.

Appendix 9.21: Comparison of effects of reformulated gasoline on emissions for catalytic cars with those with regular grade unleaded petrol

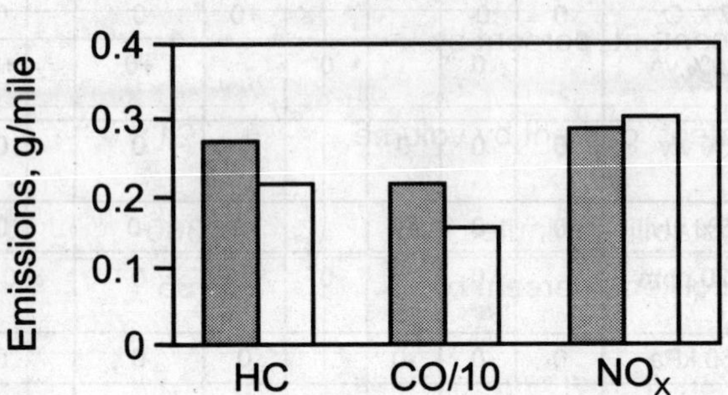

Appendix 9.22: Indian Specification for Euro IV/Bharat Stage IV Emission Norms

S. No.		Characteristics	Limits
		Gasoline 91 RON	Gasoline 95 RON
1.	Density @ 15 °C, kg/m³	720-775	720-775
2.	Distillation:		
	a) % evaporated at 70 °C (E-70), v/v	10-45	10-45
	b) % evaporated at 100 °C (E-100), v/v	40-70	40-70
	c) %evaporated at 150 °C (E-150), v/v, min	75	75
	d) Final boiling point, °C max	210	210
	e) Residue, percent by volume, max	2	2
3.	Research Octane Number (RON), min	91	95
4.	Motor Octane Number (MON), min	81	85
5.	Existent gum, g/m³, max	40	40
6.	Sulphur, total, mg/kg, max	50	50
7.	Lead content (as Pb), g/1, max	0.005	0.005

S. No.		Characteristics	Limits
		Gasoline 91 RON	Gasoline 95 RON
8.	Reid vapour pressure (RVP) at 38 °C, kPa,	60 (67)	60 (67)
9.	Benzene content, percent by volume, max	1	1
10.	Olefin content, percent by volume, Max	21	18
11.	Oxidation stability, minutes, Min	360	360
12.	Aromatic content, percent by volume, Max	35	35
13.	Oxygen content, percent by mass, Max	2.7	2.7
14.	Oxygenates, percent by volume, Max:		
	a)Methanol	Nil	Nil
	b)Ethanol	5	5
	c) Iso-propyl alcohol	10	10
	d)Iso-butyl alcohol	10	10
	e)Tertiary-butyl alcohol	7	7
	f)Ethers containing 5 or more 'C' atoms per molecule	15	15
	g)Other Oxygenates	8	8

RON- Research Octane number

Appendix 9.23: Effect of diesel cetane number (CN) increase from 50 CN to 58 CN on NO_x emissions

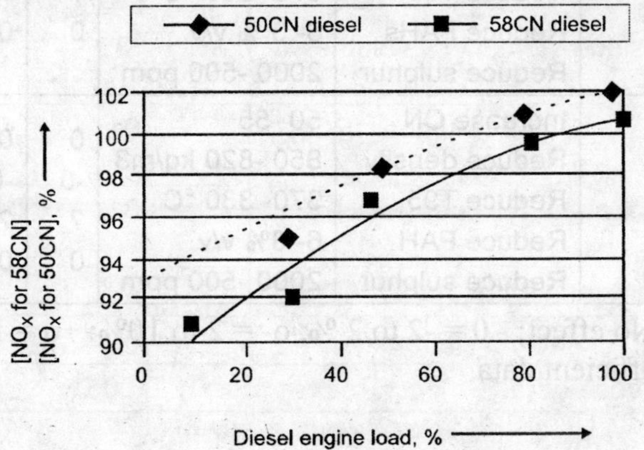

Note: Effect of CN on PM emissions depends on the engine design. For high speed, light duty engines a small reduction in PM emissions with high CN has been observed while low speed, heavy duty engines showed no significant change. With high cetane fuels the ignition delay is shorter and more fuel burns in diffusion combustion phase, which may increase soot emissions depending upon the engine operating conditions and design. Current fuel specifications are setting minimum limits of more than 50 units for the CN and minimum cetane index is also being specified as the natural high cetane fuels generally give overall better engine performance with respect to PM & HC emissions.

The temperature for 50 percent distillation temperature (mid-boiling point) , 90 percent point and the final boiling point are the important distillation parameters for diesel.

Lower the boiling point of the diesel fuel, more readily it vaporizes and mixes with air giving more complete combustion. The mid boiling volatility is also correlated to the other physico-chemical properties like, density, viscosity and ignition quality. A higher mid boiling point fuel has higher density & viscosity, and usually a lower CN. Low mid boiling point fuels give faster cold starting and hence lower HC emissions. The fuel components boiling above 350^{0}C may not burn completely, forming high soot concentrations and combustion chamber deposits. Fuels with high T90 & final boiling point (F.B.P) are seen to result in an increased injector cooling leading to poor combustion and higher smoke emission.

Appendix 9.24: Overall Effects of Diesel Fuel Property Changes on vehicular emissions

Vehicle	Fuel Property	Property Change	CO	HC	NOx	PM
Light-duty vehicles	Increase CN	50 -55			0	
	Reduce density	850- 820 kg/m^3	-0	-0	0	-0
	Reduce T95	370- 330 °C	?	?	-0	?
	Reduce PAHs	6- 3 % v/v	0	0	?	
	Reduce sulphur	2000 -500 ppm			0	
Heavy-duty vehicles	Increase CN	50 -55	0	0		
	Reduce density	850 -820 kg/m3	-0	-0	-0	-0
	Reduce T95	370- 330 °C	?	?	?	?
	Reduce PAH	6- 3% v/v	0	0	0	
	Reduce sulphur	2000 -500 ppm				

Key : 0 = No effect; - 0 = -2 to 2 %; o- = 2 to 10%; +o = 10 to 20% ,o+ = >20%; ? = Insufficient data

Appendix 9.25: Comparison of Diesel Fuel Specifications in Europe and India

Property	Europe (EN590:1999)	India (IS 1460) (BS IV – 2010)
Cetane number, min.	51	51
Cetane index, min	46	46
Density @ 15 °C, kg/m^3	820-845	820-845
Viscosity @ 40 °C, mm^2/s	2.0-4.5	2.0-5.0
Sulphur, % m/m, max.	0.035	BS III: 0.035 BS IV: 0.005
Total aromatics, % m/m, max.	-	-
Polyaromatics, (di+tri++), max.	11	11
Distillation:		-
% recovered at250 °C, max.	65	-
T 85, °C, max.	350	-
T90, °C max./range	-	-
T95, °C, max.	360	360
FBP, °C, range	-	-
Flash Point, °C, min.	55	35
Carbon residue, % m/m, max.	0.30	0.30
Water content, mg/kg, max.	200	200
Oxidation stability g/m^3, max.	25	25
Particulates, mg/l	24	-
Lubricity(HFRR scar dia. @ 60 °C),um	460	460
Oxygen content, % by mass max.	-	0.6

10

Storage and Handling of Petroleum Product

Fuel Storage Tank

Low flash point liquid fuels are stored in tanks fitted with roofs which float on the surface of the liquids while those having flash points well above room temperature (*e.g.* kerosene, diesel, fuel oils etc.) are stored in tanks provided with fixed conical shaped roofs. Fuel oil is normally stored in the storage tank erected above the ground when the capacity exceeds 70000 litres. For lower capacities, underground storage tanks are economical.

Storage tanks are normally made of welded mild steel plates with a height to diameter ratio of about 1.5. The overhead tank is mounted on concrete blocks which provide a slight gradient of 1 : 50 towards the drain valves. All fuel oils contain water which settles to the bottom of the tank and should be removed through a drain cock. In case of coal tar fuels (*e.g.* Pitch Creosote Mixture or raw tar), the water being lighter floats on the fuel and is removed by a draw off arrangement. The outlet or draw off pipe should be located at the raised tank end (not less than 15 cms from the bottom). The elimination of any air space above the fuels reduces fire hazards and evaporation losses. The vent pipe (gooseneck type) is located at the highest point of the storage tank to facilitate filling by venting air and oil vapor during liquid fuels filling operation. It should be about 5 cms in diameter and provided with a strainer at the outlet. The oil filling pipe can be located either at the top or at the bottom of the tank. When located at the top, the pipe must descend to the bottom of the tank to avoid free fall of oil from the top on to oil surface during tank filling causing frictional/impact heat generation. The drain pipe located at the lower end bottom is used periodically to remove the accumulated water.

Vertical cylindrical tanks upto 2.5 metre diameter may be erected direct on to a reinforced concrete foundation with the base having bitumen seal to prevent entry of water between tank bottom and concrete. One manhole at the top should be provided for tank upto 3.5 metre height. An additional manhole at the bottom should be provided if the tank height is more than 3.5 metres. Brass dip stick should be used to measure the tank level. Alternatively, float or remote reading gauges can be employed. Glass level indicator should not be used for level measurement in the tank. Oil draw off level should be above the oil heater mat and always > 7.5 cms above the bottom. Water accumulated at the bottom of the tank should be drained regularly, otherwise if carried to the burner with oil will cause sputtering of flame and possible flame failure. Water should be drained from main storage tank once every 15 days and from service tank once every 7 days as fuel oil contains upto 1% water which accelerates the sludge formation.

The liquid fuels are transported from the main installations to the distribution depots and to the consumers either by pipeline or in road tankers or other containers. The storage tanks in the filling stations are sited underground. They are then delivered to the consumers by road tanks or rail tanks and then stored in tanks of bulk storage capacity. A minimum safe distance from the surrounding structures need to be maintained while erecting a storage tank above the ground.

The permissible condition of safe storage & use are different for petrofuels with flash point < 23°C (*e.g.* gasoline which is dangerous and inflammable), between 23 to 66°C (*e.g.* kerosene) and above 66°C (*e.g.* fuel oils & gas oils which are safe).

Storage Tank Heating

Heating of liquid fuels in storage tank is done usually by low pressure steam coils (called mat heater) located as low as possible in the tank preferably below the level of the offtake. Hot water heater or electrical heaters are also used which are economical for small tanks. Normally, lagging/insulation of these tanks is not necessary. Hot oil pipelines are lagged with electrical or steam tracer lines within the lagging to balance the heat loss from the pipes through the lagging. Temperature of the liquid fuels in storage tank should be so maintained that it will have a viscosity of about 1000 seconds (Redwood No. I). Recommended preheat temperatures for easy pumping of some liquid fuels/oils is tabulated below.

Table: Oil Preheat Temperatures for Liquid Fuels for easy pumping.

Oil Grade	Preheating Temperature Required
Diesel/LDO (Light diesel oil)	no preheating
80 cst at 50°C (600 RWI at 100°F)	10°C
125 cst at 50C (1000 RWI at 100°F)	20°C
176 cst at 50°C (1500 RWI at 100°F)	25°C
370cst at 50°C (3500 RWI at 100°F)	35°C
PCM/Coal Tar Fuel (CTF-200)	200°F
LSHS (Low sulphur heavy stock) furnace oil	70°C

Pumps used for oil pumping

The pumps used for pumping oil are normally of following three types :

- bucket type
- reciprocating pumps
- gear or screw type rotary pumps.

If driven by electric motors, it should be ensured that all electrical insulations are sound and no loose connections are made. The safe current carrying capacity for each pump motor should be noted. High viscosity of oil means more load on motor and high pressure drop of oil. Hence, correct pumping temperature of the oil is very important.

In case pumping capacity is much more than burner oil output, then recirculation of more hot oil (at 100°C) in return line may cause cavitation in pump due to higher vapor pressure. Hence, it should be ensured that minimum quantity of hot fuel oil is returned from the burner.

Liquid Fuels Preheating

The preheating of heavier liquid fuels before pumping & before atomisation should be only upto the temperature required by the viscosity consideration for easy pumping and without excessive pressure drop. The optimum degree of preheat is decided by the plant trials. For reasons of heat economy, excessive heating of oil should be avoided. A temperature-viscosity chart for the oil may be used to get the correct preheat temperature. Normally, a low air pressure oil burner specifies a viscosity of 100 seconds Redwood No. 1 at the burner tip. Preheater capacity depends upon the oil rate and the required rise in temperature. The oil can be preheated either electrically or by steam, though the latter is preferred. Heating coils are placed at the bottom of the tank above the suction level of oil and at least 15 cms from the bottom to permit sludge cleaning. Thermostatic control is kept at a higher level (say 20 cms above the bottom) to one side of heater.

Oil preheater capacities required for 100°C rise in temperature is tabulated below.

Table: Oil Preheater Capacity required for 100°C Rise in Temperature

Fuel Oil flow, litres/hr	Preheater capacity, kw,	Steam requirements, kg/hr.
5	0.5	1
10	1	2
15	1.5	3
20	1.5	4
25	2	5
50	3.5	10
Every additional 50 litres/hr	Additional 3.5 kw is needed.	Additional 10 kg/hr capacity is needed.

Liquid Fuels Supply System

There are two systems namely gravity feed system and ring *main system* for supply of liquid fuels to furnaces/burners. In gravity feed system (which is used when firing through only one or two burners are involved), either an overhead feed tank or a ground level feed tank with a pump in delivery lines supplies the fuel to the burner through a single feed line.

The ring main system is used for medium and heavy fuel oils burnt in a number of burners supplying fuel to one or more furnaces. This ensures the minimum temperature and pressure differences between the burners. The oil pumps are located nearer to the consuming points. Oil transfer to the consuming end is through preheated pipes. Steam tracers are usually provided in these pipes. These are small diameter steam lines which are laid along sides of the oil line and then lagged in the same insulation bundle. Thus radiation losses are minimised and oil preheat temperatures are maintained at optimum level right upto the burner.

In the ring main oil supply system, the oil is normally circulated through a ring main round the plant at a rate of 2-3 times the actual consumption of the burners. Branch lines feed the individual burners. Both suction and discharge sides of the oil delivery pump are provided with oil filters for protecting the pump as well as the burner nozzles from dirt. There is an oil preheater in the ring main and no separate heating is required in the branch lines as they are having steam tracers. The return pipe of the ring main is connected to the suction end of the pump (not to the storage tank) to increase the heat economy. Standby pump, preheaters and oil filters are provided to ensure continuous plant operation. The oil in the ring main system is kept at the atomisation temperature by the steam tracer lines. Alternately, no heating is done in the ring main and the oil is kept at a pumpable temperature only. The branch line to the furnaces are in that case provided with small heaters to feed the burners with oil at required atomisation temperature. However, for highly viscous oils, the fuel handling system comprises of concentrically steam jacketted pipelines, heated pumps and heated filters.

The schematic diagram of a typical ring main system using L.S.H.S. (Low Sulphur Heavy Stock) furnace oil for burning in 6 nos. rotary Calcination kilns is shown below in figure.

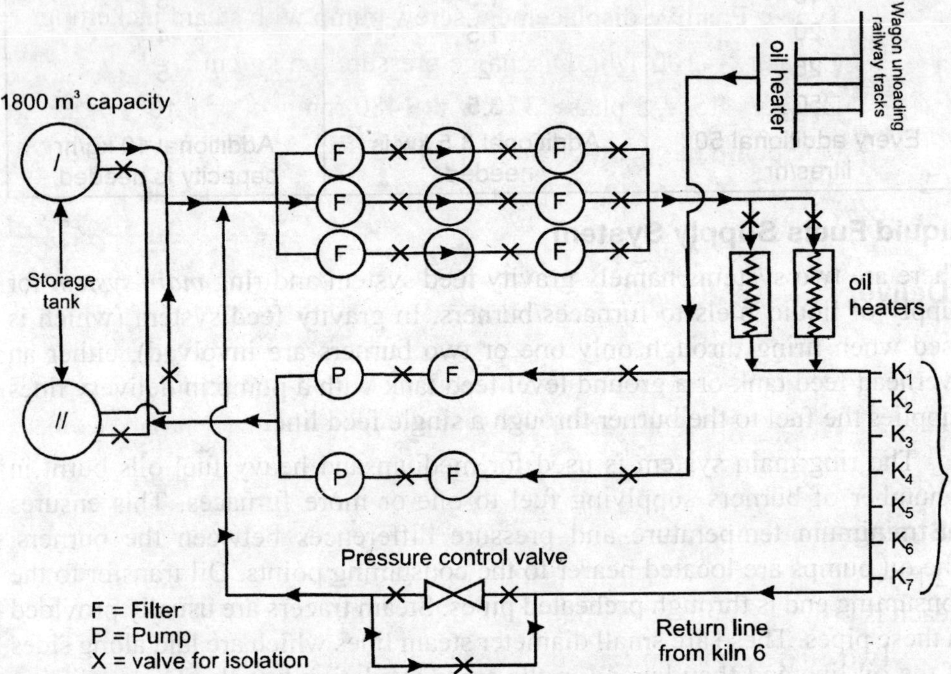

Figure : L.S.H.S. firing facility for rotary kilns (used for calcination of limestone).

The specification of LSHS and the equipments details are as follows :

Specifications & Equipment Details of Typical LSHS (Furnace Oil) Firing Facility

Specification of L.S.H.S.

Flash point = 66°C, Viscosity at 70°C, minimum = 40 centistokes, Pour point = 50°C,

Total sulphur = 1% by weight, maximum, Water content = 1.0% by volume, maximum steam availability at railway wagons is at 10 kg/cm^2& 220°C, LSHS pumpable temp. = 85°C , Steam hoses, 30 points (15 each side), branch connections = 80 mm N.B. pipes, 80 mm N.B. steam headers provided.

Total 5 pumps - 2 pumps for unloading and 3 pumps for delivery of LSHS oil to the kilns.

Features of unloading pumps (2 nos)

Type = Positive displacement screw pump with steam jacketting.

Capacity = 100 T/hr, Discharge pressure = 5 kg/cm^2

Drive = 415 V, 3 phase, 37 kW at 1480 rpm

One pump in operation and one stand by

While pumping 100 T/hr at 60-70°C, max. pressure drop = 0.1 kg/cm^2

Delivery pumps (3 nos.)

Capacity = 10 T/hr, Discharge pressure = 40 kg/cm^2

Drive = 415 V, 3 phase, 30 kW at 2940 rpm

One pump in operation, one standby and one under repair.

Filter = 60 mesh on suction side and 80 mesh on delivery side.

Storage Tank

2 Nos. of vertical cylindrical oil storage tank of effective capacity of 1800 m^3 each is provided with steam coil and oil level indicator. Oil Heater :

Two oil heaters common for both storage tanks.

Capacity of heater = 10 T/hr each from 55°C to 120°C

Super heated steam at a pressure of 10 kg/cm^2 and 220°–250°C is used for the *mat heater* which heats the oil to 60°C in tank.

The pump draws the oil from any of the two tanks and discharges through the line heaters to the kiln. Fuel oil is delivered from the oil heaters at 120°C and 40 kg/cm^2 pressure. Here also some steam is used.

Pressure control valve provided in the return line automatically controls the oil delivery pressures to the burners in the kilns.

Procedures for correctly quantifying the quantity of oil in the road tankers and railway wagons and statutory rules for handling petroleum oils are described below :

Procedure to be adopted before assessing the quantity of oil in the Road Tankers

- Ensure that the tank lorry is parked on firm level ground.
- The dipstick should be checked to ensure that it is the one calibrated by the Weights and Measures Department for the compartment concerned.
- The dip readings of all the compartments should be checked and shortfall, if any can be determined by topping up of the compartments with measured quantities of furnace oil from the main storage tank.
- If the tank of truck has been filled hot, the product in the compartment is likely to shrink to the extent of 0.7 litre/KL per degree centigrade fall in temperature. If the shortfall is higher, it may be due to pilferage.
- Stationery storage tanks should also be calibrated. Though dip readings in storage tanks are reliable, a major deviation may provide a clue to malpractices. However, the geometry of stationery tanks may not be true during construction stage, for it gets disturbed due to external damage or build-up of the tank in use.

Procedure to be adopted for handling Railway Tank Wagons Carraying furnace oil

- Dip of the tank wagon is to be taken.
- The valves are to be examined whether these are operative or not before connecting the unloading hose pipe.
- Seal also is to be examined.
- After decanting, again dip of the wagon is to be taken to make sure complete decantation.
- If there is accumulation of the oil in the sumps, that also should be pumped to the main storage tank while decanting the tank wagons by using sump pumps.
- Release memo for wagons to given only after the hose pipes are disconnected.
- Split oil is irretrievable. A little care while oil is decanted from the tank wagons to the storage tank will eliminate spillage. All oil leaks from joints, flanges and pipelines must be eliminated.

Leakage Loss of Even One Drop of Oil Every Second Can Cost/ Waste Over 4000 Litres a Year

All petroleum *oil storage installations* should conform to the following *Petroleum Rules* which is statutory/safety requirement:

- The storage tank shall be electrically connected with the earth in an efficient manner by means of not less than two separate and distinct connections placed at opposite extremities of the tank.
- Every oil installation should be surrounded by a wall at least 2 metres in height.
- Storage capacity and flash point are to be written on the body of the tank.
- Provisions of fire fighting equipment and adequate supply of dry sand in buckets with necessary implements for its convenient application in extinguishing petroleum fires should be made.
- The ground in the interior of the installation and protected area surrounding the storage tank should be kept clean & free from all inflammable materials & rubbish.
- Care should be taken to prevent any oil escaping into drain, sewer or public road.
- Petroleum having flash point 65°C and above but below 93°C comes under petroleum class 'C' *i.e.* Furnace oil and L.D.O.
- For storage capacity above 45 K.L., licence from the *Chief Inspector of Explosives, Nagpur* is to be obtained.
- Every tank or receptacle for storage of petroleum in bulk (other than a well head tank)shall be constructed of iron or steel in accordance with the relevant Indian standards or specifications approved by chief controller as per petroleum rules 1976. However underground petroleum storage tanks may be made of Fibre Reinforced Plastics (FRP) also.
- When two or more tanks are installed in one enclosure, the total capacity of the tanks in the enclosure shall not exceed 60,000 KL in case of fixed roof tanks and 1,20,000 KL in case of floating roof tanks. A combination of fixed and floating head tanks can also be located in the same enclosure/premises but the total capacity should not exceed 60,000 KL.
- Height of storage tank should not exceed 1.5 times its diameter or 20 m whichever is less.
- An air space of not less then 5% of the total storage capacity shall be considered while designing oil storage tank.
- All petroleum storage tanks installed above or below the ground shall be protected against corrosion by the use of protective coating and cathodic protection. Cast iron valves are not permitted in oil installations.

- Tanks after erection or repair will be tested with clean water.
- All storage and handling facility shall maintain the specified distance from any other facility nearby.
- No fire water/foam ring main shall pass through dyked enclosure.
- Grouping of petroleum products for storage purpose shall be based on the product classification. Class – A and /or class – B petroleum products can be stored in the same dyked enclosure. Class – C petroleum products should preferably be stored in separate enclosure. However, where class – C product is stored in the same or common dyke along with class –A and/or class – B products, all safety stipulations applicable for class **A** or **B** products should apply.
- Tanks shall be arranged in maximum two rows so that each tank is approachable from the road/area surrounding the enclosure. However, tank having capacity of 50000 m^3 and above shall be laid in a single row.
- Piping from/to any tank located in a single dyked enclosure should not pass through any other dyked enclosure. Piping within the dyked enclosure should be minimised.
- Any tank having a diameter of more than 30 metres should be separated with firewalls from other tanks.
- The minimum distance between a tank shell and inside of the dyke wall shall not be less than half of the tank diameter.
- Firewalls should be constructed by limiting the aggregate capacity of each group of tanks within 20000 m^3.

Explosives Regulations on the Storage of Petroleum Products

(*a*) Petroleum products are classified as under as per the Explosive Regulations.

Flash Point	Classification	Products
Below 23°C	A	Petrol, Naphtha
23°C—less than 65°C	B	Kerosene, Mineral Turpentine
65°C—less than 93°C	C	Light Diesel Oil, High Speed Diesel Oil
93°C and above	Excluded	(LDO), Furnace Oil LSHS, Lubricatiing Oils & Greases

(*b*) If any two classes of petroleum are stored together, the storage rules applicable for more dangerous product between the two shall be applicable *e.g.* if furnace oil and LSHS are stored adjacent to each other, even the storage of LSHS which is normally excluded from the purview of the Explosive Regulations should conform to those applicable to furnace oil. The whole installation in such case will be deemed to be that of furnace oil alone.

(*c*) If LSHS alone is being stored in the premises, any amount of it could be stored without either the approval or a license from the Dept. of Explosives, since it falls under exempted category.

11

General Appendix

Appendices

Appendix 11.1	History of oil/petroleum refining processes in chronological order
Appendix 11.2	Units and their Conversions
Appendix 11.3	Main Conversions Used in Petroleum Industry
Appendix 11.4	Conversion factors for petroleum products
Appendix 11.5	Viscometer Conversion Factors
Appendix 11.6	Table for Viscosity Conversion
Appendix 11.7	Viscosity Conversion Nomogram
Appendix 11.8	Viscosity (centistokes)of some Hydrocarbon Fuels at Various Temperatures
Appendix 11.9	Temperature Vs. Viscosity curves for Furnace oil
Appendix 11.10	All AboutGases- Natural Gas, Liquefied Petroleum Gas (LPG) and Refinery Gases
Appendix 11.11	Typical Product Yield & Uses (disposition) of Various oil Refinery Processes
Apendix 11.12	Properties of Conventional and Alternative Automotive Fuels
Appendix 11.13	Advantages and Disadvantages of Alcohol Motor Fuels Compared to petrol and Diesel
Appendix 11.14	Properties of Biodiesel Derived from Some Vegetable Oils

Appendix 11.1: History of oil/petroleum refining processes in chronological order

S.N	Year	Process name	Process objective	Process by-products
1	1862	Atmospheric distillation	Produce kerosene	Naphtha, tar, etc.
2	1870	Vacuum distillation	Lubricants (original), Cracking feedstocks (1930s)	Asphalt, residual Coker feedstocks
3	1913	Thermal cracking	Increase gasoline	Residual, bunker fuel
4	1916	Sweetening	Reduce sulphur and odour	Sulphur
5	1930	Thermal reforming	Improve octane number	Residual
6	1932	Hydrogenation	Remove sulphur	Sulphur
7	1932	Coking	Produce gasoline base stocks	Coke
8	1933	Solvent extraction	Improve lubricant viscosity index	Aromatics
9	1935	Solvent dewaxing	Improve pour point	Waxes
10	1935	Catalytic polymerization	Improve gasoline yield and octane number	Petrochemical feedstocks
11	1937	Catalytic cracking	Higher octane gasoline	Petrochemical feedstocks
12	1939	Visbreaking	Reduce viscosity	Increased distillate, tar
13	1940	Alkylation	Increase gasoline octane and yield	High-octane aviation gasoline
14	1940	Isomerization	Produce alkylation feedstock	Naphtha
15	1942	Fluid catalytic cracking	Increase gasoline yield and octane	Petrochemical feedstocks
16	1950	Deasphalting	Increase cracking feedstock	Asphalt
17	1952	Catalytic reforming	Convert low quality naphtha	Aromatics
18	1954	Hydrodesu-lphurization	Remove sulphur	Sulphur

S.N	Year	Process name	Process objective	Process by-products
19	1956	Inhibitor sweetening	Remove mercaptan	Disulphides
20	1957	Catalytic isomerization	Convert to molecules with high octane number	Alkylation feedstocks
21	1960	Hydrocracking	Improve quality and reduce sulphur	Alkylation feedstocks
22	1974	Catalytic dewaxing	Improve pour point	Wax
23	1975	Residual hydrocracking	Increase gasoline yield from residual	Heavy residuals

Note: Main processing units used in modern oil refineries include the following:

- **Crude Oil Distillation unit**: distills the incoming crude oil into various fractions for further processing.
- **Vacuum Distillation unit**: further distills the residue oil from the bottom of the crude oil distillation unit.
- **Naphtha Hydrotreater unit**: uses hydrogen to desulfurize the naphtha fraction from the crude oil distillation or other units within the refinery.
- **Catalytic Reforming unit**: converts the desulfurized naphtha molecules into higher-octane molecules to produce reformate, which is a component of the end-product gasoline or petrol.
- **Alkylation unit**: converts isobutane and butylenes into alkylate, which is a very high-octane component of the end product gasoline (petrol).
- **Isomerization unit**: converts linear molecules such as normal pentane into higher-octane branched molecules for blending into the end product gasoline. Also used to convert linear normal butane into iso-butane for use in the alkylation unit.
- **Distillate Hydrotreater unit**: uses hydrogen to desulfurize some of the other distilled fractions from the crude oil distillation unit (such as diesel oil).
- **Merox (mercaptan oxidizer) or similar units**: desulfurize LPG, kerosene or jet fuel by oxidizing undesired mercaptans to organic disulphides.
- **Amine gas treater, Claus unit, and tail gas treatment**:for converting hydrogen sulphide gas from the hydrotreaters into end product elemental sulphur.

- **Fluid catalytic cracking (FCC) unit**: upgrades the heavier, higher boiling fractions from the crude oil distillation by converting them into lighter and lower boiling, more valuable products, ,
- **Hydrocracker unit**: uses hydrogen to upgrade heavier fractions from the crude oil distillation and the vacuum distillation units into lighter, more valuable products.
- **Visbrteaker unit**: upgrades heavy residual oils from the vacuum distillation unit by thermally cracking them into lighter, more valuable reduced viscosity products.
- **Delayed coking and Fluid coker units**: convert very heavy residual oils into end product petroleum coke as well as naphtha and diesel oil by-products.

Auxiliary facilities required in oil refinieries include:

- **Steam reformer unit**: converts natural gas into hydrogen for the hydrotreaters and/or the hydrocracker.
- **Sour water stripper unit**: Uses steam to remove hydrogen sulfide gas from various wastewater streams for subsequent conversion into end product sulfur in the Claus unit.
- **Utility units**:such as cooling towers for furnishing circulating cooling water, steam generators, instrument air systems for pneumatically operated control valves and an electrical substation.
- **Wastewater collection and treating systems**: consisting of API separators, dissolved air flotation (DAF) units and some type of further treatment (such as an activated sludge biotreater) to make the wastewaters suitable for reuse or for disposals
- **Liquefied gas (LPG) storage vessels**: for propane and similar gaseous fuels at a pressure sufficient to maintain them in liquid form. These are usually spherical vessels or *bullets* (horizontal vessels with rounded ends).
- **Storage tanks**:for crude oil and finished products, usually vertical, cylindrical vessels with some sort of vapour emission control and surrounded by an earthen boundary to contain liquid spills.

Oil Refining end-products include:

The primary end-products produced in petroleum refining may be grouped into four categories: light distillates, middle distillates, heavy distillates and others.

Light distillates

- Liquid petroleum gas (LPG)
- Gasoline (also known as petrol)
- Kerosene
- Jet fuel and other aircraft fuel.

Middle distillates

- Automotive and rail road diesel fuels
- Residential heating fuel
- Other light fuel oils

Heavy distillates

- Heavy fuel oils
- Bunker fuel oil and other residual fuel oils

Others

Many of these are not produced in all petroleum refineries:

- Speciality petroleum naphthas
- Speciality solvents
- Elemental sulphur, and sometimes sulphuric acid
- Petrochemical feed stocks
- Asphalt and tar
- Petroleum coke
- Lubricating oils
- Waxes and greases
- Transformer & cable oils
- Carbon black

Appendix 11.2: Units and their Conversions

Length

[In-inch, ft-feet, m-metre, mm = millimeter, cm = centimeter, Å= Angstrom, km- kilometer, pm-micro-metre]

1Å = 10–4 micron = 10–8 cm = 10–10m, 1 µm (micrometre) = 10–6 m, 1 mm = 1000 microns

1 micron = 10–4cm ; 200 mesh size = 74 µm (micro-metre) = 0.074 mm = 74 microns

1 inch = 2.54 cms = 25.4 mm - 0.08333 ft = 0.0254 m, 1 pm = 1 micron

= 10–6 m

1 ft = 12 inches = 30.48 cms = 0.3048 m = 0.3333 yard, 1 yard = 0.9144 m

1m = 3.28 ft = 39.37 inches = 1.0936 yards = 100 cm = 1000 m = 10–3 km

1mile = 1.609 km (i.e, 5 miles = 8 km) = 5280 ft = 1760 yards = 1609 m

1 Nautical mile = 1852 m = 1.852 km = 1.15 mile; 1 mm = 1000 microns

AREA

1 barn = 10^{-28} m^2, 1 mm^2 = 10^{-6} m^2 = 0.00155 in^2 = 10.76 × 10^{-6} ft^2

1 in^2 = 6.45 cm^2, 1 ft^2 = 0.0929 m^2 = 929.03 cm^2 = 144 in^2 = 0.093 m^2

1 hectare = 10000 m^2 = 2.47 acres, 1 acre = 100 m^2, 1 cm^2 = 10^{-4} m^2

1 acre = 4840 $yard^2$ = 0.4047 hectare = 4047 m^2 = 43560 ft^2 = 4840 $yard^2$

1 m^2 = 10.76 ft^2 = 1550 in^2 = 1.196 $yard^2$ = 0.000247 hectare = 10^6 mm^2

1 km^2 = 10^{12} mm^2 = 10.76 x 10^6 ft^2 = 10^6 m^2 = 1.196 x 10^6 $yard^2$ = 247.1 acre = 0.3861

$mile^2$ 1 $mile^2$ = 27.88 x 10^6 ft^2 = 3.097 x 10^6 $yard^2$ = 2.59 x 10^6 m^2 = 640 acre = 2.59 km^2

VOLUME

[ml-millilitre, KL-kilolitre, cc-cubic centimetre (cm^3), dm-decimeter]

1 mm^3 = 10^{-9} m^3 = 16.39 cm^2, 1 ft^3 = 28.32 litres = 0.02832 m^3

1 litre = 1000 ml = 1000 cc = 10^{-3} m^3 = 0.22 U.K. gallon = 0.264 U.S. gallon = 1 dm^3

1 imperial gallon = 1 U.K. gallon = 4.546 litres (dm^3) = 1.201 U.S. gallon = 4.546 ×10^{-3} m^3

1 U.S. gallon = 3.785 litres = 0.833 U.K. gallon

1 U.S. barrel (bbl) = 42 U.S. gallon = 34.973 U.K. gallon = 0.159 m^3 = 0.159 KL = 5.615 ft^3

1 ml = 10^{-6} m^3, 1 pint = 0.5682 litre = 1/8 U.K. gallon

1 m^3 = 1000 litre = 1 KL = 35.3148 ft^3 = 219.97 U.K. gallon = 264.17 U.S. gallon

1.308 $yard^3$ = 10^3 dm^3 = 61024 $inch^3$ = 10^9 mm ^{3} = 6.29 U.S. barrel

1 ft^3 = 0.028317 m^3 = 6.023 U.K. gallon = 7.481 U.S. gallon = 1728 in^3 = 28.32 dm^3

SPECIFIC VOLUME

1 litre/kg = 1 c.c/gm = 10^{-3} m^3/kg, 1 litre/gm = 1 m^3/kg

1 ft^3/kg = 2.8317 x 10^{-2} m^3/kg, 1 ft^3/lb = 6.2428 x 10^{-2} m^3/kg

VOLUMETRIC FLOW RATE

1 ft^3/hr = 7865 mm^3/sec = 0.0166 ft^3/min = 7.865 x 10^{-6} m^3/sec, 1 ft^3/s = 1 cusec = 2.8317 × 10^{-2} m^3/s

1 ft^3/min = 60 ft^3/hr = 471.9 x 10^{-6} m^3/sec = 1 cumin = 4.7195 ×41$0^{-4}$ m^3/s

1 m^3/sec = 127.1 x 10^3 ft^3/hr = 2119 ft^3/min = 10^9 mm^3/sec

VELOCITY

1 in/sec = 60 in/min = 5 ft /min = 0.08333 ft/sec = 0.0254 m/sec = 25.4 x 10~^{6} km/sec

1ft/sec = 720 in/min = 60 ft/min = 0.3048 m/sec = 304.8 x 10~^{6} km/sec

1 m/sec = 2368 in/min = 196.9 ft/min = 39.37 in/sec = 3.281 ft/sec = 10^{-3} km/sec

1 km/sec = 2368 x 10^3 in/min = 196.9 x 10^3 ft/min = 39.37 x 10^3 in/sec= 3.281 ft/sec = 10^3 m/sec

1 km/hr = 0.6214 mile/hr, 1 mile/hr = 1.609 km/hr

FREQUENCY

1 cycle/second (cps) = 10000 Hertz (Hz)

MASS

[lb-pound, oz-ounce]

1 grain = 0.0648 gm ; 1 mg = 10~^{6} kg ; 1 gm = 15.432 grains = 0.0353 (ounce) oz = 0.002205 lb ;

1 oz = 28.35 gm = 984.2 x 10^{-9} ton (U.K/long)

1 lb = 453.6 gm = 7000 grains = 0.454 kg = 16 oz = 446.4 x 10~^{6} ton (U.K./long)

1 long ton = 1 U.K. ton = 2240 lb = 1.12 short ton = 1.12 U.S. ton = 1.016 metric ton = 1.016 x 10^6 gm = 1016 kg,

1 U.S. ton (short) = 2000 lb = 0.8929 U.K. ton (long) = 0.9072 tonne = 907.2 kg

1 tonne = 1 metric ton = 1000 kg = 2204.6 lb = 0.984 U.K. ton (long) = 1.1023 U.S. ton (short)

1 kg = 1000 gm = 2.205 lb = 984.2 x 10^{-6} ton (U.K./long)

1 slug = 14.59 kg, 1 oz = 28.35 gm, 1 tonne = 10^6 gm.

MASS FLOW RATE

1 lb/hr = 1.26 x x 10~^{4} kg/hr = 2.7778 x 10^4 kg/s ; 1 tonne/day = 1.1574 x 10^{-2} kg/s,

1 tonne/hr = 2.7778×10^{1} kg/s

1 ton (short)/day = 1.05 x 10“2 kg/s, 1 ton (short)/hr = 2.52 x lO^1 kg/s ;
1 ton (long)/day = 1.176 x 10~2 kg/s, 1 ton (long)/hr = 2.8224 x 10’1 kg/s

DENSITY AND CONCENTRATION

[ppm-parts per million]

1 ppm = 1 ml/m^3 = 1 mg/kg ;1% = 10000 ppm (for water, density = 1 gm/cc)

1 gm/ml = 1 gm/cc = 62.4 lb/ft^3 = 10.022 lb/gallon = 10^{-3} gm/m^3

1 lb/ft^3 = 16.02 kg/m^3 = 0.4536 kg/ft^3 = 578.7×10^{-6} lb/in^3 ;
1 kg/m^3 = 0.625 lb/ft^3 = 9.8 N/m^3 = 0.02832 kg/ft^3, 1 kg/litre = 10 kg/m^3

1 kg sec^2/cm^4 = 9.81 kg/m^3, 1 mg/ml = 1 kg/m^3, 1 mg/litre = 10^{-3} kg/m^3

1 m^3/kg = 16.019 lb/ft^3 ; 1 ft^3/lb = 0.062428 m^3/kg

1 kg/ft^3 = 35.31 kg/m^3 = 2.205 lb/ft^3 = 1.278×10^{-3} lb/ in^3 = 35.31×10^{-6} gm/mm^3

1 lb/in^3 = 27.68×10^{3} kg/m^3 - 1728 lb/ft^3 = 783.8 kg/ft^3 = 27.68×10^{-3} gm/mm^3

1 gm/mm^3 = 10^6 kg/m^3 = 62.43×10^3 lb/ft^3 = 28.32×10^6 kg/ft^3 = 36.05 lb/in^3

ABSOLUTE (OR DYNAMIC) VISCOSITY

[cp-centipoise, Pa-Pascal, s-second]

1 poise = 100 cp = 1 gm/cm sec = 0.01 kgf sec/m^2 = 1 dyne sec/cm^2 = 10^{-1} Pa s = 0.1 N sec/m^2 = 2.0885×10^{-3} lbf sec/ft^2

1 Pa.s = 10 poise = 0.02089 lbfs/ft^2

1 kgf sec/cm^2 = 9.81 N/m^2 ; 1 kgf sec/m^2 = 98 poise = 0.205 lbf sec /ft^2

1 lb_f sec/ft^2 = 478.8 poise = 4.88 kgf sec/m^2 = 47.88 Pa.s

1 kg/m.hr = 2.7778×10^{-4} N.S/m^2, 1 lb/ft.hr = 4.1338×10^{-4} N.S/m^2

1 mN.S/m^2 = 1 centipoise = 10^{-3} N.S/m^2, 1 kg/M.S = 1 N.S./m^2

1 lb/ft.S = 1.4882 N.S/m^2

KINEMATIC VISCOSITY AND THERMAL DIFFUSIVITY

[cst—centistoke, rpm—rotation per minute]

1 stoke = 100 cst = 1 cm^2/sec = 0.36 m^2/hr = 3.875 ft^2/hr = 10^{-4} m^2/sec

1 m^2/s = 10^4 stokes = 10.76 ft^2/s

1 m^2/hr = 2.78 stokes = 10.76 ft^2/hr = 2.78×10^{-4} m^2/s

1 ft^2/hr = 0.26 stokes = 0.09 m^2/hr = 2.5806×10^{-5} m^2/s

1 ft^2/s = 0.09290 m^2/s

1 cst = 10^{-6} m^2/sec = 1 mm^2/s ; 1 rpm = 1.047 x 10^{-1} radian/sec

FORCE

[pdl—poundal, kN—kilo Newton]

1 pdl = 0.1383 N = 0.03108 lbf = 0.01410 kgf

1 N = 7.233 pdl = 0.2248 lbf = 0.1020 kgf

1 kN = 7233 pdl = 10^3 N = 224.8 lb = 102 kgf

1 lbf = 31.17 pdl = 4.448 N = 0.4536 kgf

1 kgf = 70.90 pdl = 9.803 N = 2.205 lbf

1 dyne = 10^{-5} N

PRESSURE

[mmwg—millimeter water gauge, inwg—inch water gauge, mmHg—millimeter mercury gauge, psi—lbf/in^2, mB—mega bar, atm—atmospheric pressure, mmwc—millimeter water column, N—Newton, M—mega, Pa—Pascal, kPa—kilo pascal]

1 atm = 760 mmHg = 29.92 in Hg = 14.7 psi = 1.013 bar = 1.013 kgf/cm^2 = 34 ft water = 10033 mmwc = 10033 kgf/m^2 = 1.013 x 10^5 N/m^2 = 1013.2 mB = 2116 lbf/ft^2 = 1.013 x 10^5 Pa

1 in wg = 1.87 mm Hg = 0.07351 in wg = 0.036 psi = 1.87 torr = 1.87 mm Hg = 249.1 N/m^2 = 249.1 Pa = 2.49 mB = 25.4 kg/m^2 = 25.4 mm wg = 5.202 lbf/ft^2

1 Pa = 4.015 x 10^{-3} in water = 295.3 x 10^{-6} in Hg = 145 x 10^{-6} psi = 64.75 x 10^{-9} tors/in^2

1 kPa = 20.89 lb/ft^2 = 0.145 psi = 10^{-2} bar = 64.75 x 10^{-6} ton/in^2 = 1 kN/m^2 = 102 mmwc

1 in Hg = 13.6 in wg = 3386 mB = 25.4 torr = 25.4 mm Hg

= 0.0334 atm = 0.4912 psi = 0.03386 bar = 70.73 lb/ft^2 = 3386 Pa = 3386 N/m^2

1 ft water = 304.8 kgf/m^2 = 2986.1 N/m^2 ; 1 mB = 100 N/m^2

1 mm Hg = 1 torr = 133.3 Pa = 133.3 N/m^2 ; 1 u bar = 10^{-1} N/m^2, 1 millibar = 100 Pa

1 Pascal = 1 N/m^2, 1 bar = 1 kgf/cm^2 = 10^5 N/m^2 = 10^5 Pa = 10^6 dynes/cm^2=401.5 in H_2O = 14.5 psi =29.53 in Hg

1 psi = 0.07031 kgf/cm^2 = 68.95 mB = 6.895 x 10^3 N/m^2

= 6895 P = 0.06895 bar = 2.036 in Hg = 27.68 in water = 144 lb/ft^2

1 kgf/cm^2 = 14.7 psi = 10000 mm wc = 0.1 MN/m^2 = 9.81 N/cm^2 = 98 kPa

1 kgf = 9.8 N, 1 lbf = 4.448 N, 1 MN/m^2 = 10 kgfi/cm^2, 1 kgf/m^2 = 1 mmwc = 9.8067 N/m^2

1 lbf/ft^2 = 47.88 Pa = 47.88 N/m^2 = 0.1922 in water = 0.01414 in Hg = 6.944 x 10^{-3} psi = 478.8 x 10^{-6}bar 1 ata = 1 kgf/cm^2 = 9.8067 x 10^3 N/m^2, 1 N/mm^2 = 10^6 N/m^2

1 ton/inch^2 = 1.5477 x 10^5 N/m^2, 1 Poundal/ft^2 = 1.4881 N/m^2

HEAT, WORK, POWER AND ENERGY

[BTU—British Thermal Unit, CHU—Centigrade Heat Unit, J—Joule, kJ—kilo Joule, Cal—Calorie, Kcal—kilo calorie, kWh—kilowatt hour, W—watt, hp—horse power, min— minute, TNT—Tri-nitro-toluene, G—Giga = 10^9]

Heat

1 J = 0.7376 ft lbf = 0.2388 cal = 947.8 x 10^{-6} BTU = 277.7 x 10~^{9} kWh = 9.478 x 10^{-9} therm= 1 N.m

1 BTU = 252 calorie = 0.252 kcal = 778 ft lbf = 0.5556 CHU = 25030 pdl = 1055 J = 2.931 x 10^{-4} kWh

1 Therm = 2503 x 10^6 pdl = 105.5 x 10^6 J = 77.82 x 10^6 ft lbf = 25200 kcal = 10^5 BTU = 29.31 kWh

1 calorie = 0.003968 BTU = 4.186 J = 4.186 watt. sec. = 3.088 ft lbf = 1.163 x 10^{-6} kWh

1 kcal = 1000 calorie = 3.968 BTU = 0.00116 kWh = 3087 ft lb = 4186 J = 4.186 kJ = 2.205 CHU = 427 kg m = 0.001560 hp

1 ft.lbf = 0.3238 calorie = 0.1383 kg.m = 1.356 J = 1.356 N.m

1 kgf. m = 9.8067 J

1 hp.hr (metric) = 2.6477 x 10^6 J

1 hp.hr (British) = 2.6845 times 10^6 J

12.85 x 10^{-9} therm = 1.285 x 10^{-3} BTU = 3.776 x 10^{-7} kWh

1 ft lb/min = 0.0226 W, 1 ft lb/sec = 0.00136 kW

1 ton of TNT = 4.2 x 10^9 J = 4.2 x 10^6 kJ = 4.2 x 10^3 MJ = 4.2 GJ

Work and Heat Flow Rate:

[sec—second, MW—mega watt = 10^6 W, kW—kilowatt, = 10^3 watt, hr—hour, N-Newton]

1 BTU/s = 1055.1 J/S, 1 BTU/min = 0.02356 hp, 1 CHU = 1.8 BTU = 0.453 kcal

1 BTU/hr = 0.2931 W = 0.2162 ft lbf/s = 0.00393 hp = 0.2931 J/S

1 Therm = 100, 000 BTU = 25200 kcal = 29.31 kWh = 2503 x 10^6 pdl =

105.5 x 10^6 J = 77.82 x 10^6 ft lbf; 1 m.Kgf/S = 9.8065 J/S

1 ton of refrigeration = 3.5169 x 10^3 J/S

1 J = 1 W . sec. = 10^7 ergs = 9.5 x 10^{-4} BTU = 0.2388 calorie = 1 N.m

1 ft lbf/s = 4.628 BTU/hr = 1.356 W = 0.01819 hp = 1.356 J/S

1 kcal/hr = 1.163 J/s, 1 MW = 10^6 W = 1000 kW, 1 N = 10^5 dyne, 1 therm/hr = 2.9308 x 10^4 J/S

1 MW day = 8.64 x 10^{10} J

1 W/m^2 = 0.3170 BTU/ft^2 hr = 1 J/S.m^2; 1 BTU/ft^2 hr = 3.155 W/m^2 = 3.155 J/S.m^2

1 kcal/hr.m^2 = 1.163 J/S.m^2, 1 KW/m^2 = 1000 J/S.m^2, 1 kcal/hr.ft^2 = 12.518 J/S.m^2

Power and Energy

[W—watt, kW—kilowatt, MJ—mega joule]

1 hp = 0.746 kW = 746 W = 550 lb ft/sec. = 632.34 kcal/hr = 2545 BTU/hr

1 W = 3.412 BTU/hr = 44.24 ft.lb/min = 0.2389 cal/sec = 0.7376 ft.lbf/s = 0.01341 hp

1 kW = 13.41 hp = 1000 W = 102 kg m/sec = 3412 BTU/hr = 737.6 ft.lbf/s;

1 BTU/min = 17.58 W, 1 MW = 1000 kW, 1 Watt = 3.412 BTU/hr

1 kWh = 3.6 x 10^6 J = 2.655 x 10^6 ft lb f = 859.8 kcal = 3412 BTU = 3.6 MJ = 0.03412

therm

Joule's equivalent = 778.3 ft.lb

SPECIFIC HEAT

1 cal/gm, °C = 1 kJ/kg. °C = 1 BTU/lb . °F = 1 BTU/lb . °R = 4.1868 kJ/kg . °K

ENTHALPY

1 cal/gm = 1 kcal/kg = 4.1868 kJ/kg, 1 BTU/lb = 2.326 kJ/kg

CALORIFIC VALUE

[N—at N.T.P. *i.e.* 0°C and 760 mm Hg, S = at S.T.P. *i.e.* 15°C and 760 mm Hg]

1 kcal/kg = 1.8 BTU/lb = 0.0403 therm/ton = 4.187 kJ/kg

1 BTU/lb = 0.555 kcal/kg = 0.0224 therm/ton = 2.320 kJ/kg

1 BTU/Nft^3 = 8.9 kcal/Nm^3 = 0.038 J/N cm^3 = 0.038 MJ/Nm^3

1 kcal/Nm^3 = 0.1124 BTU/N ft^3, 1 therm/U.K. gallon = 5543.3 kcal/litre
10000 kcal/litre = 1.804 therm/U.K. gallon

1 cal/m^3 = 0.11237 BTU/ft^3

1 J/kg = 4.29 x 10“3 BTU/lb ; 1 MJ/kg = 238 kcal/kg

1 J/cm^3 = 27 BTU/ft^3 ; 1 MJ/Nm^3 = 238 kcal/Nm^3

1 MJ/sm^3 (dry) = 26.81 BTU/sft^3 (dry); 1 MJ/sm^3 (saturated) = 26.79 BTU/sft^3 (saturated)

1 BTU/sft^3 (dry) = 0.0373 MJ/sm^3 (dry);

1 BTU/sft^3 (saturated) = 0.03732 MJ/sm^3 (saturated)

1 kg coal equivalent = 7000 kcal/kg = 2.9308 x 10^4 kJ

1 kg oil equivalent = 10180 kcal/kg = 4.2622 x 10^4 kJ

TEMPERATURE

Conversion of degree Farenhite (°F) to degree centigrade/celcius (°C) and *vice versa* is done by the following formula:

$$\frac{C}{5} = \frac{F-32}{9}$$

From where, it follows that:

$$°C = (°F - 32) \times \frac{5}{9} = (°F - 32) \times 0.555$$

$$°F = °C\ \frac{5}{9} + 32 = (°C + 17.778) \times 1.8$$

°A (degree absolute) = °K (kelvin) = 273 + °C

°F absolute = °R (Rankine) = 460 + °F

Temperature difference: $1°K = 1°C = \frac{9}{5}°F = \frac{9}{5}°R$

HEAT TRANSFER RATE

1 BTU/ft^2 .hr = 2.712 kcal/m^2 hr, 1 kcal/hr = 1.163 W

1 kcal/m^2 . hr = 0.3687 BTU/ft^2 hr = 1.163 W/m^2

1 BTU/in^2 . hr = 390.6 kcal/m^2 . hr ; 1 kcal/m^3 hr = 1.163 W/m^3

HEAT TRANSFER CO EFFICIENT

1 kcal/m^2 hr °C = 0.2048 BTU/ft^2 hr °F = 1.163W/m^2 °K =1.163 J/S.m^2. °K

1 BTU/ft^2 hr °F = 4.88 kcal/m^2 hr °C = 20.44 kJ/m^2 hr °C = 5.678 J/S.m^2 °K = 5.678 W/m^2 °C = 135.6 x 10^{-6} cal/cm^2 sec °C

1 W/m² °C = 0.1761 BTU/ft² .hr. °F = 23.88 x $1O^{-6}$ cal/cm² . sec. °C = 1 J/S.m² . °K

1 kcal/hr. ft² °C = 1.2518 x 10 J/S.m² . K

1 cal/S.cm² . °C = 4.1868 x 10^4 J/S.m² . °K

Radiation co-efficient:

1 kcal/m² .hr. °K = 1.163 W/m² . °K^4

1 cal/cm². sec. °C = 41868 W/m². °C = 7373 BTU/ft² .hr .°F

THERMAL CONDUCTIVITY

1 BTU/ft² .hr °F/in thickness = 0.124 kcal/m².hr. °C/m = 0.344 cal/cm sec °C = 6.93 W/min. °K = 1.4423 x $1O^{-1}$ J/S.m. °K

1 BTU in/ft² S °F = 519.2 W/m °C = 300 BTU/ft.hr.°F

1 BTU/ft² hr °F/ft = 1.488 kcal/m² hr °C/m = 1.7308 J/S.m. °K

1 kcal/m hr °C = 0.672 BTU/ft hr°F = 8.063 BTU/ft² hr °F/inch = 1.163 W/m °K = 1.163 J/S.m. °K

1 cal/cm sec °C = 360 kcal/m hr °C = 242 BTU/ft hr °F = 2902.9 BTU in/ft² hr °F = 4.1868 x 10^2 J/S.m. °K

1 W/m °C = 0.5778 BTU/ft hr °F = 1.928 BTU in/ft² s °F

1 W/cm °C = 85.985 kcal/m.hr. °C = 57.779 BTU/ft.hr. °F

1 BTU/ft hr°F = 1.731 W/m °C = 3.333 x $1O^{-3}$ BTU in/ft²s. °F

1 kcal/hr.ft. °C = 3.8156 J/S.m. °K

SI (INTERNATIONAL SYSTEM) UNIT

SI units have been introduced to standardise the units throughout the world. India has also adopted this unit but still M.K.S. Unit is in vogue here. The base unit in SI unit system is also metre, kilogram and second.

Some of the important *SI derived units* having special names and symbols are tabulated below.

Table - Names and Symbols of SI derived units

Quantity	Name	Symbol	Expression in terms of other SI units	Expression in terms of base units
Force	Newton	N	—	kg m/sec²
Pressure	Pascal	Pa	N/m²	kg m/sec²
Energy	Joule	J	N.m	m² kg/sec²
Absolute viscosity	Pascal second	PaS	—	kg/m. sec
Heat capacity	Joule/kelvin	J/K	—	m² kg/sec².°K

Quantity	Name	Symbol	Expression in terms of other SI units	Expression in terms of base units
Thermal conductivity	Watt per metre kelvin	W/m.°K	—	m kg/sec³.°K
Molar heat capacity	Joule per mole kelvin	J/mol. °K	—	m² kg/ sec²mol °K
Moment of force	Newton-metre	N.m	—	m² kg/s²
Heat flux density	Watt per square metre	W/m²	—	kg/s³
Surface tension	Newton per metre	N/m	—	kg/s²
Specific heat capacity	Joule per kilogram kelvin	J/kg.°K	—	m²/s².K
Energy density	Joule per cubic metre	J/m³	—	kg/m.s²
Specific energy	Joule per kilogram	J/kg	—	m²/s²
Currents density	Ampere per square metre	A/m	—	A/m²
Electric field strength	Volt per metre	V/m	—	m kg/s³ .A
Magnetic field strength	Ampere per metre	A/m	—	A/m
Molar energy	Joule/mole	J/mol	—	m² kg/s² . mol
Molar heat capacity	Joule per mole kelvin	J/mol.K	—	m² kg/s² mol.k
Angular velocity	radian per second	rad/s	—	rad/s
Angular acceleration	radian per second square	rad/s²	—	rad/s²

Abbreviations used for prefixes in SI units are tabulated below.

Table - Abbreviations used for prefixes in SI units

Prefix	Symbol	Factor	Prefix	Symbol	Factor
deca	da	10^1	deci	d	1^{-1}
hecto	h	10^2	centi	c	10^{-2}
kilo	k	10^3	milli	m	10^{-3}
mega	M	10^6	micro	P-	10^{-6}
giga	G	10^9	nano	n	10^{-9}
tera	T	10^{12}	pico	P	10^{-12}
femto	f	10^{-15}	atto	a	10^{-18}

Table-Solid Fuel to Furnace Oil Conversion Ratio

Calorific value of solid fuel, Kcal/kg	Kg. of solid fuel required to substitute 1 litre furnace oil
3000	4.5
3500	3.9
4000	3.4
4500	3.1
5000	2.7
5500	2.5
6000	2.3
6500	2.1

Table-Approximate Conversion of Various Fuels to KWh

S.No.	Fuels	Conversion factor for converting into KWh
1.	Natural gas	m^3 x 10.6 KWh ft^3 x 0.3 KWh Therms x 29.3 KWh
2.	LPG (Propane)	m^3 x 25 KWh
3.	Coal	Kg x 8.05 KWh
4.	Coke	Kg x 10 KWh
5.	Gas oil	Litres x 12.5 KWh
6.	Light fuel oil	Litres x 12.9 KWh
7.	Medium fuel oil	Litres x 13.1 KWh
8.	Heavy fuel oil	Litres x 13.3 KWh

Appendix 11.3: Main Conversions Used in Petroleum Industry

S. *No.*	Items	Conversion Factor
1.	Crude Oil	1 Metric Tonne = 7.33 Barrels = 1.165 Cubic Metres (Kilolitres) 1 Barrel =0.136 Tonnes = 0.159 Cubic Metres (Kilolitres) 1 Cubic Metre = 0.858 Tonnes = 6.289 Barrels 1 Million Tonne = 1.111 Billion Cubic Metres Natural Gas = 39.2 Billion Cubic Feet natural Gas = 0.805 Million tonnes LNG = 40.4 Trillion British Thermal Units

S. *No.*	Items	Conversion Factor
2.	Natural Gas	1 Billion Cubic Metre = 35.3 Billiom Cubic Feet Natural Gas = 0.90 Million tonnes Crude Oil = 0.73 Million Tonnes LNG = 36 Trillion British Thermal Units = 6.29 Million Barrels of Oil Equivalent
3.	LNG (Liquefied Natural Gas)	1 Million Tonne = 1.38 Billion Cubic Metres Natural Gas = 48.7 Billion Cubic Feet Natural Gas = 1.23 Million tonnes Crude Oil = 52 Trillion British thermal Units = 8.68 Million Barrels of Oil Equivalent
4.	CNG (Compressed Natural Gas)	1 Kilogram = 1.244 Standard Cubic Metres Natural Gas = 1.391 Litres of Petrol = 1.399 Litres of HSDO

Appendix 11.4: Conversion factors for petroleum products

S. No.	Products of petroleum refining	TOE / Tonne	Barrel / Tonne
1.	Refinery Gas	1.150	8.00
2.	Ethane	1.130	16.85
3.	LPG	1.130	11.60
4.	Aviation Gasoline	1.070	8.90
5.	Motor Gasoline	1.070	8.53
6.	Jet Gasoline	1.070	7.93
7.	Jet Kerosene	1.065	7.93
8.	Other Kerosene	1.045	7.74
9.	Naphtha	1.075	8.50
10.	Gas/Diesel Oil	1.035	7.46
11.	Heavy Fuel Oil	0.960	6.66
12.	Petroleum Coke	0.740	5.50
13.	White Spirit	0.960	7.00
14.	Lubricants	0.960	7.09
15.	Bitumen	0.960	6.08
16.	Paraffin Waxes	0.960	7.00
17.	Non specified products	0.960	7.00

Note: TOE: Tonnes of Oil Equivalent, 1 barrel = 160 litres

Appendix 11.5: Viscometer Conversion Factors

	Factor for Conversion to Required Viscosity							
Given Viscosity	Kinematic viscosity, centistokes	Engler, ^{0}E	Redwood I seconds	Redwood II, seconds	Saybolt Universal seconds	Saybolt Furol seconds	Standard Tar Viscometer 10 mm, sec.	Standard Tar Viscometer 4 mm, sec.
Kinematic viscosity, centistokes	1	0.131	4.08	0.408	4.57	0.457	0.0025	0.074
Engler, degree (from 6° E upwards)	7.64	1	31.2	3.12	34.9	3.49	0.0191	0.565
Redwood I, sec (from 90 sec upwards)	0.245	0.0321	1	0.100	1.12	0.112	0.000612	0.0181
Redwood II, sec (from 30 sec upwards)	2.45	0.321	10.0	1	11.2	1	0.00612	0.181
Saybolt Universal, sec (from 120 sec upwards)	0.219	0.0287	0.893	0.089	1	0.10	0.000547	0.0162
Saybolt Furol, sec (from 40 sec upwards)	2.19	0.287	8.93	0.893	10.0	1	0.00547	0.162
Standard Tar Viscometer 10 mm cup, sec (from 10 sec upwards)	400	52.4	1630	163	1830	183	1	29.6
Standard Tar Viscometer 4 mm cup, sec (from 100 sec upwards)	13.5	1.77	55.1	5.51	61.6	6.16	0.034	1

Appendix 11.6: Table for Viscosity Conversion

Centis-tokes (cst)	Redwood no. 1 seconds at			Saybolt universal seconds at			°Engler
	70°F or 21.1°C	140°F or 60°C	200°F or 93.3° C	100P°F or37.7°C	150°F or 65.5° C	210°F or 98.9°C	°E
4	35.3	35.9	36.5	39.1	39.2	39.4	1.32
5	37.9	38.5	38.9	42.3	42.4	42.6	1.40
6	40.5	41.1	41.5	45.5	45.6	45.8	1.48
7	43.2	43.7	44.1	48.7	48.8	49.0	1.57
8	46.0	46.3	46.9	52.0	52.1	52.4	1.65
9	48.8	49.1	49.7	55.4	55.5	55.8	1.74
10	51.7	52.0	52.6	58.8	58.9	59.2	1.84
12	57.9	58.1	58.7	65.9	66.0	66.4	2.05
14	64.4	64.6	65.2	73.4	73.5	73.9	2.25
16	71.0	71.4	72.2	81.1	81.5	81.7	2.44
18	77.9	78.5	79.3	89.2	89.4	89.8	2.66
20	85.0	85.8	86.9	97.5	97.7	98.2	2.88
22	92.4	93.3	94.5	106.0	106.2	106.7	3.10
24	99.9	100.2	100.6	114.6	114.8	115.4	3.32
26	107.5	108.6	110.0	123.3	123.5	124.2	3.57
28	115.3	116.5	118.0	132.1	132.4	133.0	3.82
30	123.1	124.4	126.0	140.9	141.2	141.9	4.08
32	131.0	132.3	134.1	149.7	150.0	150.7	4.34
34	138.9	140.2	142.2	158.7	159.0	159.8	4.59
36	146.9	148.2	150.3	167.7	168.0	168.9	4.86
38	155.0	156.2	158.3	176.3	176.7	177.9	5.10
40	163.0	164.3	166.7	185.7	186.1	187.0	5.56
45	183.1	184.5	187.5	208.4	208.8	209.9	6.00
50	203.3	204.7	208.3	231.4	231.8	233.0	6.58
60	243.5	245.3	250.0	277.4	277.9	279.3	7.89
70	283.9	286.0	291.7	323.4	324.0	325.7	9.21
80	323.9	326.6	333.4	369.6	370.3	372.2	10.5
90	364.4	367.4	375.0	415.8	416.8	418.7	11.8
100	404.9	408.2	416.7	462.0	462.9	465.2	13.2
For viscosities above 100 centistokes multiply by							
	4.049	4.082	4.167	4.620	4.629	4.652	0.152

Note: The redwood no. 2 scale is used to define the viscosity of very thick oil. Redwood no. 1 figure for thick oil is approximately obtained by multiplying the redwood no. 2 figure by a factor of 10.

Similarly, there is an alternative saybolt scale known as saybolt furol. Saybolt universal figures is approximately obtained by multiplying the saybolt furol figure by a factor of 10.

Appendix 11.7: Viscosity Conversion Nomogram

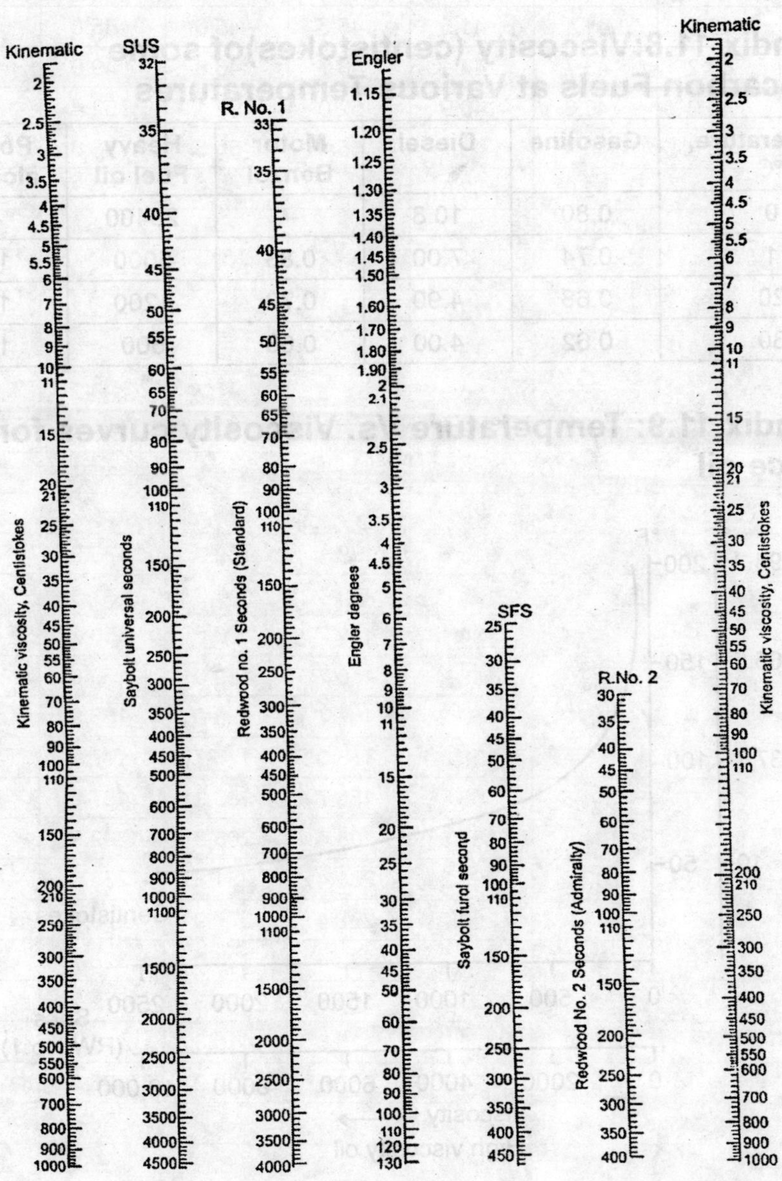

Note: To make use of it, line up the straight edge so that Centistokes value on both Kinematic scales is the same. Viscosities at the same temperature on all scales are then equivalent. To extend range of only the Kinematic, Saybolt Universal, Redwood No. 1, and Engler scales: Multiply by 10, the viscosities *on* these scales between 100 and 1000 centistokes on the Kinematic scale and the corresponding viscosities on the other three scales. For further extension, multiply these scales as above by 100 or a higher power of 10. (Example: 1500 centistokes = 150 x 10 CST = 695 x 10 SUS = 6950 SUS).

Appendix 11.8: Viscosity (centistokes) of some Hydrocarbon Fuels at Various Temperatures

Temperature, °C	Gasoline	Diesel	Motor Benzol	Heavy Fuel oil	Power alcohol
0	0.80	10.8	–	20400	–
1	0.74	7.00	0.83	5000	1.5
20	0.68	4.90	0.72	1200	1.2
30	0.62	4.00	0.66	500	1.0

Appendix 11.9: Temperature Vs. Viscosity curves for Furnace oil

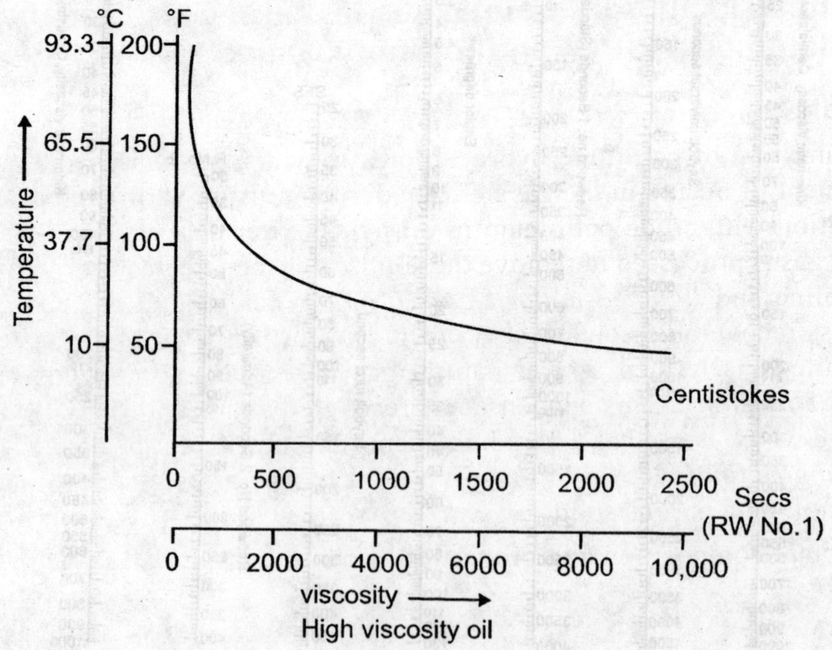

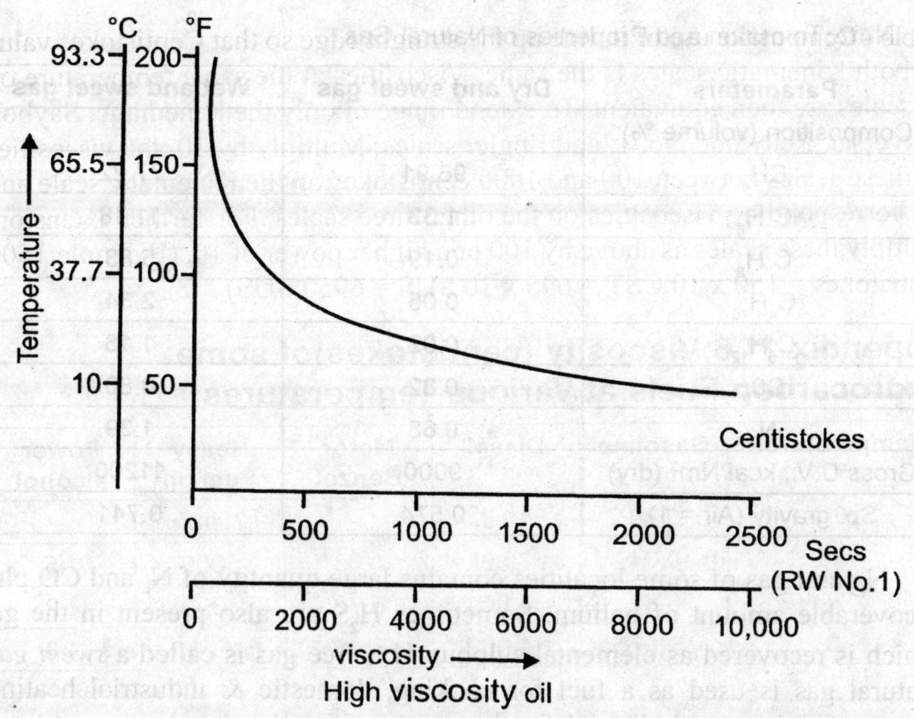

Fig. Temperature Vs. Viscosity curves for Furnace Oil

Appendix 11.10: All AboutGases- Natural Gas, Liquefied Petroleum Gas (LPG) and Refinery Gases

Natural Gas

It is a mixture of paraffinic hydrocarbons, in which methane is the main constituent. It occurs in the gas field (under the crust of earth) and also in association with crude petroleum in oil fields. After delivery from wells, natural gas is processed to remove the solids. It is then treated for recovery of gasoline and *liquified natural gas (LNG)*. When natural gas contains very less recoverable condensate (*<15 gm/m*^{3}*)*, it is known as *dry natural gas* while it is termed as wet natural gas if recoverable condensate is more (*>50 gm/ m*^{3}*)*. This condensate is recovered and is known as *natural gasoline* which is used as a blending agent for motor fuel. The condensate (*natural gasoline)* recovery from gas is done by compressing & cooling or adsorption on a porous solid.

Typical composition and properties of natural gas are given below in table.

Table - Composition and Properties of Natural Gas

Parameters	Dry and sweet gas	Wet and sweet gas
Composition (volume %):		
CH_4	96.91	77.22
C_2H_6	1.33	11.18
C_3H_8	0.19	5.83
C_4H_{10}	0.05	2.34
C_5H_{12}	0.02	1.18
CO_2	0.82	0.80
N_2	0.68	1.39
Gross C.V., kcal/Nm³ (dry)	9000	11200
Sp. gravity (Air = 1)	0.574	0.741

Natural gas of some localities contains large quantity of N_2 and CO plus recoverable amount of helium. Sometimes H_2S are also present in the gas which is recovered as elemental sulphur. H_2S free gas is called a *sweet gas.* Natural gas is used as a fuel for cooking, domestic & industrial heating, process furnace and boilers. It is also used for fertiliser production and as a source of carbon & hydrogen in chemical industries.

Uses of Natural Gas

Natural gas is used for:

- Enriching the gaseous fuels of low calorific value *e.g.* Lurgi gas.
- Internal combustion (I.C.) engine as its high anti knocking qualities permits use of high compression ratios possible resulting in improved engine efficiency.
- Gas turbines also use natural gas as a fuel.
- Producing carbon black by burning in a limited quantity of air in the presence of cooled metal surfaces. Carbon black is a pigment as well as a reinforcing agent for rubber.
- Making water gas as its methane content can be reacted with steam for this purpose.
- Making synthesis gas (by reaction with steam) which is further used for the production of methyl alcohol or gasoline (by Fischer Tropsch process) or hydrogen for ammonia synthesis.
- Partially converting into hydrogen, methane, benzene & other aromatic compounds by pyrolysis which are all valuable chemicals.

- The production of methyl alcohol & formaldehyde by controlled oxidation of methane present in it.
- Producing methyl chloride by the chlorination of methane present in it.

Natural gas is normally delivered to the consumer by a network of pipelines (dia 0.5–0.6 metres) from the oil field (after treatment) at 15-30 kg/cm^2 pressure. In cold climate, the gas must be dehydrated to avoid choking of pipelines owing to formation of crystalline hydrates of methane, ethane, propane & butane which are stable solids below **19°C.**

In USA, natural gas is widely used in domestic heating & cooking and a wide network gas pipeline has made it available to almost each & every home. In India natural gas is available from Gujarat, Assam and Bombay high oil wells. A 1700 km long pipeline (HBJ pipeline) running from Hazira to Jagdishpur *via* Bijapur supplies natural gas to fertiliser plants and thermal power stations en route.

Liquefied natural gas (LNG) is shipped also. Oil and Natural Gas Corporation Ltd. in India has set up a bottling plant at Uran (Near Bombay High) for filling LNG in cylinders.

Compressed Natural Gas (CNG)

Natural gas has a very low energy density ; hence to use it as an alternative fuel for a spark ignition (S.I)engine on board the vehicle/car, it has to be either *compressed* to a high pressure of 200 atm or cooled to –160°C to liquefy it for storage. Hence, it is used as compressed natural gas (CNG).

Advantages of CNG over petrol as a S.I. engine fuel:

- Octane number of CNG is 130, hence it does not need anti knock additive like TEL, which causes pollution.
- Its combustion is more regular and it rapidly disperses into atmosphere at normal pressure and temperature.
- It is a light weight gas with a high ignition temperature and is no more dangerous than petrol.
- Emission of carbon monoxide and unburnt hydrocarbons is comparatively less in exhaust.
- CNG is insoluble in engine oil, hence lubricating oil retains its properties for longer duration resulting in longer interval between oil changes.
- CNG fired engines have longer life, as it does not carbonise spark plugs or wash out lubricating oil films.

- It has a comparatively better cold starting characteristics, because unlike petrol, it does not first have to be vaporised and it does not condense on cold engine parts.

Disadvantages of CNG

- CNG storage tanks add weight because heavy cylinders are required for the storage of natural gas at high pressures.
- CNG storage tanks take up five times higher trunk space for comparable energy storage.
- Refuelling station cost of CNG is very high.
- Compared to petrol driven vehicles, CNG driven vehicles have shorter operating range.

Use of ***catalytic converter*** to reduce exhaust emissions like carbon monoxide, unburnt hydrocarbons and nitrogen oxides helps in controlling the pollution from vehicles. Catalytic converter fitted in the exhaust system of vehicle comprises of honeycomb ceramic or metallic disc coated with platinum, palladium and rhodium, which catalyses the conversion of carbon monoxide into carbon dioxide, hydrocarbons into carbon dioxide & water and nitrogen oxides into nitrogen. There are two types of catalytic converter namely *(i)* ***two way system*** employing platinum & palladium as a catalyst which remove only carbon monoxide & unburnt hydrocarbons leaving nitrogen oxides unchanged, and ***three way system*** employing platinum, palladium & rhodium, which remove all the three pollutants simultaneously.

Presence of lead in the petrol poisons the catalysts coated on the honeycomb disc of the catalytic converter thereby reducing its potency.

Liquified Petroleum Gas (LPG)

It is mainly a mixture of propane and butane used mainly as cooking gas in domestic ovens as well as for industrial heating in furnaces (*e.g.* heating of annealing furnaces). C_3(propane etc.) and C_4(butane etc.) hydrocarbons are easily liquified at room temperature with the application of very low pressure. Hence, propane & butane is liquified, stored and transported in light cylinders. Generally, a mixture of about 80% butane and 20% propane is used for filling in LPG cylinders (also called bottled gas). LPG is prepared from wet natural gas and refinery gases. Even pure butane & propane are also liquified and used as fuels.

LPG is a highly volatile liquid, which expands 247 times its volume as vapor (1litre LPG liquid = 247 litres of vapor). It is non toxic, does not support life and is heavier than air.

LPG have high calorific value, high specific gravity, high air requirement and low flame speeds. Since, LPG is odorless, hence odorants like mercaptans

(50 ppm) or sulphides are added to detect its leakage from cylinders by smell as mercaptans have very pungent smell. The lighter grade LPG commercially available contains 90% propane and 10% butane.

Properties of main components of LPG *i.e.* butane and propane are given below in table.

Table - Properties of main components of LPG i.e., butane and propane

	Properties	Unit	Butane	Propane
1.	Gross calorific value	Kcal/Nm^3	30680	23670
2.	Specific gravity of gas (Air = 1)	—	2	1.52
3.	Specific gravity of liquid	—	0.5844	0.5079
4.	Theoretical air requirement	Nm^3 air/Nm^3 of LPG	30.94	23.80
5.	Theoretical amount of flue gas generated:	Nm^3/Nm^3 of LPG		
	CO_2	–do–	3.9	3.1
	H_2O	–do–	4.9	4.1
	N_2	–do–	24.4	18.8
	Total	–do–	33.2	26.0
6.	Sulphur content	%	0.02	0.02
7.	Liquefaction pressure at room temperature	Kg/cm^2	3	10
8.	Limit of inflammability in air	% LPG vapor in vapor-air mixture	1.9–8.5	2.2–9.5
9.	Ignition temperature	°C	500	480
10.	Maximum flame temperature in air	°C	1925	1995
11.	Flash point	°C	–60	– 105
12.	Boiling point	°C	–0.5	–42
13.	Critical temperature	°C	–96.8	–152

LPG is a popular domestic fuel because of the fact that a large concentration of thermal energy is contained in a small portable cylinder. LPG is quite popular fuel for industrial heating also where alternative fuel is not available. There are more than 100 LPG bottling plant in the country. It is also used as a fuel in I.C. engine as well as for welding & cutting.

Propane and butane are also used for enriching the low C.V. gases in a town gas supply system abroad upto limits of permissible specific gravity. Butane is also diluted with air to get a suitable fuel with a controlled heat release rate. A mixture of 20% butane and 80% air has a C.V. of 6136 kcal/$Nm^3$and is outside the limit of inflammability of butane in air. In fact, the LPG used in domestic heating abroad is diluted by adding inert nitrogen or air to reduce its high calorific value so as to obtain a controlled heat release rate in domestic ovens which do not require very high temperature. The amount of air/inert gas added is such that it is well outside the limit of inflammability (explosion limit) of LPG which is about 1.8–8.5% of LPG in LPG/air mixture.

Density of liquid LPG, which is non-corrosive to steel, copper, copper alloys and aluminium is, half that of water. LPG is slightly toxic & anaesthetic, if inhaled in very high concentration. It has no lubricating property and is colorless in either liquid or vapour phase. It is used also for welding, cutting, soldering, brazing and portable blow lamps. Yield (weight %) of LPG is 1 – 2% in straight run distillation of crude oil; 5 - 10% in catalytic reforming of naphtha; 12% in catalytic cracking of vacuum gas oil; 3 – 14% in hydrocracking of vacuum gas oil and 5 – 10% in coking of residuum.

Propane and butane are also used in the production of petrochemicals. LPG being heavier than air has a tendency to settle down while leaking from cylinder. At times methane rich natural gas is also liquified and sold as bottled gas in cylinders. In that case, it is called *Liquified Natural Gas* (LNG).

For *industrial grade* LPG (80% butane + 20% propane):

- C.V. = 29275 kcal/Nm^3(11200 Kcal/Kg)
- Sp. gravity =1.9 (Air =1), Theoretical combustion air requirement = 29 $Nm^3/Nm^3$of LPG

And for *domestic grade* LPG (70% propane + 30% butane);

- C.V. = 25775 kcal/Nm^3
- Sp. gravity = 1.65 (Air = 1)
- Theoretical Combustion air requirement = 26 Nm^3/Nm^3 of LPG

Use of LPG as a S.I. engine fuel.

Use of LPG in a *spark ignition engine* (S.I)offers following *advantages:*

- LPG is readily miscible with air and can burn at leaner air/fuel ratios.
- Being resistant to pre ignition, it leaves less combustion chamber deposits.
- It has an excellent cold starting & warm up performance and gives lower emission of pollutants like unburnt hydrocarbons & carbon monoxide.

- Its anti-knocking properties are excellent provided large amount of unsaturates are not present in it.

However, LPG suffers from the following *disadvantages*:

- LPG has to be stored & transported in bulky tanks capable of withstanding very high pressure.
- Some of the unsaturated compounds present in LPG have a comparatively poor octane number and are prone to causing high speed knock.

The performance of LPG as a S.I. engine fuel with respect to exhaust emissions, power developed & economy depends upon its composition (C_3/C_4 ratio & the level of unsaturates).

Refinery Gases

Gases obtained during distillation, cracking and other processing of petroleum and its fractions which contain paraffins (*e.g.* methane, ethane, propane & butane), olefins (*e.g.* ethylene, propane & butane), hydrogen sulphide and hydrogen are called refinery gases. A possible range of volumetric composition and properties of refinery gases is given below in table .

Table - Composition and Properties of Refinery Gas

Propene + Propane	40-55%
Butene + Butane	upto 30%
Ethylene + Ethane	8-12%
Methane	8-10%
Hydrogen	6-8%
Hydrogen sulphide	6-8%
Calorific value, kcal/Nm^3 (dry)	20500
Specific gravity (Air = 1)	1.35

Normally, C_3 & C_4 hydrocarbons (*i.e.* propane, propene, butane, butene etc.) are separated from refinery gases by liquifaction and sold as LPG/bottled gas. Similarly, sulphur is recovered from hydrogen sulphide. Hydrogen and higher hydrocarbons are utilised in the production of chemicals, fertilizer& plastics. These may be used to enrich town gas also. Surplus refinery gas (after using as refinery fuel and for above uses) is burnt off in atmosphere.

The yield of refinery gases (in the form of LPG) during various petroleum refinery conversion processes is tabulated below.

Table - Yield of refinery gases in the form of LPG in various refinery processes

Refinery process	Feedstock	Main products	Yield of LPG (weight % on feed)
Visbreaking/ coking	Residuum, heavy gas oil	Coke	5—10
Catalytic reforming	Light virgin naphtha	Aromatics, gasoline blend stocks	5—10
Alkylation/ Polymerisation	Butane-unsaturated gases	Gasoline	10—15
Catalytic cracking	Gas oil	Ethylene, propylene (for petrochemicals)	15—20

Appendix 11.11: Typical Product Yield & Uses (disposition) of Various oil Refinery Processes

(a) Typical Yields & Dispositions (uses) of crude distillation unit (CDU)process are tabulated below:

Product of CDU process	Yield, Wt% of crude oil	Disposition/ uses of products
Light Ends	2.3	LPG
light Naphtha	6.3	Naphtha Hydrotreating
Medium Naphtha	14.4	Naphtha Hydrotreating
Heavy Naphtha	9.4	Distillate Hydrotreating
Kerosene	9.9	Distillate Hydrotreating
Atmospheric Gas Oil	15.1	Fluid Catalytic Cracking
Reduced Crude	42.6	Vacuum Distillation Unit

Note: CDU distill & separates valuable distillates (naphtha, kerosene, diesel) and atmospheric gas oil (AGO) from the crude oil feedstock .

Process Steps include:

- Preheat the crude feed utilizing recovered heat from the product streams
- Desalt and dehydrate the crude using electrostatic enhanced liquid/ liquid separation (desalter)
- Flash the crude in the atmospheric distillation column

- Utilize pump around cooling lopps to create internal liquid reflux
- Product & draws are from the top, sides, and bottom

(b) Typical Yields &Dispositions (uses) of vacuum distillation unit (VDU) is tabulated below:

Products OfVDU Process	Yield, wt% of Crude	Disposition/ uses of products
Light Ends	<1	LPG
Light VGO	17.6	Distillate Hydrotreating
Heavy VGO	12.7	Fluid Catalytic Cracking
Vacuum residue (Residuum)	12.3	Coking

Note: VDU process recovers valuable gas oils from reduced crude via vacuum distillation.

It Reduces the hydrocarbon partial pressure via vacuum and stripping steam.

Process steps include:

- Heat the reduced crude to the desired temperature using fired heaters
- Flash the reduced crude in the vacuum distillation column.
- Utilize pumparound cooling loops to create internal liquid reflux.
- Product & draws are from top, sides, and bottom.

(C) Typical Yields & Dispositions / uses of delayed & fluid coking processes are tabulated below:

Products	Yield, wt% of feed	Diposition/uses of products
Light ends	12.5 – 20	LPG
Naphtha	10 – 15	Naphtha Hydrotreating
Light Coker Gas Oil	18 – 24	Distillate Hydrotreating
Heavy Coker Gas Oil	30 – 40	Fluid Catalytic Cracking
Petroleam Coke	20 –35	Sponge – carbon anodes; Needle – graphite electrodes; Any coke – power generation

Note: Objective of **delayed coking process** is to convert low value residue to valuable products (naphtha & diesel) and coker gas oil. Thermocracking increases H/C ratio by carbon rejection in a semi-batch process.

Process steps include:

- Preheat residue feed, and provide primary condensing of reactor vapors by introducing the feed to the bottom of the main fractionators.

- Heat the coke drum feed by fired heaters.
- Flash superheated feed in a large coke drum where the coke remains and vapors leave the top and goes back to the fractionator.
- Off-line coke drum is drilled and the petroleum coke is removed via hydrojetting.

Objective of **fluid coking process** is to convert low value residue to valuable products (naphtha & diesel) and coker gas oil. Thermocracking increases H/C ratio by carbon rejection in a continuous process.

Process steps include:

- Preheat residue feed, scrub coke particles, and provide primary condensing of reactor vapors by introducing the feed to the scrubber.
- Residue atomized into a fluid coke bed and thermocracking occurs on the particle surface
- Coke particles leaving the reactor are steam stripped to remove remaining liquid hydrocarbons
- Sub-stoichiometric air is introduced to burner to burn some of the coke and provide the necessary heat for the reactor
- Reactor vapors leave the scrubber and go to the fractionator.

(d) Typical Yields & Dispositions/uses of fluid catalytic cracking (FCC) process is tabulated below:

PRODUCTS OF FCC Process	Yield, wt% of feed	Disposition/uses of products
Light Ends	16.5 – 22	LPG; Alkylation
Nephtha	44 –56	Naphtha Hydrotreating
Light Cycle Oil	13 – 20	Distillate Hydrotreating
Medium Cycle Oil	10 – 26	Hydrocracking
Slurry Oil	4 – 12	Heavy fuel oil; carbon black processing
Coke	5 – 6	Flue gas to CO boiler

Note: Objective of FCC process is to convert low value gas oils to valuable products (naphtha & diesel) and slurry oil.Catalytic caracking increases H/C ratio by carbon rejection in a continuous process.

Process steps include:

- Gas oil feed is dispersed into the bottom of the riser using steam .
- Thermal cracking occurs on the surface of the catalyst.
- Disengaging drum separates spent clatalyst from product vapors .

- Steam strips residue hydrocarbons from spent catalyst.
- Air burns away the carbon film from the catalyst in either a"partial-burn" or "full-burn"mode of operation•
- Regenerated catalyst enters bottom of riser-reactor.

(e) Typical Yields & Dispositions/ uses of catalytic reforming process is tabulated below:

Products of Catalytic Reforming	Yield, vol% feed	Disposition/uses of products
Light ends	5– 8	LPG
Reformate	84 –85	Gasoline; Petrochemical Plants
Hydrogen	650 – 1100 scf/bbl	Hydrotreating

Note: Objective of catalytic reforming process is to convert low octane naphtha into a high octane, reformate for gasoline blending and/or to provide aromatics (benzene, toluene, & xylene) for petrochemical plants. Reforming also produces high purity hydrogen for hydrotreating processes.

Reforming reactions occur in chloride promoted fixed catalyst beds or continuous catalyst regeneration beds where the catalyst is transferred from one stage to another, through a catalyst regenerator and back again. Desired reactions include: dehydrogenation of naphthenes to form aromatics isomerization of naphthenes; dehydrocyclization of paraffins to form aromatics and isomerization of paraffins. Hydrocracking of paraffins is undesirable due to increased light-ends make.

Process steos include:

- Naphtha feed and recycle hydrogen are mixed, heated and sent through successive reactor beds .
- Each pass requires heat input to drive the reactions.
- Final pass effluent is separated with the hydrogen being recycled or purged for hydrotreating.
- Reformate product can be further processed to separate aromatic components or be used for gasoline blending.

(f) Typical Yields & Dispositions/ uses of catalytic hydrocracking process is tabulated below:

Products of Hydrocarcking	Yield, vol% feed	Disposition
Light ends	Varies depending upon objectives	LPG

Products of Hydrocarcking	Yield, vol% feed	Disposition
naphtha	-do-	Gasoline; Catalytic Reformer
Diesel	-do-	Diesel
Total volume yield	130 – 140	–
Gas Oil Conversion	60 – 99%	–

Note: Objective of hydrocracking process is to remove feed contaminants (nitrogen, sulfur, metals) and to convert low value gas oils to valuable products (naphtha, middle distillates, & ultra-clean lube base stocks) .Hydrogenation occurs in fixed hydrotreating catalyst beds to improve H/C ratios and to remove sulfur, nitrogen, & metals. This is followed by one or more reactors with fixed hydrocracking catalyst beds to dealkylate aromatic rings, open naphthene rings, and hydrocrack paraffin chains.

Process steps include:

- Preheated feed is mixed with hot hydrogen and passes through a multi-bed reactor with inter stage hydrogen quenches for hydrotreating .
- Hydrotreated feed is mixed with additional hot hydrogen and passes through a multi-bed reactor with quenches for first pass hydrocracking.
- Reactor effluents are combined and pass through high & low pressure separators and are fed to the fractionator where valuable products are drawn from the top, sides, & bottom.
- Fractionator bottoms may be recycled to a second pass hydrocracker for additional conversion all the way up to full conversion.

Objective of hydroterating process is to remove contaminants (sulfur, nitrogen, metals) and saturate olefins & aromatics to produce a clean product for further processing or finished product sales.Hydrogenation occurs in a fixed catalyst bed to improve H/C ratios and to remove sulfur, nitrogen, & metals.

Process steps include:

- Feed is preheated using the reactor effluent.
- Hydrogen is combined with the feed and heated to the desired hydrotreating temperature using a fired heater.
- Feed and hydrogen pass downward in a hydrogenation reactor packed with various types of catalyst depending upon reactions desired.
- Reactor effluent is cooled and enter the high pressure separator which separates the liquid hydrocarbon from the hydrogen/hydrogen sulfide/ammonia gas.

- Acid gases are absorbed from the hydrogen in the amine absorber.
- Hydrogen minus purges, is recycled with make-up hydrogen .
- Further separation of LPG gases occurs in the low pressure separator prior to sending the hydrocarbon liquids fractionation.

Various hydrotreating processes include:

- **Naphtha Hydrotreating**:
  - ♦ Primary objective is to remove sulfur contaminant for downstream processes; typically < 1 ppmw.
- **Gasoline Hydrotreating:**
  - ♦ Sulfur removal from gasoline blending components to meet clean fuels specifications.
- **Mid-Distillate Hydrotreating**:
  - ♦ Sulfur removal from kerosene for home heating .
  - ♦ Convert kerosene to jet via mild aromatic saturation .
  - ♦ Remove sulfur from diesel for clean fuels.
  - ♦ Ultra-low sulfur diesel requirements are leading to major unit revamps .
- **FCC Feed Pretreating**:
  - ♦ Nitrogen removal for better FCC catalyst activity .
  - ♦ Sulfur removal for SO_X reduction in the flue gas and easier post FCC treatment.
  - ♦ Aromatic saturation improves FCC feed "crackability".
  - ♦ Improved H/C ratios increase FCC capacity & conversion

(g) Typical Yields & Dispositions/uses of HF(Hydrogen fluoride)alkylation process is tabulated below:

Product of HF Alkylation	Yield, vol% of olefinfeed	Disposition uses of products
Propane	20-30	LPG
N-Butane	40-52	LPG; Gasoline LPG; Gasoline
Alkylate	150-170	Gasoline
Acid Oils	<1	Furnace
Isobutane consumption	67-75	–

Note: Objective of HF alkylation process is to combine light olefins (propylene & butylene) with Isobutane to form a high octane gasoline (alkylate).Alkylation occurs in the presence of a highly acidic catalyst hydroflouric acid or sulfuric acid).

Process steps include:

- Olefins from FCC are combined with Iso-Butane and fed to the HF Reactor where alkylation occurs.
- Acid settler separates the free HF from the hydrocarbons and recycles the acid back to the reactor.
- A portion of the HF is regenerated to remove acid oils formed by feed contaminants or hydrocarbon polymerization.
- Hydrocarbons from settler go to the Deisobutanizer for fractionating the propane & isobutene from the n-butane and alkylate .
- Propane is then fractionated from the Isobutane propane as a product and the Isobutane to be recycled to the reactor.
- N-Butane and alkylate are deflourinated in a solid adsorbent and fractionated as separate products.

(h) Typical Yields & Dispositions os Isomerisation is process is tabulated below:

Product	–	Disposition uses of products
Hydrogen and Light ends	–	LPG, Fuel gas
Isomerate	–	Gasoline; iso-butane for Alkylation LPG; Gasoline
Conversion	Up to 97%	–

Note: Objective of isomerization process is to convert low octane n-paraffins to high octane iso paraffins.Isomerization occurs in a chloride promoted fixed bed reactor where n-paraffins are converted to iso-paraffins. The catalyst is sensitive to incoming contaminants (sulfur and water).

Process Step include:

- Desulfurrzed feed and hydrogen are dried in fixed beds of solid desiccant prior to mixing together .
- The mixed feed is heated and passes through a hydrogenation reactor to saturate olefins to paraffins & saturate benzene .
- The hydrogenation effluent Is cooled and passes through *an* isomerization reactor.
- The final effluent is cooled & separated as hydrogen & LPG which typically go to fuel gas, and isomerate product for gasoline blending .

Appendix 11.12: Properties of Conventional and Alternative Automotive Fuels

S.No,	Property	Petrol (Gasoline)	Diesel	Methyl Alcohol (Methanol)	Ethyl Alcohol (Ethanol)	Propane (LPG)	Methane (CNG)
1.	H/C ratio	1.9	1.88	4.0	3.0	2.7	4.0
2.	Engery content (LHV) (MJ/Kg)	44.0	42.5	20.0	26.	46.4	50.0
3.	Liquid density (kg/l)	0.72–0.78	0.84–0.88	0.792	90.785	0.51	0.422
4.	Liquid energy density (MJ/l)	3300	36.55	15.84	21.12	23.66	21.13
5,	Boiling point	37-205	140–360	65	79	42.15	161.6
6.	Research Octane Numbers(RON)	92-98	–25	106	107	112	120
7.	Motor Octane Numbers(MON)	80-90	–	92	89	97	120
8.	Cetane Numbers	0-5	45-55	5	5	o	0
9.	Stoichiometric air-fuel ratio	14.7	14.6	6.5	9.0	15.7	17.2
10.	Reid Vapor Pressure (psi)	8–15	0.2	4.6	2.3	208	2.400

Note: LHV - lower heating value: LPG = liquefied petroleum gas: CNG = compressed natural gas.

a. Energy density of propane at standard temperature and pressure: 0.093 MJ/I.

b. Energy density of methane at standard temperature and pressure: 0.036MJ/I ; at 200 bar pressure: 70.4MJ/I.

Appendix 11.13: Advantages and Disadvantages of Alcohol Motor Fuels Compared to petrol and Diesel

S.No.	Alcohol fuel Performance Parameter	Compared to gasoline and diesel fuels	Advantages/Disadvantages
I.	Flame temperature	Lower	Potentially lower NO emissions and lower heat losses
2.	Vapour pressure	Lower	Poor cold Starting and warm up performance, higher unburned fuel emissions during starting/ warm up phase
3.	PM emissions	Lower	Due to clean burning characteristics; PM emissions are even lower than the gasoline engines
4.	Air Toxic Emissions	Lower	Lower benzene and 1-3 butadiene
5.	CO and NO_X Emissions	Similar	No definite trend is observed. So. no advantage over petroleum fuels have been noted
6.	Aldehyde emissions	Higher	Formaldehyde and acetaldehyde emissions are higher
7	Nature of sources	Renewable especially of ethanol	Sources more widespread around the world, hence better energy security, Lower net CO emissions.

Note:

- Alcohols are a preferred alternative for SI engines only due to their high octane number. A higher engine compression ratio could be used to obtain a higher engine thermal efficiency.
- Cetane number of methanol & ethanol are close to 5 & 8, respectively.
- Direct injection alcohol engine prototypes operating similar to diesel/ CI engines have also been developed using either a positive source of ignition or high-dosage of ignition quality improvers (3 to 7% by volume).
- Alcohols are not easily miscible in the diesel fuels. To prepare alcohol-diesel blends high amounts of emulsifiers or solublizers are required.
- Heating value of ethanol is approximately 60 percent and that of methanol is only 45 % of gasoline. The stoichiometric air-fuel ratio

due to presence of oxygen in the molecule is much lower than the gasoline.

- The volumetric energy content of stoichiometric mixture(gaseous state) of alcohols, and gasoline however, are not very different. Thus, engine specific power output that may be obtained with alcohols and gasoline is nearly the same.
- The latent heat of vaporization of methanol & ethanol is nearly 4 & 2.7 times respectively compared to gasoline. Cold starting performance with neat alcohol is therefore, poor compared to gasoline.
- Flames of neat alcohols in air are not easily visible to the naked eye. Hence 15% gasoline is mixed to alcohol for making the flame visible in case of an accidental fire.

Lower flame temperature with alcohol may result in lower NO_x emissions. With neat alcohol as the emissions of most of the volatile organic compounds in the exhaust would consist of alcohol or aldehydes only the photo chemical reactivity and the ozone forming potential is lower with alcohol fuel operation.

Appendix 11.14: Properties of Biodiesel Derived from Some Vegetable Oils

S.No.	Properties of biodiesel	Rapeseed methyl ester	Jatropha methyl ester
1.	Molecular weight	300	293
2.	Hydrogen/carbon ratio, m/m	0.15	0.157
3.	Oxygen content, % m/m	9–11	10.9
4.	Relative density @ 15°C	0.882	0.88
5.	Kinematic viscosity @ 40˚C, mm²/s	4.57	4.4
6.	Cetane number	51.6	57.1
7.	Lower heat of combustion, MJ/kg	37.7	38.45
8.	Sulphur content, % m/m	<0.002	<0.020

Note: The Vegetable oil esters are practically free of sulphur and have a high cetane number ranging from 46 to 60 depending upon the feedstock. Due to presence of oxygen, biodiesels have a lower calorific value than the diesel fuels. The influence of biodiesel on emissions varies depending on the type of biodiesel (soybean, rapeseed, or animal fats) and on the type of conventional diesel to which the biodiesel is added due to differences in their chemical composition & properties.Use of biodiesel results in reduction of CO, HC &PM, but slight increase in NO_x emissions is obtained.

- Reduction in CO emissions is attributed to presence Of oxygenate in the fuel molecule.
- A slight increase in NO_x emissions results perhaps due to advancement of dynamic injection timing with biodiesel. The methyl esters have a lower compressibility, which results in advancement of dynamic injection timing with biodiesel compared to diesel.
- Lower SOF(Soluble Oil Fraction) with biodiesel and advanced injection timing also results in lower PM emissions.
- Volumetric fuel consumption with biodiesel is higher than diesel due to its lower heating value. An increase of 10-11 % in fuel consumption compared to diesel may be expected when comparing their heating values. An increase in volumetric fuel consumption by 0.9-2.1% with 20% blends has been obtained, more bioavailable than their precursors. The cancer potency of diesel vehicles is more than two times than that of petrol vehicles in India.

Appendix 11.15: All about alcohols - a blending component of automotive fuels petrol & diesel

Alcohols are the simple class of organic compounds containing hydroxyl (-OH) group attached to a carbon atom. This (OH) group is the functional group of alcohols and is called alcoholic group, e.g.,

(i) Methanol: CH_3 . OH; also called methyl alcohol

(ii) Ethanol: CH_3 CH_2OH; also called ethyl alcohol

Methanol and Ethanol are among the two commercially important alcohols. **Methanol** is commercially prepared by hydrogenation of carbon monoxide in presence of a metal oxide catalyst. It is also called wood alcohol as it can be obtained by destructive distillation of wood.

Ethanol is commercially prepared by fermentation process which is the process of slow decomposition of complex organic compounds into similar substances, in presence of enzymes. The term 'alcohol', unless specified otherwise, refers to ethanol or ethyl alcohol.In common language alcohol and spirit are same but technically spirit is obtained from distilling the alcohol.

Types of Ethanol/Ethyl alcohol/Spirit:

- **Rectified Spirit or Rectified Alcohol:** This is the constant boiling mixture of ethanol and water. It contains 94.8 to 95.5% of ethyl alcohol (by v/v) with water. A rectified spirit or rectified alcohol is highly concentrated ethanol (drinking alcohol) which has been purified by means of rectification (repeated distillation). It is used in mixed drinks, in the production of liquors, for medicinal purposes, and as a household solvent. Rectified spirits are illegal for non medical purposes in the country because of high ethanol content.

- **Neutral Spirit/Neutral Grain Spirit/Neutral Alcohol: They** are rectified spirit made from grain. However, rectified spirits are made from other kinds of plant materials, like sugar beets or potatoes also .
- **Extra Neutral Alcohol (ENA)** is the alcohol produced by taking extraordinary precaution during process of redistribution of rectified spirit after the addition of chemicals, and treatment by activated carbon. The spirit is made neutral. The term neutral refers to the spirit's lacking the flavor that would have been present if the mash ingredients were distilled to a lower level of alcoholic purity, and also lacking flavor added to it during distillation. Extra Neutral Alcohol is used as volatile carriers in Flavour & Fragrances, raw material of potable liquor for human consumption. It contains 95 to 96% ethanol by v/v.
- **Potable Alcohol/Potable Spirit:** Alcohol for human consumption is called potable spirit. It should conform to minimum quality standards laid down under the prevention of Food Adulteration Act (P.F.A.), 1954.
- **Absolute alcohol:** The rectified spirit is dehydrated to produce Absolute, Ethanol (Fuel Ethanol) suitable for blending with petrol and diesel. The rectified spirit or industrial alcohol is converted into absolute alcohol.
- **Power Alcohol:** When ethanol is used as an additive to motor fuels to act as a fuel for internal combustion of engines, it is called "'power alcohol" .Power alcohol is a mixture of petrol (75 to 95%) +ethyl alcohol + water. Fuel alcohol contains 99.5 to 99.90% alcohol by v/v.
- **Industrial Alcohol:** Industrial alcohol is distilled ethyl alcohol, produced and sold for other than beverage purposes. Pure-ethyl alcohol is used in laboratories and in industry. Many medicines, food products, flavorings and cosmetics can not be produced without it.
- **Pure Undenatured Ethanol:** Pure Ethanol is Ethyl Alcohol with no other additives or denaturants. Pure Alcohol (Ethanol) is also commonly referred to as Beverage Grade Alcohol.
- **Fusel oil:** Fusel oil is a mixture of higher boiling alcohol, such as butyl alcohol(C_4H_9OH) and amyl alcohol($C_5H_{11}OH$) and other substances. This is used industrially as solvents and in preparing artificial essence etc.

Properties of Alcohols

(i) Both methanol and ethanol are colorless liquids at room temperature. They are acidic in nature;

(ii) They mix with water in all proportions,

(iii) Ethanol is flammable and burns in air with a blue flame; and

(iv) Methanol is highly poisonous in nature. Ingestion of even small quantity of methanol can cause blindness and large quantities cause even death.

Uses of Alcohols

(i) Both methanol and ethanol are used as fuel: mainly as blending component of automotive fuels petrol & diesel.

(ii) Methanol is used as a solvent for fats, oils and gums. It is used chiefly for making formaldehyde.

(iii) Ethanol is used as a solvent for drugs, perfumes, dyes etc;

(iv) Both are used as an anti-freeze for automotive radiators;

(v) Methanol is used in industrial purposes, in chemical laboratory, as home cleaning products, skin lotions, as industrial solvents and commonly available as methylated spirit . The end product of methanol after it is digested by the human body is formaldehyde, which is responsible for "alcohol Poisoning". Methanol Poisoning leading to blindness has been known to occur on consuming even small amounts.

(v) Ethanol; This has been consumed by human beings for its intoxicating and mind altering effects. It is usually consumed in diluted concentrations of absolute (e.g., 100%) ethyl alcohol. It is also used as a reagent in industry.

Spurious alcohol/Denatured spirit: The commercial alcohol is made unfit for drinking by mixing in it some copper sulphate (to give colour) and pyridine(a foul smelling liquid). It is known as denaturation of alcohol. It is the methylated spirit, which is ethyl alcohol (ethanol), rendered unfit for drinking by adding poisonous substance like methanol, acetone or pyridine to prevent theft for human consumption. Biological oxidation of methanol and ethanol in the body produces the corresponding aldehyde.

Use of oxygenated fuels to reduce unhealthy levels of carbon monoxide necessitated ethanol as an oxygenated blend gasoline. In U.S.A alcohols are added to gasoline to increase octane no. and improve the emission quality of gasoline. In some areas ethanol is blended with gasoline to form an B 10 blend (10% ethanol and 90% gasoline), but it can be used in higher concentrations such as E85 or in its pure form well depending on the availability.

The production phases in Ethanol production are shown in the following diagram

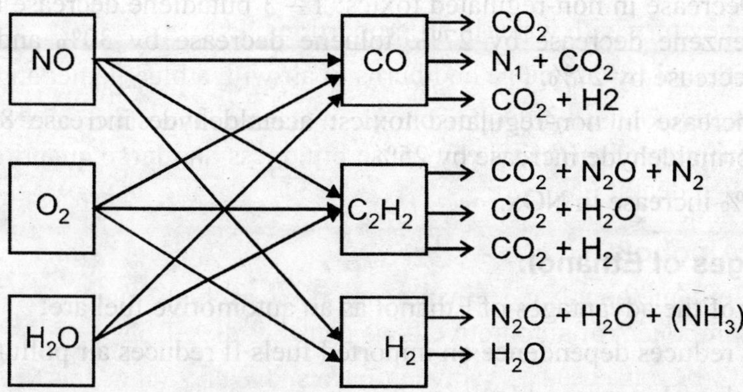

Ethanol production Process

Fuel ethanol is denatured with small amount (2–5%) of some product such as gasoline, to make it unfit for human consumption. Two main co-products of ethanol production are CO_2 and distillers grain. Many ethanol manufacturers collect the CO_2, clean it of any residual alcohol, compress it and sell it for use in carbonate beverages or to flash freeze meat.

Emissions Characteristics of ethanol (ethyl alcohol)

Emissions results of a test conducted by National Renewable Energy laboratory (NREL). USA are given in the following table. The comparative emissions from ethanol and gasoline fuelled vehicle are tabulated below.

Comparative emissions: Ethyl alcohol Vs. Gasoline (Petrol)

Emissions, gm/mile	Ethyl alcohol	Gasoline
NMHC	0.10	0.10
CO	1.48	1.13
NOx	0.12	0.09
CO_2	396.4	439.7

NMHC – Non methane hydrocarbons

Emissions of total potency weighted toxics (including benzene, 1–3, butadiene, formaldehyde, and acetaldehyde) for the E85 were 55% lower than that tested on gasoline. (E85 = 85% ethanol + 15% gasoline blend)

A recent Australian study with E1O gives the following emission results: (E10 = 10% ethanol + 90% gasoline)

- Decreased emissions of CO by 32%.
- Decreased emissions of HC by 12% .
- Decrease in non-regulated toxics: 1 – 3 butadiene decrease by 19%, benzene decrease by 27%, toluene decrease by 30% and xylene decrease by 27%.
- Increase in non-regulated toxics: acetaldehyde increase 80% and formaldehyde increase by 25%.
- 1% increase in NOx.

Advantages of Ethanol:

Some of the advantages of Ethanol as an automotive fuel are:

- It reduces dependence on imported fuels It reduces air pollution.
- Ethanol is renewable.
- Refueling is akin to that of gasoline or diesel .
- Is applicable for both light & heavy duty vehicles.
- More energy density compared to gasoline with optimized compression ratio.
- Maintenance assistance required is more or less identical to that of conventional fueled vehicles.

Disadvantages of Ethanol:

Some of the disadvantages of Ethanol are:

- Demands frequent refueling keeping the volume of the tank unaltered.
- Use of special compatible lubricants required .
- Ethanol, especially E85 requires special replacement parts.

Operation and Performance:

- High energy density compared to gasoline.
- No loss in power, acceleration and payload.
- Special lubricants are required.
- Special parts required.

Storage & Distribution:

Storage and distribution of ethanol is quite similar to that of gasoline and diesel.

Indian initiatives on Ethanol:

In the year 1980. IIP and IOC, R&D conducted a study with ethanol as the fuel in some 13-passenger cars including army vehicles. The test included city driving, highway driving and hill driving conditions.Some of the findings of this study are:

- Loss in volumetric fuel economy of 1% & 3.9% with E10 & E20 fuels respectively under city driving conditions and 3.5% & 4.3% under highway driving conditions.
- Improvement in fuel economy In Ambassador and standard cars under hill driving conditions ranged from 4% to 13% .
- Cold starting at ambient temperatures from 0°C to 30°C remained unaffected.
- Hot startability and driveabilitv demerits found higher with ethanol blends.
- No compatibility problems observed with metallic & non-metallic components.

Government of India has taken a decision to introduce petrol blended with 5% ethanol for use in motor vehicles all over the country in a phased manner. In the first phase, the 5% ethanol blended petrol will be introduced in the states of Andhra Pradesh, Gujarat, Haryana, Karnataka, Maharashtra, Punjab, Tamil Nadu and Uttar Pradesh. Rest of the States/Union Territories will be taken up in the second phase.

On the other hand, India is also working to blend 10% ethanol in petrol. Amendment in BIS specification to accommodate this 10% blend is being pursued.

METHANOL (Methyl alcohol)

Methanol (CH_3 OH) is an alcohol fuel. Methanol is also made into ether, MTBE, which is blended with gasoline to enhance octane number and to create oxygenated gasoline. Methanol contains no sulphur or complex organic species. (MTBE- Methyl Tertiary Butyl Ether)

Production of methyl alcohol:

Methanol can be produced from a veriety of feedstocks, including natural gas, coal, biomass and cellulose.It is predominantly produced by steam reforming of natural gas to create a synthesis gas, which is then fed into a reactor vessel in the presence of a catalyst to produce methanol and water vapour. Today's economics favor its production from natural gas.

Emissions

Methanol perhaps is not the cleanest gasoline alternatives but it has a distinct advantage in controlling ozone formation. USA is focused to methanol and methanol blends as it promises significant ozone improvements & control smog formation at a reasonable cost.

The following table gives emissions comparison between gasoline, M85 & (M100, M85= 85%methanol + 15% gasoline), (M100 = 100% Pure methanol).

Emissions form gasoline, M85 & M100 in a FTP Cycle

Emissions, mg/km FTP Cycle	Gasoline	M85	M100
THC	161.59	111.87	124.30
CO	733.37	683.65	870.11
NOx	490.99	379.12	285.89
Evaporative emission (mg/test) FTP test	1720.00	680.00	880.00
Benzene	7.79	4.38	0.32
Toluene	33.66	8.66	2.11
Buta 1-3-diene	.019-0.50	0.44	2.05
Formaldehyde	4.78	13.87	2.1.76
Acetaldehyde	0.94	10.02	0.27

THC - Total hydrocarbons

Advantages of Methanol:

Some of the advantages of methanol when used as a fuel are:

- Methanol has very low ozone forming potential.
- Emissions of sulphur and sulphur compounds are virtually negligible.
- Very low evaporative emissions due to its low vapor pressure.
- Easy refueling.
- Methanol is the most practical carrier of hydrogen to run fuel cells.
- Methanol has high-octane quality .

Disadvantages of Methanol:

This disadvantages of methanol include the following:

- High formaldehyde emissions .
- Acute toxicity.
- Availability is much dependent on the availability of natural gas.
- Low energy content compared to gasoline.
- Demands special lubricants and spare parts.
- May be costly.

Operation & Performance with methanol:

- Because of low energy content, mileage will be slightly lower.

- Power, acceleration and payload are comparable to those of equivalent internal combustion engines. Methanol needs special lubricants.
- Compatible replacement parts are required.Methanol is mostly used in light-duty vehicles. More than 20, 000 M85 flexible fuel vehicles are in operation in USA.

Safety Issues:

Methanol may not give rise to too much safety concerns when used as automotive fuel but it is inherently toxic. Adequate training is required to operate and maintain methanol-fueled vehicles. Moreover neat methanol (M-100) also presents a special safety hazard as it burns without a visible flame and even alcohol water wastes may be flammable.

Storage & Distribution:

Toxicity' & solubility of methanol in water raises concern for safe storage and distribution. Adequate training is required to store, maintain & operate methanol fueled vehicles though it has the refueling advantages like gasoline or diesel.

Indian Initiative on Methanol:

Indian Institute of Petroleum (IIP) conducted a fleet trial with M12 gasoline blend. IIP also carried out some experimental studies on the two wheelers with methanol- gasoline blends upto 20%(M12 = 12% methanol+88% petrol)

Findings of the studies are summarized below:

- Operation beyond or over 15% methanol - gasoline blends was erratic and the engine started misfiring and hunting.
- With M15 it showed marginally better output and 3-4% improvement in fuel consumption .
- It established that upto 15% blends can be used without any engine modifications. Substantial reduction in carbon monoxide was recorded.
- No hot or cold driveability problems with M12 blends was found .
- Engine performance of M12 vehicles was found comparable to that of gasoline vehicles.
- Increase of wear with cast iron rings were observed.

In the year 1992, BIS standard was amended to facilitate the use of methanol in gasoline. Indian oil companies initiated a pilot scale project in November 1993 to market a methanol-petrol blend of 3% methanol called Petrol-M. This product was supplied from 10 selected retail outlets in the city of Baroda in Gujarat. Initially this project was taken up for a period of one year. The key findings of this project are:

- Blending, transportation and quality wise Petrol-M trial marketing was successful.
- A total quantity of 376 kl of the product was sold.
- It was comfortably used by cars, two-wheelers and three-wheelers.
- To tackle the apprehended problem of corrosive effect of methanol on engine parts, the oil industry used corrosion inhibitors.

CPCB,India conducted a study to evaluate the emission performance of methanol gasoline blends . CPCB estimated that if all the petrol driven vehicels in Delhi use a methanol-gasoline blend of 3% methanol and 97% gasoline, It may be possible to have 11% reduction in hydrocarbons emissions, 7% CO reduction and 30% NOx reduction compared to pure gasoline driven vehicles.

Gasoline + alcohol blends are called 'gasohol' in Brazil which uses a blend of 30% alcohol & 70% gasoline in their vehicles.

Use of ALCOHOL + DIESEL BLENDS Called"(DIESOHOL)

Ethanol-diesel blend projects are under trial in Brazil and Sweden. Unlike ethanol-gasoline blend, ethanol-diesel blend has some concerns regarding lubricity, reduced flash point and startability problems.

International Experience of use of diesohol

Several technologies are currently under trials in different parts of the world. An Australian organization has developed a ethanol-diesel emulsion agent which is under trial in Thailand, Chile, Malawi, Australia, Germany and Sweden. This diesohol technology claims successful blends up to 15% of ethanol in diesel and also claims that this emulsion, which allows use of hydrated ethanol with diesel, gives improvements in NOx, PM, (including $PM_{2.5}$), & hydrocarbons and also increases the thermal efficiency of the engine. Vehicles can use diesohol and diesel fuels interchangeably. The fire point of diesohol is higher than flash point and the magnitude of the difference depends upon the composition of the diesel fuel. Australian Government is 'currently in the process of developing fuel quality and operability standards for dieshol. There are still several issues that need to be resolved before diesohol can be introduced commercially.

Some of the important issues are:

- Vapor lock-use of ethanol changes the vapour lock characteristics of the fuel.
- Material compatibility of pump seals, timing belts and some nitrile rubber seals used- in the fuel injection systems of a vehicle is still a concern.
- Toxicity of the emulsifier.

- Dosage quantities of the emulsifier which may vary from base fuel to base fuel .
- Blend stability, especially at low temperature.
- Establishment of diesohol test standards.
- Stability under water addition.

Indian initiative on Diesohol:

Government of India has sanctioned money for R&D Studies on ethanol-diesel blends. MOP&NG (Ministry of Oil ,Petroleum & Natural Gas) is also working on to introduce 5% ethanol-diesel blend. For this purpose Indian oil Corporation has selected some vehicle models to carry out trial tests on these vehicles. IOC (R&D) has also recommended to carry out the tests trials with 5% ethanol-diesel blend which are to be supplied by IOC.

Sufficient field trials in Indian conditions need to be carried out and its benefits on emissions, material compatibility and drivability, etc. may be assessed before trying this fuel in India

There are some field trails data on methanol-diesel blends. IIP has developed a retrofit kit for duel-fuel operation of diesel vehicles with alcohols. IIP used the fumigation concept for this purpose and successfully demonstrated this system on MSRTC and DTC diesel buses under actual commercial passenger service. The methanol fuel consisted of 10% gasoline also; to impart flame luminosity from the safety angle as the methanol flame is almost invisible in sunlight.

The findings of the study are tabulated below:

Results of IIP fuel study on Methanol-Diesel Blend

Number of Buses	60 Numbers
Total Operation	42, 00, 000 kms
Diesel Replacement	15-20%
Fuel Consumption	Comparable
Energy Efficiency	Better
Smoke Reduction	25-40%
Drivability	Good-to-Better
Oil Consumption & Degradation	Comparable
Engine Wear	Lower-to-Com parable
Engine Deposits	Lower-to-Comparable
Sludge Deposits	Marginally Lower
Material Compatibility	Adequate, Rubber components partially hardened

IIP - Indian Institute of Petroleum, Dehradun

From the above work it was concluded that 15-20% substitution of diesel by alcohols is possible by a simple retrofit fumigation system. Even though more field trials are required on new generation vehicles to assess the technical feasibility of the fuel.

Appendix 11.16: Selected fuel properties for petrodiesel & Biodiesel

S.No.	Fuel Property	Petrodiesel	Biodiesel
1.	Lower Heating Value, Btu/gal	131, 295	117, 093
2.	Kinematic Viscosity, @40°C, cst	1.3-4.1	1.9-6.0
3.	Specific Gravity, kg/l @60°F	0.85	0.88
4.	Density, lb/gal @ 15° C	7.079	7.328
5.	Water, ppm by wt.	161	0.05% max:
6.	Carbon, wt%	87	77
7.	Hydrogen, wt%	1.3	12
8.	Oxygen by difference, wt%	0	11
9.	Sulphur, wt%	0.05 max	0.00-0.0024
10.	Boiling Point, °C	188-343	182-338
11.	Flash Point, °C	60-80	100-170
12.	Cloud Point, °C	-15 to 5	- 3 to 12
13.	Pour Point, °C	-35 to -15	-15 to 10:
14.	Cetane Number	40-55	48-65
15.	Stoichometric Air/Fuel Ratio	15	13.8

Toxicity of Biodiesel

Biodiesel is not toxic and there is no hazards anticipated from ingestion incidental to industrial exposure. The acute oral LD_{50} (lethal dose) is greater than 17.4 g/kg-body weight, which by comparison is far safer then even table salt. A 96-hr. lethal concentration of biodiesel for bluegills was greater than 1000 mg/T and this aquatic toxicity is deemed as insignificant. Other related effects of biodiesel are given below:

- Very mild human skin irritation. It is less than the irritation produced by 4% soap and water solution.
- It is biodegradable. It degrades at least 4 to 5 times faster then conventional diesel fuel.
- Biodiesel has a flash point of about 150°C well above that of conventional diesel fuel (70°C) .

- Spills of biodiesel can decolorize any painted surface if left for long.
- There is no tendency for the mutagenicity of exhaust gas to increase for a vehicle running on biodiesel (20% biodiesel + 80% diesel)

Storage of biodiesel

In general, the standard storage and handling procedures used for petroleum diesel may be used for biodiesel. It is preferable to store the fuel in clean, dry and dark environment. Biodiesel may gel at low temperatures and care, needs to be taken to avoid temperature extremes. Acceptable storage tank materials include mild steel, stainless steel, fluorinated polyethylene and fluorinated polypropylene. Biodiesel has a solvent effect, which releases the deposits accumulated on tanks and pipes, which previously have been used for diesel. These deposits can be expected to clog filters initially and precautions should be taken for this.

Materials Compatibility of biodiesel

Biodiesel over time will soften and degrade certain types of elastomers & natural rubber compounds.. Materials like bronze, brass, copper, lead, tin & zinc may oxidize the diesel or biodiesel fuels and create sediments. Moreover, lead solders and zinc linings should be avoided, as should copper pipes, brass regulators and copper fittings, It is desirable to change all components, which are not biodiesel compatible to aluminum or stainless steel. The effect of B20 (20% biodiesel +80% diesel) on vulnerable materials is diluted compared to higher blends. It may also be noted that most of the new generation vehicles can take biodiesel without any materials compatibility problems as they are already tuned to using low sulphur diesel, biodiesel etc .

Compatibility of some materials with biodiesel is given below:

Material Compatibility with Biodiesel Fuels

Material	Biodiesel Type	Effect compared to petrodiesel
Teflon	B100	Little change
Nylon 6/6	B100	Little Change
Nitrile	B100	Hardness reduced 20%
	B100	Swell increased 18%
Viton A401-C	B100	Little Change
Viton GFLT	B100	Little Change
Fluorosilicon	B100	Little change in hardness
	B100	Swell increased 7%
Fluoroethane	B100	Little change in hardness

Material	Biodiesel Type	Effect compared to petrodiesel
	B100	Swell increased 6%
Polypropylene Polyvinyl chloride	BI00	Hardness reduced 10%
	B100	Swell increased **15%**
	B100	Much worse
	B50	Worse
	B40	Worse
	B30	Worse
	B20	Comparable
	B10	Comparable
Tygon	B100	Worse

B100 - 100% pure biodiesel

B50 - 50% biodiesel + 50% diesel

B40 - 40% biodiesel + 60% diesel and so on

Solvency of Biodiesel

Biodiesel is a mild solvent. On prolonged contact with painted surfaces, if may deface some paints. Always wipe lip spills and dispose of rags in a safe manner. Biodiesel soaked rags may self-combust if not banded properly. The most commonly encountered problem with solvency is biodiesel's tendency to "clean out" storage tanks, including the vehicle fuel tanks and systems. Some type of diesel tends to form sediments that stick to and accumulate in storage tanks, forming layers of sludge or slime in the fuel systems. The older the system, and poorer the maintenance, the thicker the accumulated sediments become. Biodiesel which dissolve these sediments carry the dissolved solids into the fuel systems of the vehicles. Fuel filters will catch most of it, but in severe cases, the dissolved sediments have caused fuel injector failure. The solvency effect of the biodiesel in B20 is sufficiently diluted so that most problems encountered are minor and in general the problem goes away after the first few tanks of fuel.

Appendix 11.17: Comparative percentage contribution of pollution load from petrol and diesel vehicles

Type of vehicle	CO	NOx	SO_2	HC	PM	Pb
Petrol vehicle	67	5	4	84	–	100
Diesel vehicle	33	95	96	16	100	-
Total = 100%						

PM-particulate matter, HC-Hydrocarbons

Appendix 11.18: All about some gaseous alternate fuels (eg, CNG/LNG, LPG and DME) for automobiles

Compressed Natural Gas (CNG)/ Liquefied Natural Gas (LNG)

Natural gas is a mixture of hydrocarbons-mainly methane (CH_4) and is produced either from gas wells or in conjuction with crude oil production. Due to its low energy density for use as vehicular fuel, it is compressed to a pressure of 200-250 bars to facilitate storage in cylinders mounted in vehicle and so it is called **Compressed natural gas (CNG).**

India's recoverable resources of more than 800 billion cubic meters make it long term substituted fuel for use in petrol & diesel engines. Low exhaust emissions, low noises, less maintenance, not prone to adulteration, driver's comfort, etc, are some of the attractive features of CNG as an automotive fuel.

Natural gas is produced from reservoirs or wells that contain natural gas, liquids and other materials. Processing is required to separate the gas from petroleum liquids and to remove contaminants. First the gas is separated from free liquids such as crude oil, hydrocarbon condensate, water and entrained solids. The separated gas is further processed to meet specified requirements, Natural gas when stored and distributed in the liquid phase is called Liquefied Natural Gas (LNG). LNG essentially facilities storage and transportation of natural gas. This LNG is again reheated to CNG and filled on board a vehicle for use as fuel. Vehicles running on LNG are also available in many parts of the world.

CNG qualifies to be one of the most prominent alternative fuels. It stands substantially than conventional fuels both in life cycle and vehicle exhaust emissions.

Advantages of CNG as an automotive fuel:

- No visible tail pipe emissions.
- Eliminates sulphur and lead from the external emissions.
- Reduction in CO, NOx and Particulate emissions. Significant reduction in benzene and other toxic emissions.
- Higher octane value of CNG reduces knocking problems of a vehicle.
- Reduces noise from running vehicles.
- CNG cannot be adulterated.
- Reduced noise in operation.

Disadvantages of CNG

CNG is now established as a very successful alternate fuel for automobiles throughout the world. The disadvantage of this fuel. If any is easily overruled by advantages associated with the fuel. Nevertheless, infrastructure, on board storage and issues on safety need proper attention for this fuel.

Natural gas is neither corrosive nor toxic, its ignition temperature is high, it is lighter than air, and it has a narrow flammability range, making it an inherently safe fuel compared to other fuel sources. Natural gas cannot contaminate soil water. It will always rise to the atmosphere out of doors, unlike other fuels, which are heavier than air can pool, either as a liquid or a vapor, upon the ground. Natural gas contains a distinctive odorant (mercaptan), which allows natural gas to be detected at 0.5% concentration that can support combustion. Due to high ignition temperature of natural gas (540 °C) s, imple exposure to a hot surface (such as exhaust manifold) is unlikely to lead a fire.

India too has initiated to introduce CNG as automobile fuel. Large number of buses, autos and taxis in Delhi have been switched over to CNG. Problems of infrastructure and dispensing stations are getting back to normal. CNG vehicles are also present in Mumbai and Baroda. Acknowledging the impact of CNG on the air quality of Delhi other cities of India are also taking up plans for introducing CNG in the phased manner.

Liquified Petroleum Gas (LPG)

LGP is a by-product of natural gas processing or a product that comes from crude oil refining and is composed primarily of propane & butane with smaller amounts of propylene & butylenes. Since LPG is largely propane, the characteristics of propane sometimes are taken as a close approximation to those of LPG.

Typical composition of LPG and CNG is tabulated below:

% composition	CNG	LPG
Methane	84.50	–
Ethane	7.70	0.2
Propane	2.40	57.3
Butane	0.58	41.1
Pentane	0.37	1.4

LPG is a by-product of two sources; natural gas processing and crude oil refining. When natural gas is produced, it contains methane and other light hydrocarbons that are separated in gas processing plant. Because propane boils at –42 °C and ethane boils at –88 °C, separation from methane is accomplished by combining increasing pressure & decreasing temperature. The natural gas liquid components, recovered during processing include ethane, propane, butane and other haeavier hydrocarbons.Propane and butane along with other gases are also produced during crude refining as by-productof the process that rearrange or break down the molecular structure to obtain more desirable petroleum compounds.

The main constituent of LPG is propane. Lower carbon-to-hydrogen ratio, higher octane rating and its ability to form a homogeneous mixture inside the combustion chamber enable to produce lesser emissions compared to conventional fuels. Non-regulated emissions like aldehydes and poly aromatic hydrocarbons(PAHs) are much lower from CNG vehicles compared to that of diesel vehicles.

Some of the advantages of LPG as automotive fuel are:

- Reduction in emissions.
- Very less carbon build up increase life to engine parts like spark plugs.
- Fueling is akin to that of conventional diesel or gasoline and time needed is also similar.
- Service life of vehicle increases with LPG.
- Little or no damage to soil water, if it is spilled, due to its rapid evaporation.
- Higher octane number.

Disadvantages of LPG:

Many countries in the world are using LPG as automotive fuel. Many years of experience of successful use of LPG in vehicles and the emission benefits it offers really support it to outperform the associated disadvantages of the fuel, if any. Moreover, familiarity with LPG in household application like cooking makes it more acceptable to the users. However, in the initial stages of introduction, of this fuel, issues like safety, storage & handling, extreme volatility of the fuel, etc. needs proper attention.

Safety is an issue with LPG, however, standards for application, storage and distribution already exists. The hazard associated with on board storage should be similar to that of gasoline, especially if reinforced cylinders similar to that of CNG tanks are used. LPG fuel systems have many built in safety features and they generally maintain their integrity in severe collisions & do not permit massive leaks. LPG is about twice as heavier than air and unlike CNG does not disperse easily in air. Its flammability limits (2.1-9.5 vol %) & auto ignition temperature (450 ^{0}C) are also lower than natural gas and as such should not be handled carelessly. LPG or propane is as safe as gasoline, if handled carefully.

In India, initiative for evaluating the performance of LPG as automotive fuel was carried out by IIP, Dehradun. Manufacturer in India already acquired the technology for producing LPG version of their products, especially two wheelers but huge demand of LPG for domestic cooking seems to discourage its lunch in automobiles sector in a big way.

Government of India has already passed legislation enabling use of LPG as automotive fuel.

DIMETYL ETHER (DME)

Dimethyl Ether (DME) is an ether compound having a molecular structure in which one oxygen atom and two methyl groups are bonded. It is chemically synthesized from natural gas, coal- associated etc. DME is a gas looking colourless and smelling etheric at room temperature & under normal pressure like LPG. DME is liquefied by applying a pressure of approximately 0.5Mpa. A DME engine, although having an equal energy efficiency level to that of a diesel engine, emits extremely low NOx & particulate matters. DME bus easily meets the Euro-IV diesel standards. A DME vehicle outperforms diesel vehicles for regulated emissions like NOx, total hydrocarbons and particulate matters. Peak acceleration smoke of a DME vehicle is almost nil and it also performs better with regard to combustion noise when compared to that a diesel vehicle is almost nil and also performs better with to achieve severe emission reduction targets of as low as 50% of the proposed Euro-IV levels. Fuels economy of a DME vehicle is also found to be comparable to that of conventional diesel vehicle. Volvo is working on to further reduce emissions from their DME buses, which at present easily meets equivalent Euro-IV standards.

Advantage of DME:

The greatest advantage of DME when used as an automobile fuel is that its high cetane number makes it applicable to a diesel engine and being an oxygen containing fuel, it prevents a diesel engine from emitting soot & particulate matter to a greater extent than diesel fuel does.

Disadvantage of DME:

- DME has viscosity lower than that of diesel.
- Lubricity is also low causing wearing of engine parts.
- It is found to react with some rubber components

DME Vehicles are under developmental stage in many countries of the world. Several developmental projects are being promoted for commercializing this fuel.

In the short term, synthetic diesel fuel like DME is not commercially very viable but it holds promise for the future. Some projects on DME are also going on in Japan.

Comparison of CNG/LPG/DME:

Compressed Natural Gas (CNG), Liquified Petroleum Gas (LPG) and Dimethyl Ether (DME) are the three prominent gaseous alternative to conventional fuels. CNG and LPG have already penetrated as automotive fuels in many parts of the world and DME is also fast emerging as a promising alternative.

Indian Initiative on DME:

India's Ministry of Petroleum & Natural Gas invited British Petroleum (BP) in 1997 to develop DME supplies in an exclusive partnership with India's leading energy marketers and research organizations. In 1998, a collaboration agreement was signed between BP, TOCL, GAIL and LLP for development, production & marketing of DME as a multi-purpose fuel for India. DME will be produced through Gas-to-Liquid (GTL) technology and for this gas will be sourced from the Middle East. Intention of this project was to establish DME as a multi-purpose fuel for India but in the initial stages the project identified some southern Indian states to supply DME as a feedstock for power plants.

Appendix 11.19:Automobile Fuel (Gasoline & Diesel) Aduteration & Potential Adulterants

Blending or mixing of adulterants into the base transport fuels exists in various forms and both the type and quantiy of adulterants vary from place to place. Moreover, profitability, availability & blendability are the prominent factors governing the choice of adulterants.

Specific types of adulteration may be broadly classified as follows:

- Blending relatively small amount of distillate fuels like diesel or kerosene into automotive gasoline.
- Blending variable amount (as much as 30%) of the gasoline boiling range hydrocarbons such as industrial solvents into automotive gasoline.
- Blending small amounts of spent waste industrial solvent such as used lubricants, which would be costly to dispose of in an environmentally approved manner-into gasoline and diesel.
- Blending kerosene into diesel, often as much as 20-30 percent.
- Blending small amount of heavier fuel oils into diesel.

There are several petroleum products in our country or abroad, which are close substitutes of petrol (Motor Spirit) & high-speed diesel (HSD), and are available at considerable lower prices. The consequences is that these products are widely used as adulterants.

Adulterants available in Indian markets are listed in Table below:

Potential Adulterants for Gasoline(Petrol):

S.N	Solvent/Chemicals used as petrol adulterants
1	Naphtha
2	Solvents
3	Pentane
4	Hexane
5	Resol
6	Raffinate
7	Kerosene
8	Oxygenates
9	Benzene
10	Toluene

Potential Adulterants for High Speed Diesel (HSD):

S.N	Solvent/Chemicals used as diesel adulterants
1	Aromex
2	Iomex
3	C9 Raffinate
4	Kerosene

Financial incentives arising from differential taxes are generally the primary cause of fuel adulteration. In South Asia, gasoline carries a much higher tax than diesel, which in turn is taxed more than kerosene. Industrial solvents and recycled lubricants are other materials with little or no tax. Adulteration of gasoline & diesel is indulged primarily due to the significant price difference between these products and the adulterant.

A study concluded that more than 30% of Kerosene distribution intended for household consumption through PDS outlets flowed back as adulterants of automotive fuel in India.

Fuel adulteration causes an increased tailpipe emissions of harmful & sometimes carcinogenic pollutants. While indirectly in the form of diversion of PDS kerosene to the diesel sector for adulteration, thus prompting the use of biomass as domestic fuel which in turn leads to health problems of various types due to indoor air pollution. However, all forms of adulteration are not harmful to public health. Some adulterants increase emission of harmful pollutants significantly, whereas others have little or no effect on air quality.

Adulteration results in significant loss of tax revenues to the government. The nation is losing at least 20, 000 crores annually directly as a result of adulteration of fuel, besides social costs as a result of environmental pollution and damage to vehicles & engines.

Fuel adulteration causes marked effect jn the tailpipe emissions of vehicles, as adulterants alter the chemistry of the base fuel rendering its quality inferior to the required commensurate fuel quality for the vehicles. This in turn affects the combustion dynamics inside the combustion chamber of vehicles increasing the emissions of harmful pollutants significantly. In some cases, the effects of adulteration are indirect-for example, large scale diversion of rationed kerosene subsidized for household use to the diesel sector for mixing with diesel not only hamper engine performance of diesel vehicles, but also deprives the poor of kerosene which can otherwise be used for cooking and as a consequence of lack of availability of subsidized kerosene force the poor to continue to use biomass which expose them to high levels of indoor pollution.

In general fuel adulteration can increases the tailpipe emissions of hydrocarbons (HC), carbon monoxide (CO), Oxides of nitrogen (NOx) and particulate matter (PM). Adulteration of fuels can also cause emissions of air toxins like benzene and polyaromatic hydrocarbons (PAHs); both well-known carcinogens.

Adulterating gasoline with kerosene causes increase in emissions, as kerosene is more difficult to burn than gasoline and this results in higher levels of HC, CO & PM(Particulate Matter). High sulphur contents of the kerosene can deactivate the catalyst and lower conversion of engine out pollutants. Kerosene addition may also cause fall in octane quality, which can lead to engine knocking. When gasoline is adulterated with diesel fuels, the same effects occurs but usually at lower levels of added diesel fuel. Both diesel and kerosene added to gasoline will increase engine deposit formation.

Gasoline may also be adulterated with gasoline boiling range solvent like toluene, xylene and other aromatics. With the 'judicious" adulteration, the gasoline would not exhibit drivability problems in motor vehicles. Larger amounts of toluene and /or mixed with xylene cause some increase in HC, CO, NOx emissions, and significant increase in the level of air toxins, especially benzene in the tailpipe exhaust. The adulterated gasoline itself could have increased potential human toxicity, if frequent skin contact is allowed.

Extremely high levels of toluene (>45 %) could cause premature failure of neoprene, styrene butadiene rubber and butyl rubber components in the fuel system. This has caused vehicle fires in some cases, especially in older vehicle.

Adulteration of gasoline by waste industrial solvents is especially problematic as the adulterants are so varied in composition. They will cause increased emissions, may even cause vehicle breakdown. Even low levels of these adulterants can be injurious and costly to vehicle operation.

For gasoline, any adulterant that changes its volatility can effect drivability. High volatility (resulting from addition of light hydrocarbons) in hot weathers can cause vapour lock and stalling. Low volatility in cold weather can cause starting problems and poor warm-up.

The blending of kerosene with automotive diesel is generally practiced by oil industry worldwide as a means of adjusting the low temperature operability of the fuel. This practice is not harmful or detrimental to tailpipe emissions, provided the resulting fuel continues to meet engine manufacturer's specifications (especially for viscosity & cetane number). However, high level adulteration of low sulphur diesel fuel with higher level sulphur kerosene can cause the fuel to exceed the sulphur maximum. The addition of heavier fuel oils to diesel is usually easy to detect because the resultant fuel will be darker than normal. Depending on the nature of these heavier fuel oils & the possible presence of additional PAHs, there could be some increase in both exhaust PM & PAHs emissions.

Adulteration of transport fuels at the point of sale and transportation has become a routine problem in India. There are several petroleum products available in our country, which are close substitute of gasoline and diesel, but are available at considerable lower prices. The price differential is usually in the range of Rs. 15 to 30 per litre in case of petrol and Rs. 10 to 15 per litre in case of diesel. Since kerosene is usually considered as poor man's fuel, Govt. of India has been subsidizing it for public distribution for several years. Significant portion of this subsidized kerosene is being diverted for adulterating diesel.

Tata Consultancy study has concluded that more than 30% of Kerosene distribution intended for household consumption flows back in industry in one form or the other.

In India, Naphtha is a commonly used adulterant for gasoline. The modus operandi id to import the product in guge quantity and divert it for adulteration.

Foreign car manufacturers set up manufacturing facilities in India are concerned about the quality of petrol that gets supplied to the users cars and they have found that fuel supplies are heavily adulterated and particurlaly the octane number of petrol is much lower than the specification value of 87.

A number of analytical techniques are available to detect adulteration. Some of the approaches for detecting adulteration are outlines below:

- **Full fuel specification tests of the standards:** This may be quite time consuming and many parameters, even if the fuel is adulterated. In fact fuel standards or specifications are framed to ensure that the fuel corresponds to certain level of quality commensurate to technology requirements of the vehicles. Parameters in the fuel standards may not necessarily stand as check point for any sort of fuel adulteration.
- **Testing selected parameters:** This asks for testing some critical parameters, which are likely to be affected or altered by adulteration and adversely affect engine performance & emissions and can be evaluated. Lists of such parameters are given below. In general, many of the selected parameters may already be included in the fuel specification standards. However, dosage of multifunctional additive and cetane number improver are intended for following the adulterants by dilution.

Selected Parameters for Gasoline Testing for detecting adulteration:

1. Density
2. Distillation

3. Hydrocarbon Composition:
   Aromatic, Vol%
   Olefins, Vol%
   Benzene, Vol%
   Sulphur, ppm
4. Stability:
   Existing gum, Potential gum
5. Octane Number:
   Research
   Motor
6. Multifunctional additives dosage

Selected Parameters for Diesel Testing for detecting adulteration:-

1. FlashPoint
2. Density
3. Distillation
4. Sulphur
5. Polycyclic aromatics (+2 rings)
6. Total sediment
7. Cetane number
8. Cetane index
9. Multifunctional additives dosage
10. Cetane improver presence

For diesel fuel without cetane improver, cetane index can be utilized. If dosage can be determined, it may be used for detection of adulteration based on depletion from original dosage.

Appendix 11.20: Use of Different types of alternative fuels for automobiles

Various types of alternative fuels used for automobiles are given below:

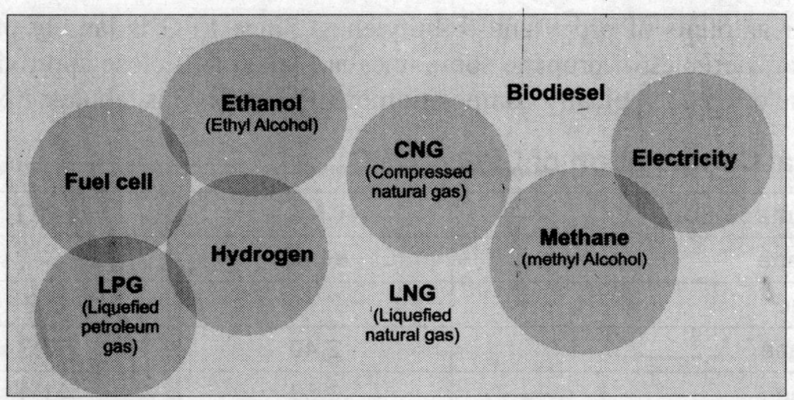

Liquefied petroleum gas (LPG)

- Widely used in light-duty vehicles in the world for many years, current population is about 20 million vehicles worldwide in use.
- LPG is mainly propane with a small amount of butane. It burns in I.C engines with little air pollution & solid residue. Besides, it has a high octane rating and does not dilute lubricants.
- It is a non-toxic, colorless & odorless gas.
- It is a clean-burning fuel:
- Produce fewer emissions than gasoline & diesel engine.
- There is 90% reduction in toxic emissions as compared to use of reformulated gasoline and it also reduces CO2 emission by 22–24% compared to gasoline.
- It gives longer service life & reduced maintenance costs.
- No cold starting problem.
- Engine performance is almost the same as gasoline.
- LPG is compatible with petrol and many vehicles can run on either fuel (dual-fuel vehicles); hence the limited distribution facility for LPG is not a problem.
- It has no spillage problem:
- "Closed" filling system, so hardly contributes to "filling pollution" which is a problem with both petrol & diesel.
- Cost: LPG vehicle is more expensive than an equivalent gasoline-powered vehicle (fuel cost per km run is cheaper than diesel & petrol).
- It is not renewable as it is petroleum based product and does not qualify for tackling climate change problem.

LPG is a by-product of natural gas processing or a product that comes from crude oil refining and is composed primarily of propane & butane with smaller amounts of propylene & butylenes. Since LPG is largely propane, the characteristics of propane sometimes are taken as a close approximation to those of LPG. Typically composition of LPG & CNG is tabulated below:

Typical Composition of LPG & CNG

% Composition	CNG	LPG
Methane	84.50	–
Ethane	7.70	0.2
Propane	2.40	57.3
Butane	0.58	41.1
Pentane	0.37	1.4

LPG is a by-product of two sources: natural gas processing and crude oil refining. When natural gas is produced, it contains methane and other light hydrocarbons that are separated in a gas processing plant. Because propane boils at –44 °F and ethane boils at –127 °F, separation from methane is accomplished by combining increasing pressure and decreasing temperature. The natural gas liquid components recovered during processing include ethane, propane & butane and other heavier hydrocarbons. Propane and butane along with other gases are also produced during crude refining as a by-product of the process that rearrange or break down the molecular structure to obtain more desirable petroleum compounds.

The main constituent of LPG is propane. Lower carbon-to-hydrogen ratio, higher octane rating and its ability to form a homogeneous mixture inside the combustion chamber enable it to produce lesser emissions compared to conventional fuels. LPG outperforms conventional fuels in both regulated and non-regulated emissions. Non-regulated components of emissions like aldehydes and poly aromatic hydrocarbons are much lower from CNG vehicles compared to that of diesel vehicles as shown below in table.

Test results of Comparative Emissions (in gm/km) from Euro-II Diesel & LPG run Bus is tabulatedbelow:

S.No.	Pollutants	Euro-II	LPG	%Change
1.	HC	0.143	0.027	–81
2.	CO	0.212	0.013	–94
3.	NOx/10	1.254	0.54	–57
4.	Particulates	0.028	0.017	–39
5.	CO_2/100	1.344	1.309	–4

Test results of Comparison of non-regulated Emissions (Diesel Vs. LPG) is tabulated below:

Emission Component	Euro-II Diesel Bus	LPG Bus
Toxic Effects (%):		
• PAH	100	< 15 to 35%
• BTX	100	20 to 40%
• Lower Aldehydes	100	
Summer Smog Potential (Ethene Equivalent) (%)	100	40%
Acidification (%)	100	20 to 50%
Winter Smog Potential (%)	100	< 15%

Advantages of LPG:

Some of the benefits of LPG as automotive fuel are:

- Reduction in emissions.
- Very less carbon build-up increases life of engine parts like spark plugs.
- Fueling is akin to that of conventional diesel or gasoline and time needed is also similar.
- Service life of a vehicle increases with LPG.
- Little or no damage to soil & water if it is spilled, due to its rapid evaporation.
- Higher octane number.

Disadvantages of LPG:

Many countries in the world are using LPG as automotive fuel. Many years of experience of successful use of LPG in vehicles and the emission benefits it offers really support it to outperform the associated disadvantages of the fuel, if any. Moreover, familiarity with LPG in household application like cooking makes it more acceptable to the users. However, in the initial stages of introduction of this fuel, issues like safety, storage & handling, extreme volatility of the fuel, etc. needs proper attention.

Safety Issues:

Safety is an issue with LPG, however, standards for application, storage and distribution already exists. The hazard associated with on-board storage should be similar to that of gasoline, especially if reinforced cylinders similar to that of CNG tanks are used. LPG fuel systems have many built-in safety features and they generally maintain their integrity in severe collisions and do not permit massive leaks. LPG is about twice as heavier than air and unlike CNG does not disperse easily in air. Its flammability limits (2.1-9.5 vol.%)

and auto ignition temperature (450°C) are also lower than natural gas and as such should not be handled carelessly.

CNG & LPG

CNG: Compressed Natural Gas is composed of Methane, Ethane, Propane, Butane and other contaminants. The main constituent of CNG is Methane which will be up to 90%. Due to its low energy density, it is compressed to a pressure of 200–250 kg/cm^2 and the name is Compressed Natural Gas.

Comparative emission from diesel and CNG run buses are tabulated below:

Fuel	Pollutant emissions, gm/km		
	CO	NOx	PM
Diesel	2.4	21	0.38
CNG	0.4	8.9	0.012
%Reduction	84	58	97

It reduces CO to 97%, HC to 20 to 25% as compared to gasoline fuel.

In practice, LNG and CNG are replacement for petrol/diesel and suitable for heavier freight vehicles. Natural gas is intrinsically cleaner than petrol/diesel, but since it is a fossil fuel it is not renewable. It also contributes to global warming. At a local level it produces much less pollution than petrol or diesel and its use attracts financial incentives. The fuel tanks are specially designed ;intense refrigeration (for LNG) or high pressure (for CNG) which makes them larger and heavier.

Another test result of the comparative pollutants emissions from CNG & Diesel run I.C Engine is tabulated below:

Emissions in g/km					
Fuel	CO	NMVOC	NOx	PM	PM Emissions relative to CNG
Low sulphur diesel (500 ppm)	1.32	0.50	14.72	0.22	340% higher
ULSD (50 ppm)	1.41	0.52	14.32	0.16	220% higher
CNG	0.66	2.75	9.87	0.05	-

NMVOC – Non-methane volatile organic compounds

ULSD – Ultra low sulfur diesel (S < 50 ppm)

Pollutants emission results for Bajaj Three Wheelers run on petrol & CNG is tabulated below . (unit:gm/kg)

Type	CO	THC	NMHC	NOx
2-stroke (Petrol RE)	0.29	1.45	–	0.1
2-stroke (CNG RE)	0.8	5.74	0.291	0.02
4-stroke (Petrol RE)	0.45	0.68		0.56
4-stroke (CNG RE)	0.29	2.4	0.17	.75

THC – Total hydrocarbons, NMHC – Non-methane hydrocarbons

Test result for catalytic converter equipped Bus (TELCO) Emission Results for Euro-II Diesel Vs CNG is tabulated below .(unit-gm/kWh):-

Emissions (g/Kwh)	Diesel (Euro-II + Catalytic Converter)	CNG (with Catalytic Converter)
Carbon monoxide	1.06	1.68
Total Hydrocarbons	0.36	1.64
Nitrogen Oxides	5.89	3.42
Particulates	0.113	0.03

Pollutants Emission results for Ashok Leyland Diesel Buses & CNG Buses is tabulated below. (unit-gm/Kwh):

Test results (gm/Kwh)		
Pollutant	CNG	Diesel
Oxides of Nitrogen	3.24	7.721
Carbon monoxide	3.12	1.820
Hydrocarbons	1.30	0.262
Non-methane hydrocarbons	0.04	–
Particulate Matter	0.014	0.31

Biodiesel

- Produced from vegetable oil, animal fats, recycled oil or plants (Jetropha).
- Motor vehices can be very efficiently run by bio-diesel. Internal combustion engines are common in motor vehicles and are traditionally fuelled by diesel derived from petroleum fuels which is a compound which can be replaced with bio-diesel, which is an organically based product and is renewable. It is relatively easily produced from plant & animal oils, fats and greases. Environmentally bio-diesel run vehicles also gives benefits in reduction of pollution.

- To contribute towards protection of earth from the threat of environmental degradation, bio-fuels can be good alternative for diesel for most of developing countries. Biodiesel produced from vegetable oils have considerable potential to be considered as appropriate alternative, as they possess fuel properties similar to that of diesel. There are more than 300 different species of trees in India which produce oil bearings for the production of biodiesel.

Advantages of biodiesel:

- Biodiesel is a clean burning fuel:-
  - ♦ Reduces emissions of unburned HC, CO, SO_2 & PM.

The following table shows the reduction in various /pollutants by the use of Bio-Diesel blends.

Average Bio-Diesel Emissions Compared to Conventional Diesel

Emission Type	B100	B20
Total Unburnt Hydrocarbons	-67%	-20%
Carbon Monoxide	-48%	-12%
Particulate Matter	-47%	-12%
NOx	+ 2 to 2 %	+10%

If B100 i.e.l 00% of Bio-Diesel is in use, then the Unburnt Hydrocarbons will be reduced to 67%, Carbon Monoxide to 48%, Particulate Matter to 47% and Nitrogen Oxide may be minimized by two percent or it may increase.

If B20 i.e. 20% blend of Bio-Diesel is in utilization, then the Unburnt Hydrocarbons will be reduced to 20%, Carbon Monoxide to 12%, Particulate Matter to 12% and Nitrogen Oxides may increase by 10%. But here the Nitrogen oxide problem may be minimized by using the catalytic convertors.

- It is safer in use:-
  - ♦ Bio-diesel exhaust has a less harmful impact on human health than petroleum diesel.
  - ♦ It is Biodegradable & non-toxic.
- It is a renewable energy source:-
  - ♦ Produced from natural, renewable sources such as new & used vegetables oils & animal fats.
- Feasible in CI (diesel) engines with no major engine modifications.
- No special requirement for biodiesel.
- Already well developed and used since last decade:-

- ♦ Popular in Europe & the USA
- Blended with diesel and used mainly as:
  - ♦ B20 (20% biodiesel + 80% diesel)
  - ♦ B100 (pure biodiesel)

Disadvantages of Biodiesel

- Biodiesel is more expensive than petro diesel fuel due to the high cost of raw vegetables oils.
- It damages natural rubber:
  - ♦ Can damage natural rubber hoses/seals.
  - ♦ Requires to replace to synthetic rubber type.
- Clogging may occur due to use of biodiesel:
  - ♦ Cleans the scale from the engines. This dirt then collects in the fuel filter, which can clog it.
  - ♦ Biodiesel has been used as a substitute for diesel for more than 3 decades in abroad countries. Its use in Indian vehicles is in nascent stage.
  - ♦ Consumed more than 500 million gallon/year outside India.
  - ♦ Have 500 million miles on road usage record.

Countries using bio-diesel fuel include:

- Austria (B100) • Italy
- Belgium • Slovakia (B30)
- Brazil • Sweden (B20-B100)
- Canada (B2-B100) • Switzerland (B100)
- Czechoslavia (B30) • Thailand
- France (B5-B30) • U.K.
- Germany (B100) • USA (B20)

B100 → 100% biodiesel, B20 → 20% biodiesel + 80% petrodiesel

Ethyl Alcohol fuel (Ethanol):

- Most widely used alternative fuel abroad.

Advantages of ethyl alcohol as fuel:

- Low pollution (improve air quality):
  - ♦ low emissions of toxic compounds: emit almost no PM & much less NOx than diesel-fueled vehicles.

- ♦ combust more completely than petrol & diesel.
- ♦ highly soluble & will disperse rapidly; biodegradable, & will evaporate quickly, if spilled on land.

- **Fire safety (much less flammable than gasoline):**
  - ♦ less likely to ignite compared to gasoline.
- **Fuel supply options**:
  - ♦ can be manufactured from a variety of carbon based feedstocks such as natural gas, coal, & biomass (molasses)
  - ♦ could diversify the country's fuel supply & reduce its dependence on imported petroleum.
- Economically attractive
  - ♦ with advances in technologies, could be produced, distributed & sold to consumers at a competitive price

Disadvantages of ethtyl alcohol (ethanol) fuel:

- Requires modifications in engine:
  - ♦ corrosive to certain materials – some modifications have to be made to engines & the fuel delivery systems.
  - ♦ less energy content: for E85 (85% ethanol + 15% gasoline) consume 15–30% more fuel per km than gasoline-powered cars.
  - ♦ for blends containing less than 15% ethanol, fuel economy is comparable to gasoline.
- Higher costs:
  - ♦ an ethanol fueled vehicle is more expansive than a comparable gasoline vehicle.
  - ♦ fuel price depends on prices of agricultural crops and pricing policy of the government.

Current status of ethanol fuel usage:

- Blended with gasline: E10 (10% ethanol + 90% petrol) and E85 blend → feasible for all types of gasoline vehicles & engines.
- Countries using ethanol as automobile fuel:
- Brazil (E22-E100); Sweden (E100; USA (E10-E85); Malawi (E10); Mexico (E10); China & Thailand (under trial tests). India also plans to use 5 to 10% ethanol blend in petrol in coming future by 2016.
- Widely used in Brazil & the USA:
  - ♦ ~40% of the cars in Brazil operate on E100 (>5 million cars).

The remaining cars run on a E22 (22% ethanol + 78% gasoline); also called "GASOHOL".

♦ The USA uses ethanol in 12% of its fuel, mostly as E10 called gasohol.

Electricity as alternative Automobile fuels:

- Characteristics of electricity as automobile fuels:
  - ♦ Zero-emission of air pollutant.
  - ♦ Noiseless, odorless, and free of smoke.
  - ♦ High performance.
- More efficient than internal combustion engine (ICE)
- Three types of electricity source.
  - ♦ Rechargeable batteries
  - ♦ Rechargeable batteries with IC engine (Hybrid vehicles)
  - ♦ Fuel cells

Appendix 11.21: Research octane number (RON) and motor octane number (MON) of some hydrocarbons

S.N	Hydrocarbon	Octane number[a]		Boiling point (K)
		RON	MON	
1	*n*-Pentane	62	—	309
2	2-Methylbutane	90	—	301
3	Cyclopentane	85	—	322
4	*n*-Hexane	26	—	342
5	2-Methylpentane	73	—	333
6	2 .2-Dimethylbutane	93	—	323
7	1-Hexene	63	—	337
8	2-Hexene	81	—	341
9	Benzene	>100	—	353
10	Cyclohexane	77	—	354
11	*n-Heptane*	0[b]	—	362
12	2-methylheptane	13	—	381
13	2, 2, 4-Trimethylpentane	100[b]	—	372
14	1 -Octene	35	—	395
15	2-Octene	56	—	398

S.N	Hydrocarbon	Octane number[a]		Boiling point (K)
		RON	MON	
16	3-Octene	68	—	396
17	Xylenes	>100	—	≈415
18	Ethylbenzene	98	—	410
19	1.2-Dimethylcyclohexane	79	—	403
20	Ethylbenzene	41	—	403
21	Methyl-tertiary-butyl-ether (MTBE)	118	—	328
22	Ethyl-tertiary-butyl-ether (ETBE)	118	—	345
23	Tertiary-amyl-methyl-ether (TAME)	111	—	359
24	Light straight-run gasoline	68	67	—
25	Isomerate	85	82	—
26	FCC light gasoline	93	82	—
27	FCC heavy gasoline	95	85	—
28	Alkylate	95	92	—
29	Reformate (OCR)	99	88	—

[a] *Research octane number (RON). The motor octane number (MON) is generally lower, depending on the particular compound.*

The difference is particularly large for alkenes and aromatics.

[b] *By definition.*

Appendix: 11.22 All about utilities and various auxiliary operations in a petroleum refinery

A. Heat Exchangers, Coolers, and Process Heaters

- **Heating Operations.** Process heaters and heat exchangers preheat feedstock in distillation towers and in refinery processes to reaction temperatures. Heat exchangers use either steam or hot hydrocarbon transferred from some other section of the process for heat input. The heaters are usually designed for specific process operations, and most are of cylindrical vertical or box-type designs. The major portion of heat provided to process units comes from fired heaters fueled by refinery or natural gas, distillate, and residual oils. Fired heaters are found on crude and reformer preheaters, coker heaters, and large-column reboilers.

- **Cooling Operations.** Heat also may be removed from some processes by air & water exchangers, fin fans, gas & liquid coolers, and overhead condensers, or by transferring heat to other systems. The basic mechanical vapor compression refrigeration system, which may serve one or more process units, includes an evaporator, compressor, condenser, controls, and piping. Common coolants are water, alcohol/water mixtures, or various glycol solutions.

B. Steam Generation

- **Heater and Boiler Operations.** Steam is generated in main generation plants, and/or at various process units using heat from flue gas or other sources. Heaters (furnaces) include burners and a combustion air system, the boiler enclosure in which heat transfer takes place, a draft or pressure system to remove flue gas from the furnace, soot blowers, and compressed air systems that seal openings to prevent the escape of flue gas. Boilers consist of a number of tubes that carry the water steam mixture through the furnace for maximum heat transfer. These tubes run between steam distribution drums at the top of the boiler and water-collecting drums at the bottom of the boiler. Steam flows from the steam drum to the superheater before entering the steam distribution system.
- **Heater Fuel**
  a) Heaters may use any one or combination of fuels including refinery gas, natural gas, fuel oil, and powdered coal. Refinery off-gas is collected from process units and combined with natural gas and LPG in a fuel-gas balance drum. The balance drum provides constant system pressure, fairly stable Btu content fuel, & automatic separation of suspended liquids in gas vapors, and it prevents carryover of large slugs of condensate into the distribution system. Fuel oil is typically a mix of refinery crude oil with straight-run & cracked residues and other products. The fuel-oil system delivers fuel to process unit heaters and steam generators at required temperatures & pressures. The fuel oil is heated to pumping temperature, sucked through a coarse suction strainer, pumped to a temperature control heater, and then pumped through a fine mesh strainer before being burned.

  b) In one example of process unit heat generation, carbon monoxide boilers recover heat in catalytic cracking units as carbon monoxide in flue gas is burned to complete combustion. In other processes, waste-heat recovery units use heat from the flue gas to make steam.

- **Steam Distribution.** The distribution system consists of valves, fittings, piping, and connections suitable for the pressure of the steam transported. Steam leaves the boilers at the highest pressure required by the process units or electrical generation. The steam pressure is then reduced in turbines that drive process pumps and compressors. Most steam used in the refinery is condensed to water in various types of heat exchangers. The condensate is reused as boiler feedwater or discharged to wastewater treatment. When refinery steam is also used to drive steam turbine generators to produce electricity, the steam must be produced at much higher pressure than required for process steam. Steam typically is generated by heaters (furnaces) and boilers combined in one unit.
- **Feedwater Supply**
  a) Feedwater supply is an important part of steam generation. There must always be as many kg of water entering the system as there are kg of steam leaving it. Water used in steam generation must be free of contaminants including minerals and dissolved impurities that can damage the system or affect its operation. Suspended materials such as silt, sewage, & oil, which form scale & sludge, must be coagulated or filtered out of the water. Dissolved gases, particularly carbon dioxide & oxygen, cause boiler corrosion and are removed by deaeration & treatment. Dissolved minerals including metallic salts, calcium, carbonates, etc., that cause scale, corrosion, & turbine blade deposits are treated with lime or soda ash to precipitate them from the water. Recirculated cooling water must also be treated for hydrocarbons and other contaminants.
  b) Depending on the characteristics of raw boiler feedwater, some or all of the following six stages of treatment will be applicable:
     - Clarification;
     - Sedimentation;
     - Filtration;
     - Ion exchange;
     - Deaeration; and
     - Internal treatment.

C. Pressure-Relief and Flare Systems

- **Pressure-Relief Systems.** Pressure relief systems control vapors & liquids that are released by pressure relieving devices and blowdowns. Pressure relief is an automatic, planned release when operating pressure reaches a predetermined level. Blowdown normally refers to the intentional release of material, such as blowdowns from process unit startups, furnace blowdowns, shutdowns, and emergencies. Vapor depressuring is the rapid removal of vapors from pressure vessels in case of fire. This may be accomplished by the use of a rupture disc, usually set at a higher pressure than the relief valve.

- **Safety Relief Valve Operations.** Safety relief valves, used for air, steam, and gas as well as for vapor & liquid, allow the valve to open in proportion to the increase in pressure over the normal operating pressure. Safety valves designed primarily to release high volumes of steam usually pop open to full capacity. The overpressure needed to open liquid relief valves where large volume discharge is not required increases as the valve lifts due to increased spring resistance. Pilot operated safety relief valves, with up to six times the capacity of normal relief valves, are used where tighter sealing and larger volume discharges are required. Non volatile liquids are usually pumped to oil water separation & recovery systems, and volatile liquids are sent to units operating at a lower pressure.
- **Flare Systems.** A typical closed pressure release & flare system includes relief valves and lines from process units for collection of discharges, knockout drums to separate vapors & liquids, seals, and/ or purge gas for flashback protection, and a flare & igniter system which combusts vapors when discharging directly to the atmosphere is not permitted. Steam may be injected into the flare tip to reduce visible smoke.

D. Wastewater Treamtment:

Wastewater treatment is used for process, runoff, and sewerage water prior to discharge or recycling. Wastewater typically contains hydrocarbons, dissolved materials, suspended solids, phenols, ammonia, sulfides, and other compounds. Wastewater includes condensed steam, stripping water, spent caustic solutions, cooling tower & boiler blowdown, wash water, alkaline & acid waste neutralization water, and other process associated water.

- **Pretreatment.** Pretreatment is the separation of hydrocarbons and solids from wastewater. API separators, interceptor plates, and settling ponds remove suspended hydrocarbons, oily sludge, & solids by gravity separation, skimming, & filtration. Some oil-in-water emulsions must be heated to assist in separating the oil and water. Gravity separation depends on the specific gravity differences between water& immiscible oil globules and allows free oil to be skimmed off the surface of the wastewater. Acidic wastewater is neutralized using ammonia, lime, or soda ash. Alkaline wastewater is treated with sulfuric acid, hydrochloric acid, carbon dioxide rich flue gas, or sulfur.

- **Secondary Treatment.** After pretreatment, suspended solids are removed by sedimentation or air flotation. Wastewater with low levels of solids may be screened or filtered. Flocculation agents are sometimes added to help separation. Secondary treatment processes biologically degrade and oxidize soluble organic matter by the use of activated sludge, unaerated or aerated lagoons, trickling filter methods, or anaerobic treatments. Materials with high adsorption characteristics are used in fixed-bed filters or added to the wastewater to form a slurry which is removed by sedimentation or filtration. Additional treatment methods are used to remove oils and chemicals from wastewater. Stripping is used on wastewater containing sulfides and/or ammonia, and solvent extraction is used to remove phenols.
- **Tertiary Treatment Operations.** Tertiary treatments remove specific pollutants to meet regulatory discharge requirements. These treatments include chlorination, ozonation, ion exchange, reverse osmosis, activated carbon adsorption, etc. Compressed oxygen is diffused into wastewater streams to oxidize certain chemicals or to satisfy regulatory oxygen content requirements. Wastewater that is to be recycled may require cooling to remove heat and/or oxidation by spraying or air stripping to remove any remaining phenols, nitrates, & ammonia.

E. Cooling Towers

Cooling towers remove heat from process water by evaporation and latent heat transfer between hot water & air. The two types of towers are crossflow and counterflow. Crossflow towers introduce the airflow at right angles to the water flow throughout the structure. In counterflow cooling towers, hot process water is pumped to the uppermost plenum and allowed to fall through the tower. Numerous slats or spray nozzles located throughout the length of the tower disperse the water and help in cooling. Air enters at the tower bottom and flows upward against the water. When the fans or blowers are at the air inlet, the air is considered to be forced draft. Induced draft is used when the fans are at the air outlet.

Recirculated cooling water must be treated to remove impurities and dissolved hydrocarbons. Because the water is saturated with oxygen from being cooled with air, the chances for corrosion are increased. One means of corrosion prevention is the addition of a material to the cooling water that forms a protective film on pipes and other metal surfaces.

F. Electric Power

Refineries may receive electricity from outside sources or produce their own power with generators driven by steam turbines or gas engines. Electrical substations receive power from the utility or power plant for distribution throughout the facility. They are usually located in non classified areas, away from sources of vapor or cooling-tower water spray. Transformers, circuit breakers, and feed-circuit switches are usually located in substations. Substations feed power to distribution stations within the process unit areas. Distribution stations can be located in classified areas, providwd that classification requirements are met. Distribution stations usually have a liquid filled transformer and an oil-filled or air-break disconnect device.

G. Gas & Air Compressors:

Both reciprocating & centrifugal compressors are used throughout the refinery for gas and compressed air. Air compressor systems include compressors, coolers, air receivers, air dryers, controls, and distribution piping. Blowers are used to provide air to certain processes. Plant air is provided for the operation of air-powered tools, catalyst regeneration, process heaters, steam-air decoking, sour water oxidation, gasoline sweetening, asphalt blowing, and other uses. Instrument air is provided for use in pneumatic instruments & controls, air motors and purge connections.

H. Marine, Tank Car, and Tank Truck Loading & Unloading

Facilities for loading liquid hydrocarbons into tank cars, tank trucks, and marine vessels & barges are usually part of the refinery operations. Product characteristics, distribution needs, shipping requirements, and operating criteria are important when designing loading facilities. Tank trucks & rail tank cars are either top or bottom loaded, and vapor recovery systems may be provided where required.

I. Turbines

Turbines are usually gas or steam powered and are typically used to drive pumps, compressors, blowers, and other refinery process equipment. Steam enters turbines at high temperatures & pressures, expands across and drives rotating blades while directed by fixed blades.

J. Pumps, Piping & Valves:

Centrifugal and positive displacement (i.e., reciprocating) pumps are used to move hydrocarbons, process water, fire water, and wastewater through piping within the refinery. Pumps are driven by electric motors, steam turbines, or internal combustion engines. The pump type, capacity, and construction materials depend on the service for which it is used.

Process and utility piping distribute hydrocarbons, steam, water & other products throughout the facility. Their size and construction depend on the type of service, pressure, temperature, and nature of the products. Vent, drain, and sample connections are provided on piping, as well as provisions for blanking.

Different types of valves are used depending on their operating purpose. These include gate valves, bypass valves, globe & ball valves, plug valves, block & bleed valves, and check valves. Valves can be manually or automatically operated.

K. Storage tank

Atmospheric storage tanks and pressure storage tanks are used throughout the refinery for storage of crudes, intermediate hydrocarbons (during the process), and finished products. Tanks are also provided for fire water, process & treatment water, acids, additives, and other chemicals. The type, construction, capacity & location of tanks depends on their use and materials stored.

M. Sweetening & treating Processes

Treating is a means by which contaminants such as organic compounds containing sulfur, nitrogen, & oxygen; dissolved metals & inorganic salts; and soluble salts dissolved in emulsified water are removed from petroleum fractions or streams. Petroleum refiners have a choice of several different treating processes, but the primary purpose of the majority of them is the elimination of unwanted sulfur compounds. A variety of intermediate & finished products, including middle distillates, gasoline, kerosene, jet fuel, & sour gases are dried and sweetened. Sweetening, a major refinery treatment of gasoline, treats sulfur compounds (hydrogen sulfide, thiophene & mercaptan) to improve color, odor, and oxidation stability. Sweetening also reduces concentrations of carbon dioxide.

Treating can be accomplished at an intermediate stage in the refining process, or just before sending the finished product to storage. Choices of a treating method depend on the nature of the petroleum fractions, amount & type of impurities in the fractions to be treated, the extent to which the process removes the impurities, and end-product specifications. Treating materials include acids, solvents, alkalis, oxidizing, and adsorption agents.

- **Acid, Caustic, or Clay Treating.** Sulfuric acid is the most commonly used acid treating process. Sulfuric acid treating results in partial or complete removal of unsaturated hydrocarbons, sulfur, nitrogen, & oxygen compounds, and resinous & asphaltic compounds. It is used to improve the odor, color, stability, carbon residue, and other properties of the oil. Clay/lime treatment of acid refined oil removes traces of asphaltic materials and other compounds improving product color, odor, and stability. Caustic treating with sodium (or potassium) hydroxide is used to improve odor & color by removing organic acids (naphthenic acids, phenols) and sulfur compounds (mercaptans, H_2S) by a caustic wash. By combining caustic soda solution with various solubility promoters (e.g., methyl alcohol and cresols), up to 99% of all mercaptans as well as oxygen & nitrogen compounds can be dissolved from petroleum fractions.

- **Drying and Sweetening.** Feedstocks from various refinery units are sent to gas treating plants where butanes and butenes are removed for use as alkylation feedstock, heavier components are sent to gasoline blending, propane is recovered for LPG, and propylene is removed for use in petrochemicals. Some mercaptans are removed by water-soluble chemicals that react with the mercaptans. Caustic liquid (sodium hydroxide), amine compounds (diethanolamine) or fixed-bed catalyst sweetening also may be used. Drying is accomplished by the use of water absorption or adsorption agents to remove water from the products. Some processes simultaneously dry & sweeten by adsorption on molecular sieves.
- **Sulfur Recovery.** Sulfur recovery converts hydrogen sulfide in sour gases and hydrocarbon streams to elemental sulfur. The most widely used recovery system is the Claus process, which uses both thermal and catalytic-conversion reactions. A typical process produces elemental sulfur by burning hydrogen sulfide under controlled conditions. Knockout pots are used to remove water and hydrocarbons from feed gas streams. The gases are then exposed to a catalyst to recover additional sulfur. Sulfur vapor from burning and conversion is condensed & recovered.
- **Hydrogen Sulfide Scrubbing.** Hydrogen sulfide scrubbing is a common treating process in which the hydrocarbon feedstock is first scrubbed to prevent catalyst poisoning. Depending on the feedstock and the nature of the contaminants, desulfurization methods vary from ambient temperature-activated charcoal absorption to high-temperature catalytic hydrogenation followed by zinc oxide treating.

N. Unsaturated Gas Plants

Unsaturated (unsat) gas plants recover light hydorcarbons (C_3 & C_4 olefins) from wet gas streams from the *FCC*, *TCC*, and delayed coker overhead accumulators or fractionation receivers. In a typical unsat gas plant, the gases compressed and treated with amine to remove hydrogen sulfide either before or after, they are sent to a fractionating absorber where they are mixed into a concurrent flow of debutanized gasoline. The light fractions are separated by heat in a reboiler, the off gas is. sent to a sponge absorber, and the bottoms are sent to a debutanizer. A portion of the debutanized hydrocarbon recycled with the balance is sent to the splitter for separation. The overhead gases go to a depropanlzer for use as alkylation unit feedstock.

O. Amine Plants

Amine plants remove acid contaminants from sour gas and hydrocarbon streams. In amine plants, gas & liquid hydrocarbon streams containing carbon dioxide and/or hydrogen sulfide are charged to a gas absorption tower or liquid contactor where the acid contaminants are absorbed by counter flowing amine solutions (i.e., MEA, DEA, MDEA). The stripped gas or liquid is removed overhead, and the amine is sent to a regenerator. In the regenerator, the acidic components are stripped by heat & reboiling action and disposed of, and the amine is recycled.

P. Saturate Gas Plants:

Saturate (sat) gas plants separate refinery gas components including butanes for alkylation, pentanes for gasoline blending, LPG's for fuel, and ethane for petrochemicals. Because sat gas processes depend on the feedstock and product demand, each refinery uses different systems, usually absorption fractionation or straight fractionation. In absorption-fractionation, gases & liquids from various refinery units are fed to an absorber deethanizer where C_2 and lighter fractions are separated from heavier fractions by oil absorption and removed for use as fuel gas or petrochemical feed. The heavier fractions are stripped and sent to a debutanizer, and the lean oil is recycled back to the absorber deethanizer. C_3/C_4 is separated from pentanes in the debutanizer, scrubbed to remove hydrogen sulfide, and fed to a splitter where propane & butane are separated. In fractionation sat gas plants, the absorption stage is eliminated.

Q. Asphalt Plants:

Asphalt is a portion of the residual fraction that remains after primary distillation operations. It is further processed to impart characteristics required by its final use. In vacuum distillation, generally used to produce road-tar asphalt, thc residual is heated to about 750° F and charged to a column where vacuum is applied to prevent cracking.

Asphalt for roofing materials is produced by air blowing. Residual is heated in a pipe still almost to its flash point and charged to a blowing tower where hot air is injected for a pre determined time. The dehydrogenization of the asphalt forms hydrogen sulfide, and the oxidation creates sulfur dioxide. Steam, used to blanket the top of the tower to entrain the various contaminants, is then passed through a scrubber to condense the hydrocarbons.

A third process used to produce asphalt is solvent deasphalting. In this extraction process, which uses propane (or hexane) as a solvent, heavy oil fractions are separated to produce heavy lubricating oil, catalytic cracking feedstock, and asphalt. Feedstock and liquid propane are pumped to an extraction tower at precisely controlled mixtures temperatures (65–120° F), and pressures of 25-42 kg/cm^2. Separation occurs in a rotating disc contactor, based on differences in solubility. The products are then evaporated and steam stripped to recover the propane, which is recycled. Deasphalting also removes some sulfur & nitrogen compounds, metals, carbon residues, and paraffins from the feedstock.

R. Hydrogen Production

High-purity hydrogen (95–99%) is required for hydrodesulfurization, hydrogenation, hydrocracking, and petrochemical processes. Hydrogen, produced as a by-product of refinery processes (principally hydrogen recovery from catalytic reformer product gases), often is not enough to meet the total refinery requirements, necessitating the manufacturing of additional hydrogen or obtaining supply from external sources.

In steam methane reforming, desulfurized gases are mixed with superheated steam (595 – 870°C) and reformed in tubes containing a nickel base catalyst. The reformed gas, which consists of steam, hydrogen, carbon monoxide & carbon dioxide, is cooled and passed through converters containing an iron catalyst where the carbon monoxide reacts with steam to form carbon dioxide & more hydrogen. The carbon dioxide is removed by amine washing. Any remaining carbon monoxide in the product stream is converted to methane.

Steam-naphtha reforming is a continuous process for the production of hydrogen from liquid hydrocarbons and is, in fact, similar to steam methane reforming. A variety of naphthas in the gasoline boiling range may be employed, including fuel containing up to 35% aromatics. Following pretreatment to remove sulfur compounds, the feedstock is mixed with steam and taken to the reforming furnace (675 -815°C) where hydrogen is produced.

S. Blending:

Blending is the physical mixture of a number of different liquid hydrocarbons to produce finished product with certain desired characteristics. Products can be blended in line through a manifold system, or batch blended in tanks and vessels. In line blending of gasoline, distillates, jet fuel, and kerosene is accomplished by injecting proportionate amounts of each component into the main stream where turbulence promotes thorough mixing. Additives including octane enhancers, metal deactivators, antioxidants, anti-knock agents, gum & rust inhibitors, detergents, etc. are added during and/or after blending to provide specific properties not inherent in hydrocarbons.

Appendix 11.23: Sources and Control of Environmental Pollutants in a Petroleum Refinery

Petroleum refining is the physical, thermal, and chemical separation of crude oil into its major distillation fractions, which are then further processed through a series of separation & conversion steps into finished petroleum products. The primary products of the industry fall into three major categories: fuels (motor gasoline, diesel & distillate fuel oil, liquefied petroleum gas, jet fuel, residual fuel oil, kerosene, and coke); finished nonfuel products (solvents, lubricating oils, greases, petroleum wax, petroleum jelly, asphalt, & coke); and chemical industry feedstocks (naphtha, ethane, propane, butane, ethylene, propylene, butylenes, butadiene, benzene, toluene, & xylene). These petroleum products are used as primary input to a vast number of products, including: fertilizers, pesticides, paints, waxes, thinners, solvents, cleaning fluids, detergents, refrigerants, antifreeze, resins, sealants insulations, latex, rubber compounds, hard plastics, plastic sheeting, plastic foam, and synthetic fibers.

Process Operations and Sources of Air Pollutants Emissions

Crude oil is a mixture of many different hydrocarbons and small amounts of impurities. The composition of crude oil can vary significantly depending on its source. Petroleum refineries are a complex system of multiple operations and the operations used at a given refinery depend upon the properties of the crude oil to be refined and the desired products. For these reasons, no two refineries are alike. Portions of the outputs from some processes are re-fed back into the same process, fed to new processes, fed back to a previous process, or blended with other outputs to form fished products. The major unit operations typically-involved at petroleum refineries are described briefly below.

Refining crude oil into useful petroleum products can be separated into two phases and a number of supporting operations. The first phase is desalting of crude oil and the subsequent distillation into its various components or "fractions." The second phase is made up of three different types of "downstream" processes: combining, breaking, and reshaping.

Downstream processes convert some of the distillation fractions into petroleum products (residual fuel oil, gasoline, kerosene, etc.) through any combination of different cracking, coking, reforming, and alkylation processes. Supporting operations may include wastewater treatment, sulfur recovery, additive production, heat exchanger cleaning, blowdown systems, blending of products, and storage of products.

Various air pollutants emitted and their sources in a petroleum refinery is tabulated below.

S. No.	Air pollutants emitted	Sources of air pollutants
1	Hydrocarbons (HC)	Air blowing, barometric condensers, blind changing, blow-down system, boilers, catalyst regenerators, compressors, cooling towers, decoking operations, flare, heaters, incinerators, loading facilities, processing vessels, pumps, sampling operations, tanks turn around operations, vacuum jets, effluent-handling equipment
2	Sulphur oxides	Boilers, catalyst regenerators, decoking operations, flares, heaters, incinerators, treaters, acid sludge disposal
3	Carbon monoxide	Catalyst regenerators, compressor engines, coking operations, incinerators
4	Nitrogen oxides	Boilers, catalyst regenerators, compressor engines, flares
5	Particulate matter	Boilers, catalyst regenerators, coking operations, heaters, incinerators
6	Odours	Air blowing, barometric condensers, drains process vessels, steam blowing tanks, treaters, effluent-handling equipment
7	Aldehydes	Catalyst regenerators, compressor engines
8	Ammonia	Catalyst regenerators

Air pollution control methods employed in different petroleum refining processes is tabulated below.

S.N	Oil refinery sources	Air pollutants control methods employed.
1	Storage vessels	Vapour recovery systems; floating roof tanks; pressure tanks; vapour balance; painting tanks white
2	Catalyst regenerators	Cyclones-precipitator-CO boiler; cyclories-water scrubber; multiple cyclones.
3	Accumulator vents	Vapour recovery; vapour incineration
4	Blowdown systems	Smokeless flares-gas recovery
5	Pumps and compressors	Mechanical seals; vapour recovery: sealing glands by oil pressure; maintenance
6	Vacuum jets	Vapour incineration
7	Equipment valves	Inspection and maintenance
8	Pressure relief valves	Vapour recovery; vapor incinceration; rupture disks; inspection and maintenance.

S.N	Oil refinery sources	Air pollutants control methods employed.
9	Effluent waste disposal	Enclosing separators; covering sewer boxes and using liquid seals; liquid seals on drains
10	Bulk loading facilities	Vapour collection with recovery or incineration; submerged or bottom loading
11	Acid treating	Continuous type agitators with mechanical mixing; replace with catalytic hydrogenation units; incinerate all vented cases; stop sludge burning
12	Acid sludge storage and shipping	Caustic scrubbing; incineration, vapour return system.
13	spent caustic handling	Incineration; scrubbing.
14	Doctor solution treating	Steam strip spent doctor solution to hydrocarbon recovery before air regeneration; replace treating unit with other, less objectionable units (Merroes)
15	Sour water treating	Use sour water oxidizers and gas incineration; conversion to ammonium sulphate
16	Mercaptan disposal	Conversion to disulfides; adding to catalytic cracking charge stock; incineration; using material in organic synthesis.
17	Asphalt blowing	Incineration; water scrubbing (nonrecirculating type)
18	Shutdowns, turnarounds	Depressure and purge to vapour recovery

Crude Oil Desalting and Distillation:

Distillation involves the heating, vaporization, fractionation, condensation, and cooling of feedstocks. Before separation into fractions, crude oil must first be treated to remove corrosive salts. The desalting process also removes some of the metals and suspended solids which cause catalyst deactivation. Desalting involves the mixing of heated crude oil with water (about 3 to 10% of the crude oil volume) so that the salts are dissolved in the water. The water must then be separated from the crude oil in a separating vessel by adding demulsifier chemicals to assist naphtha from the FCC & coking units, the by-product from steam cracking of naphtha or light hydrocarbons during the production of ethylene & propylene, catalytic dehydrogenation of isobutane, and conversion of tertiary butyl alcohol recovered as a by-product in the manufacture of propylene oxides. Several different processes are currently in use to produce MTBE and TAME from isobutylene and methanol. Most processes use a two-stage acidic ion exchange resin catalyst. The reaction is exothermic and cooling to the proper reaction temperature is critical in obtaining the optimal conversion efficiency. The process usually produces an MTBE or TAME stream and a relatively small stream of unreacted hydrocarbons and methanol. The methanol is extracted in a water wash and the resulting methanol-water mixture is distilled to recover the methanol for recycling.

Blending

Blending is the final operation in petroleum refining. It consists of mixing the products in various proportions to meet requirement usually produces an MTBE or TAME stream and a relatively small stream of unreacted hydrocarbons and methanol. The methanol is extracted in a water wash and the resulting methanol-water mixture is distilled to recover the methanol for recycling.

Blowdown System

Most refinery process units and equipment are manifolded into a collection unit, called the blowdown system. Blowdown systems provide for the safe handling and disposal of liquids & gases that are either automatically vented from the process units through pressure relief valves, or that are manually drawn from units. Recirculated process streams and cooling water streams are often manually purged to prevent the continued buildup of contaminants in the stream. Part or all of the contents of equipment can also be purged to the blowdown system prior to shutdown before normal or emergency shutdowns.

Blowdown systems utilize a series of flash drums and condensers to separate the blowdown into its vapor and liquid components. The liquid is typically composed of mixtures of water and hydrocarbons containing sulfides, ammonia, and other contaminants, which are sent to the wastewater treatment plant.

The gaseous component typically contains hydrocarbons, hydrogen sulfide, ammonia, mercaptans, solvents, & other constituents, and is either discharged directly to the atmosphere or is combusted in a flare. The major air emissions from blowdown systems are hydrocarbons in case of direct discharge to the separator which may contain disulfides.

Dewaxing

Dewaxing of lubricating oil base stocks is necessary to ensure that the oil will have the proper viscosity at lower ambient temperatures. Two types of dewaxing processes are used: selective hydrocracking and solvent dewaxing. In selective hydrocracking, one or two zeolite catalysts are used to selectively crack the wax paraffins. Solvent dewaxing is more prevalent. In solvent dewaxing the oil feed is diluted with solvent to lower the viscosity, chilled until the wax is crystallized, and then filtered to remove the wax. Solvents used for the process include propane and mixtures of methyl ethyl ketone (MEK) with methyl isobutyl ketone (MIBK) or MEK with toluene.

Solvent is recovered from the oil and wax through heating, two-stage flashing, followed by steam stripping. The solvent recovery stage results in solvent contaminated water which typically is sent to the wastewater treatment plant. The wax is either used as feed to the catalytic cracker or is deoiled and sold as industrial wax. Air emissions may arise from fugitive emissions of the solvents.

Propane Deasphalting

Propane deasphalting produces lubricating oil base stocks by extracting asphaltenes and resins from the residuals of the vacuum distillation unit. Propane is usually used to remove asphaltenes due to its unique solvent properties. At lower temperatures (40–60°C), paraffins are very soluble in propane and at higher temperatures (about 90°C) all hydrocarbons are almost insoluble in propane. The propane recovery stage results in propane contaminated water which typically is sent to the wastewater treatment plant. Air emissions may arise from fugitive propane emissions and process vents.

Supporting Operations

Many important refinery operations are not directly involved in the production of hydrocarbon fuels but serve in a supporting role. Some of the major supporting processes are:

Wastewater Treatment

Relatively large volumes of water are used by the petroleum refining industry. Four types of wastewater are produced: surface water runoff, cooling water, process water, and sanitary wastewater. Surface water runoff is intermittent and will contain constituents from spills to the surface, leaks in equipment, and any materials that may have collected in drains. Runoff surface water also includes water coming from crude and product storage tank roof drains. Cooling water typically does not come into direct contact with process oil streams and therefore contains less contaminants than process wastewater. Most cooling water is recycled over and over with a bleed or blowdown stream to the wastewater treatment unit to control the concentration of contaminants and the solids content in the water. Cooling towers within the recycle loop cool the water using ambient air. Some cooling water, termed "once through, " is passed through a process unit once and is then discharged directly without treatment in the wastewater treatment plant. The water used for cooling often contains chemical additives such as chromates, phosphates, and antifouling biocides to prevent scaling of pipes & biological growth.

Although cooling water usually does not come into direct contact with oil process streams, it also may contain some oil contamination due to leaks in the process equipment. Water used in processing operations also accounts for a significant portion of the total wastewater. Process wastewater arises from desalting crude oil, steam stripping operations, pump gland cooling, product fractionator reflux drum drains and boiler blowdown. Because process water often comes into direct contact with oil, it is usually highly contaminated.

Petroleum refineries typically utilize primary and secondary wastewater treatment technologies. Primary wastewater treatment consists of the separation of oil, water, and solids in two stages. During the first stage, an API separator, a corrugated plate interceptor, or other separator design is used. Wastewater moves very slowly through the separator allowing free oil to float to the surface and be skimmed off, and solids to settle to the bottom and be scraped off to a sludge collecting hopper. The second stage utilizes physical or chemical methods to separate emulsified oils from the wastewater. Physical methods may include the use of a series of settling ponds with a long retention time, or the use of dissolved air flotation (DAF). In DAF, air is bubbled through the wastewater, and both oil and suspended solids are skimmed off the top; Chemicals, such as ferric hydroxide or aluminum hydroxide, can be used to coagulate impurities into a froth or sludge which can be more easily skimmed off the top. Some wastes associated with the primary treatment of wastewater at petroleum refineries may be considered hazardous and include: API separator sludge, primary treatment sludge, sludges from other gravitational separation techniques, float from DAF units, and wastes from settling ponds. After primary treatment, the wastewater can be discharged to a publicly owned treatment works or undergo secondary treatment before being discharged directly to surface.

In secondary treatment, dissolved oil & other organic pollutants may be consumed biologically by microorganisms. Biological treatment may require the addition of oxygen through a number of different techniques, including activated sludge units, trickling filters, and rotating biological contactors. Secondary treatment generates bio-mass waste which is typically treated anaerobically, and then dewatered.

Some refineries employ an additional stage of wastewater treatment called polishing to meet discharge limits. The polishing step can involve the use of activated carbon, anthracite coal, or sand to filter out any remaining impurities, such as biomass, silt, trace metals, and other inorganic chemicals, as well as any remaining organic chemicals.

Certain refinery wastewater streams are treated separately, prior to the wastewater treatment plant, to remove contaminants that would not easily be treated after mixing with other wastewater. One such waste stream is the sour water drained from distillation reflux drums. Sour water contains dissolved hydrogen sulfide and other organic sulfur compounds and ammonia which are stripped in a tower with gas or steam before being discharged to the wastewater treatment plant.

Wastewater treatment plants are also a significant source of refinery air emissions and solid wastes. Air releases arise from fugitive emissions from the numerous tanks, ponds, and sewer system drains. Solid wastes are generated in the form of sludges from a number of the treatment units.

Gas Treatment and Sulfur Recovery

Sulfur is removed from a number of refinery process off-gas streams (sour gas) in order to meet the SO_2 emissions limits and to recover saleable elemental sulfur. Process off-gas streams, or sour gas, from the coker, catalytic cracking unit, hydrotreating units, and hydroprocessing units can contain high concentrations of hydrogen sulfide mixed with light refinery fuel gases. Before elemental sulfur can be recovered, the fuel gases (primarily methane and ethane) need to be separated from the hydrogen sulfide. This is typically accomplished by dissolving the hydrogen sulfide in a chemical solvent.

Blending

Blending is the final operation in petroleum refining. It consists of mixing the products in various proportions to meet specifications such as vapor pressure, specific gravity, sulfur content, viscosity, octane number, initial boiling point, and pour point. Blending can be carried out inline or in batch blending tanks. Air emissions from blending are fugitive VOCs from blending tanks, valves, pumps, and mixing operations.

Storage Tanks

Storage tanks are used to store crude oil & intermediate process feeds for cooling and further processing. Finished petroleum products are also kept in storage tanks before transport off site. Storage tank bottoms are mixtures of iron rust from corrosion, sand, water, and emulsified oil & wax, which accumulate at the bottom of tanks, liquid tank bottoms (primarily water and oil emulsions) are periodically drawn off to prevent their continued buildup. Tank bottom liquids and sludge are also removed during periodic cleaning of tanks for inspection, Tank bottoms may contain amounts of tetraethyl or tetramethyl lead (although this is increasingly rare due to the phaseout of leaded products), other metals, and phenols. Solids generated from leaded gasoline storage tank bottoms are hazardous waste. Even if equipped with floating tops, storage tanks account for considerable VOC emissions at petroleum refineries.

Raw material input to petroleum refineries is primarily crude oil; however, petroleum refineries use and generate an enormous number of chemicals, many of which leave the facilities as discharges of air emissions, wastewater, or solid waste. Pollutants generated typically include VOCs, carbon monoxide (CO), sulfur oxides (SO_x), nitrogen oxides (NO_x), particulates, ammonia (NH_3), hydrogen sulfide (H_2S) metals, spent acids, and numerous toxic organic compounds.

The pollutant outputs from the refining facilities, however, are modest in comparison to the pollutant outputs realized from the consumption of petroleum products by the transportation sector, electric utilities, chemical manufacturers, and other industrial and commercial users.

Air emissions from refineries include fugitive emissions of the volatile constituents in crude oil and its fractions, emissions from the burning of fuels in process heaters, and emissions from the various refinery processes themselves.

Fugitive emissions occur throughout refineries and arise from the thousands of potential fugitive emission sources such as valves, pumps, tanks, pressure relief valves, flanges, etc. While individual leaks are typically small, the sum of all fugitive leaks at a refinery can be one of its largest emission sources. Fugitive emissions can be reduced through a number of techniques, including improved leak resistant equipment, reducing the number of tanks and other potential sources, and, perhaps the most effective method, an ongoing Leak Detection and Repair program.

The numerous process heaters used in refineries to heat process streams or to generate steam (boilers) for heating or steam stripping, can be potential sources of SO_2, NO, CO, particulate matter, and hydrocarbon emissions. When operating properly and when burning cleaner fuels such as refinery fuel gas, fuel oil, or natural gas, these emissions are relatively low. If, however, combustion is not complete, or heaters are fired with refinery fuel pitch or residuals, emissions can be significant. The majority of gas streams exiting each refinery process contain varying amounts of refinery fuel gas, hydrogen sulfide, and ammonia. These streams are collected and sent to the gas treatment and sulfur recovery units to recover the refinery fuel gas and sulfur. Emissions from the sulfur recovery unit typically contain some H_2S, SO_2, and NO_2. Other emissions sources from refinery processes arise from periodic regeneration of catalysts. These processes generate streams that may contain relatively high levels of carbon monoxide, particulates, and VOCs. Before being discharged to the atmosphere, such off-gas streams may be treated first through a carbon monoxide boiler to bum carbon monoxide & any VOCs, and then through an electrostatic precipitator or cyclone separator to remove particulate matter.

A good target for simple refineries (i.e., refineries with distillation, catalytic reforming, hydrotreating, and offsite facilities) is that the total quantity of oil consumed as fuel and lost in production operations should not exceed 3.5% of the throughput. For refineries with secondary conversion units (i.e., hydrocrackers or lubricating oil units), the target should be 5 to 6% (and, in some cases, up to 10%) of the throughput. Fugitive VOC emissions from the process units can be reduced to 0.05% of the throughput, with total VOC emissions of less than 1 kg per ton of crude (or, 0.1% of throughput).

Plant operators should aim at using fuel withless than 0.5% sulfur. High sulfur fuels should be directed to units equipped with SO_2, controls. Fuel blending is another option. A sulfur recovery system that achieves at least 97% sulfur recovery should be used when the hydrogen sulfide concentration in tail gases exceeds 230 mg/Nm^3. The total release of sulfur dioxide should be below 0.5 kg per ton for a hydroskimming refinery and below 1 kg per ton for a conversion refinery.

A wastewater generation rate of 0.4 m^3/t of crude processed is achievable with good design & operation, and new refineries should achieve this target as a minimum.

Areas where efforts can be concentrated in reducing air emissions include the following:

- Minimize losses from storage tanks & product transfer areas by methods such as vapor recovery systems and double seals.
- Minimize SO_2, emissions either through desulfurizatzon of fuels, to the extent feasible, or by directing the use of high-sulfurfuels to units equipped with SO_2 emissions controls.
- Recover sulfur from tail gases in high-efficiency sulfur recovery units.
- Recover nonsilica-based/metallic catalysts and reduce particulate emissions.
- Use low-NO_X, burners to reduce nitrogen oxide emissions. Avoid and limit fugitive emissions by proper process design and maintenance.
- Keep fuel usage to a minimum.

The following provides a summary of some pollution prevention techniques found to be effective at petroleum refineries:

Place secondary seals on storage tanks- One of the largest sources of fugitive emissions from refineries is storage tanks containing gasoline and other volatile products. These losses can be significantly reduced by installing secondary seals on storage tanks.

Establish leak detection and repair program - Fugitive emissions are one of the largest sources of refinery hydrocarbon emissions. A leak detection and repair (LDAR) program consists of using a portable VOC detecting instrument to detect leaks during regularly scheduled inspections of valves, flanges, and pump seals. Leaks are then repaired immediately or are scheduled for repair as quickly as possible.

Install vapor recovery for barge loading- Loading of tanker barges is an important VOCs fugitive emissions source for many facilities.

It is estimated that these emissions could be reduced 98% by installing a marine vapor loss control system or VOC destruction in a flare.

Eliminate use of open ponds- Open ponds used to cool, settle out solids and store process water can be a significant source of VOC emissions. Wastewater from coke cooling and cake VOC removal is occasionally cooled in open ponds where VOCs easily escape to the atmosphere. In many cases, open ponds can be replaced with closed storage tanks.

Remove unnecessary storage tanks from service - Since storage tanks are one of the largest sources of VOC emissions, a reduction in the number of these tanks can have a significant impact.The need for certain tanks can be eliminated through improved production planning and more continuous operations. By minimizing the number of storage tanks, tank bottom solids and decanted wastewater may also be reduced.

Replace old boilers- Older refinery boilers can be a significant source of SO2, NOx, and particulate emissions. It is possible to replace a large number of old boilers with a single new cogeneration plant with emissions controls.

Modify the FCCU to allow the use of catalyst fines- Some FCCUs can be modified to recycle some of the catalyst fines generated.

Install rupture discs and plugs- Rupture discs on pressure relieve valves and plugs in open ended valves can reduce fugitive emissions.

Install high pressure power washer- Chlorinated solvent vapor degreasers can C be replaced with high pressure power washers which do not generate spent solvent hazardous wastes or toxic fugitive emissions.

Recycle and regenerate spent caustics- Caustics used to absorb and remove hydrogen sulfide and phenol contaminants from intermediate and final product streams can often be recycled. Spent caustics may be saleable to chemical recovery companies if concentrations of phenol or hydrogen sulfide are high enough. Process changes in the refinery may be needed to raise the concentration of phenols in the caustic to make recovery of the contaminants economical.

Caustics containing phenols can also be recycled on-site by reducing the pH of the caustic until the phenols become insoluble thereby allowing physical separation. The caustic can then be treated along with the refiery waste waters.

Control and reuse FCCU & coke fines Significant quautities of catalyst fines are present around the FCCU catalyst hoppers and reactor & regeneration vessels. Coke fines are present around the coker unit and coke storage areas. The fines can be collected and recycled before being washed to the sewers or migrating off-site via the wind. Collection techniques include dry sweeping the catalyst and coke fines and sending the solids to be recycled or disposed of as nonhazardous waste. Coke fines can also be recycled for fuel use. Another collection technique involves the use of vacuum ducts in dusty areas (and vacuum hoses for manual collection) which run to a small baghouse.

Use non hazardous degreasers. Spent conventional degreaser solvents can be reduced or eliminated through substitution with less toxic and/or biodegradable products. Commonly used are amines, such as diethanolamine (DEA). Dry adsorbents such as molecular sieves, activated carbon, iron sponge, and zinc oxide are also used, sulfide is dissolved in the solution. The fuel gases are removed for use as fuel in process furnaces in other refinery operations. The amine-hydrogen sulfide solution is then heated and steam stripped to remove the hydrogen sulfide gas.

Current methods for removing sulfur from the hydrogen sulfide gas streams are typically a combination of two processes: the Claus Process followed by the Beaven Process, SCOT Process, or the Wellman-Lord Process.

The Claus process consists of partial combustion of the hydrogen sulfide-rich gas stream (with one-third the stoichiometric quantity of air) and then reacting the resulting sulfur dioxide and unbrned hydrogen sulfide in the presence of a bauxite catalyst to produce elemental sulfur.

Since the Claus process by itself removes only about 90% of the hydrogen sulfide in the gas stream, the Beaven, SCOT, or Wellman-Lord processes are often used to further recover sulfur. In the Beaven process, the hydrogen sulfide in the relatively low concentration gas stream from the Claus process can be almost completely removed by absorption in a quinone solution.

The dissolved hydrogen sulfide is oxidized to form a mixture of elemental sulfur and hydro-quinone. The solution is injected with air or oxygen to oxidize the hydro-quinone back to quinone. The solution is then filtered or centrifuged to remove the sulfur and the quinone is then reused.

The Beaven process is also effective in removing small amounts of sulfur dioxide, carbonyl sulfide, and carbon disulfide that are not affected by the Claus process. These compounds are first converted to hydrogen sulfide at elevated temperatures in a cobalt molybdate catalyst prior to being fed to the Beaven unit.

Air emissions from sulfur recovery units will consist of hydrogen sulfide, SO_2 & NO, in the process tail gas as well as fugitive emissions and releases from vents.

The SCOT process is also widely used for removing sulfur from the Claus tail gas. The sulfur compounds in the Claus tail gas are converted to hydrogen sulfide by heating and passing it through a cobalt-molybdenum catalyst with the addition of a reducing gas. The gas is then cooled and contacted with a solution of di-isopropanol amine which removes all but trace amounts of hydrogen sulfide. The sulfide-rich solution is sent to a stripper where hydrogen sulfide gas is removed and sent to the Claus plant. The solution is returned to the absorption column.

Additive Production

A number of chemicals (mostly alcohols and ethers) are added to motor fuels to either improve performance or meet environmental requirements. Alcohols (methanol & ethanol) and ethers are added to gasoline to increase octane levels and reduce carbon monoxide generation in place of the lead additives which are being phased out. Alcohol additives are increasingly supplemented or replaced with a number of different ethers which are better able to meet both the new oxygen requirements and the vapor pressure limits.

The most common ethers being used as additives are methyl tertiary butyl ether (MTBE), and tertiary amyl methyl ether (TAME). Many of the larger refineries manufacture their own supplies of MTBE and TAME by reacting isobutylene and/or isoamylene with methanol. Smaller refineries usually buy their supplies from chemical manufacturers or the larger refineries.

The desalting process creates an oily desalter sludge and a high temperature salt water waste stream which is typically added to other process waste waters for treatment in the refinery wastewater treatment facilities. The water used in crude desalting is often untreated or partially treated water from other refining process water sources.

Atmospheric Distillation - The desalted crude oil is than heated in a heat exchanger and furnace to about 450 °C and fed to a vertical, distillation column at atmospheric pressure where most of the feed is vaporized and separated into its various fractions. Fractions obtained from atmospheric distillation include fuel gas naphtha, gasoline, kerosene, light fuel oil, diesel oils, gas oil, lube distillate, and heavy bottoms. The fuel gas also contains hydrogen sulfide and ammonia gases. The mixture of these gases is known as “sour gas” or “acid gas.” The sour gas is sent to the refinery sour gas treatment system which separates the fuel gas so that it can be used as fuel in the refinery heating furnaces. Air emissions during atmospheric distillation arise from the combustion of fuels in the furnaces to heat the crude oil and to process vents and fugitive emissions. Oily sour water (condensed steam containing hydrogen sulfate and ammonia) and oil are also generated in the fractionators.

Vacuum Distillation – Heavier fractions from the atmospheric distillation unit that cannot be distilled without cracking under its pressure and temperature conditions are vacuum distilled. Vacuum distillation is simply the distillation of petroleum fractions at a very low pressure (0.2 to 0.7 psia) to increase volatilization and separation. In most systems, the vacuum inside the fractionator is maintained with steam ejectors and vacuum pumps, barometric condensers, or surface condensers.

The injection of superheated steam at the base of the vacuum fractionator column further reduces the partial pressure of the hydrocarbons in the tower, facilitating vaporization and separation. The heavier fractions from the vacuum distillation column are processed downstream into more valuable products through either cracking or coking operations.

A potential source of emissions from distillation of crude oil is the combustion of fuels in the furnace and some light gases leaving the top of the condensers on the vacuum distillation column. A certain amount of noncondensable light hydrocarbons & hydrogen sulfide pass through the condenser to a hot well, and then are discharged to the refinery sour fuel system or are vented to a process heater, flare, or another control device to destroy hydrogen sulfide.

If barometric condensers are used in vacuum distillation, significant amounts of oily wastewater can be generated. Vacuum pumps and surface condensers have largely replaced barometric condensers in many refineries to eliminate this oily wastewater stream. Oily sour water is also generated in the fractionators.

Downstream Processing – Certain fractions from the distillation of crude oil are further refined in thermal cracking (visbreaking), coking, catalytic cracking, catalytic hydrocracking, hydrotreating, alkylation, isomerization, polymerization, catalytic reforming, solvent extraction, Merox, dewaxing, propane deasphalting, and other operations. These downstream processes change the molecular structure of hydrocarbon molecules either by breaking than into smaller molecules, joining them to form larger molecules, or reshaping them into higher quality molecu les.

Thermal Cracking/ Visbreaking – Thermal cracking, or visbreaking, uses heat & pressure to break large hydrocarbon molecules into smaller, lighter molecules. The process has been largely replaced by catalytic cracking and some refineries no longer employ thermal cracking. Both processes reduce the production of less valuable products such as heavy fuel oil and cutter stock and increase the feed stock to the catalytic cracker & gasoline yields. Air emissions from thermal cracking include emissions from the combustion of fuels in the process heater, vents, and fugitive emission.

Coking – Coking is a cracking process used primarily to reduce refinery production of low-value residual fuel oils to transportation fuels, such as gasoline and diesel. As part of the upgrading process, coking also produces petroleum coke, which is essentially solid carbon with varying amounts of impurities, and is used as a fuel for power plants if the sulfur content is low enough.

A number of different processes are used to produce coke; both delayed coking and fluid coking are the most widely used processes. Fluid coking produces a higher grade of coke. In delayed coking operations, the same basic process as thermal cracking is used except feed streams are allowed to react longer without being cooled.

Air emissions from coking operations include the process heater flue gas emissions, fugitive emissions, and emissions that may arise from the removal of the coke from the coke drum.

Wastewater is generated from the coke removal & cooling operations and from the steam injection.

In addition, the removal of coke from the drum can release particulate emissions and any remaining hydrocarbons to the atmosphere.

Catalytic Cracking – Catalytic cracking uses heat, pressure, and a catalyst to break larger hydrocarbon molecules into smaller, lighter molecules. Catalytic cracking has largely replaced thermal cracking because it is able to produce more gasoline with a higher octane and less heavy fuel oils and light gases. Feed stocks are light and heavy oils from the crude oil distillation unit which are processed primarily into gasoline as well as some fuel oil and light gases. Most catalysts used in catalytic cracking consist of mixtures of crystalline synthetic silica-alumina, termed "zeolites, and amorphous synthetic silica alumina. The catalytic cracking processes, as well as most other refinery catalytic processes, produce coke which collects on the catalyst surface and diminishes its catalytic properties. The catalyst, therefore, needs to be regenerated continuously or periodically essentially by burning the coke off the catalyst at high temperatures.

Emissions from catalytic cracking operations include: the process heater flue gas emissions, fugitive emissions, and emissions generated during regeneration of the catalyst. Relatively high concentrations of carbon monoxide can be produced during regeneration of the catalyst which is typically converted to carbon dioxide either in the regenerator or further downstream in a carbon monoxide waste heat boiler.

In addition, a significant amount of fine catalyst dust is produced in fluid catalytic craking unit as a result of the constant movement of the catalyst grains against each other. Much of this dust, consisting primarily of alumina and relatively small amounts of nickel, is carried with the carbon monoxide stream to the carbon monoxide burner.

The catalyst dust is then separated from the resulting carbon dioxide stream via cyclones and/or electrostatic precipitators and is sent off-site for disposal or treatment. Generated wastewater is typically sour water from the fractionator containing some oil and phenols. Wastewater containing metal impurities from the feed oil can also be generated from the steam used to purge and regenerate catalysts.

Catalytic Hydrocracking – Catalytic hydrocracking normally utilizes a fixed-bed catalytic cracking reactor with cracking occurring under substantial pressure in the presence of hydrogen. Feedstocks to hydrocracking units are often those fractions that are the most difficult to crack and cannot be cracked effectively in catalytic cracking units. The feedstocks include: middle distillates, cycle oils, residual fuel oils, & reduced crudes. The hydrogen suppresses the formation of heavy residual material and increases the yield of gasoline by reacting with the cracked products. However, this process also breaks the heavy, sulfur and nitrogen bearing hydrocarbons and releases these impurities to where they could potentially foul the catalyst. For this reason, the feedstock is often first hydrotreated to remove impurities before being sent to the catalytic hydrocracker.

Hydrocracking feedstocks are usually first hydrotreated to remove the hydrogen sulfide and ammonia that will poison the catalyst. Sour gas and sour water streams are produced at the fractionator; however, if the hydrocracking feedstocks are first hydrotreated to remove impurities both streams will contain relatively low levels of hydrogen sulfide and ammonia. Hydrocracking catalysts are typically regenerated off-site after two to four years of operation. Therefore, few or no emissions are generated from the regeneration processes. Air emissions arise from the process heater vents, and fugitive emissions.

Hydrotreating/Hydroprocessing – Hydrotreating and hydroprocessing are similar processes used to remove impurities such as sulfur, nitrogen, oxygen, halides, & trace metal impurities that may deactivate process catalysts. Hydrotreating also upgrades the quality of fractions by converting olefins and diolefins to paraffins for the purpose of reducing gum formation in fuels. Hydroprocessing, which typically uses residuals from the crude distillation units, also cracks these heavier molecules to lighter more saleable products. Both hydrotreating & hydroprocessing units are usually placed upstream of those processes in which sulfur and nitrogen could have adverse effects on the catalyst, such as catalytic reforming and hydrocracking units. Air emissions from hydrotreating may arise from process heater flue gas, vents, and fugitive emissions.

Alkylation – Alkylation is used to produce a high octane gasoline blending stock from the isobutane formed primarily during catalytic cracking and coking operations, but also from catalytic reforming, crude distillation, and natural gas processing. Alkylation joins an olefin and an isoparaffin compound using either a sulfuric acid or hydrofluoric acid catalyst. The products are alkylates including propane and butane liquids. When the concentration of acid becomes less than 88%, some of the acid must be removed and replaced with stronger acid. In the hydrofluoric acid process, the slip stream of acid is redistilled. Dissolved polymerization products are removed from the acid as a thick dark oil. The concentrated hydrofluoric acid is recycled.

In the sulfuric acid process, the sulfuric acid removed must be regenerated in a sulfuric acid plant which is generally not a part of the alkylation unit and may be located off-site. Spent sulfuric acid generation is substantial. Air emissions from the alkylation process may arise from process vents and fugitive emissions.

Isomerization – Isomerization is used to alter the arrangement of a molecule without adding or removing anything from the original molecule. Typically, paraffins (butane or pentane from the crude distillation unit) are converted to isoparaffins having a much higher octane. Isomerization reactions take place at temperatures in the range of 100–200 °C in the presence of a catalyst that usually consists of platinum on a base material. Two types of catalysts are currently in use. One requires the continuous addition of small amounts of organic chlorides which are converted to hydrogen chloride in the reactor. In such a reactor, the feed must be free of oxygen sources including water to avoid deactivation and corrosion problems. The other type of catalyst uses a molecular sieve base and does not require a dry & oxygen free feed. Both types of isomerization catalysts require an atmosphere of hydrogen to minimize coke deposits; however, the consumption of hydrogen is negligible. Catalysts typically need to be replaced about every two to three years or longer. Platinum is then recovered from the used catalyst off site. Light ends are stripped from the product stream leaving the reactor and are then sent to the sour gas treatment unit. Some isomerization units utilize caustic treating of the light fuel gas stream to neutralize any entrained hydrochloric acid. This will result in a calcium chloride (or other salts) waste stream. Air emissions may arise from the process heater, vents, and fugitive emissions.

Polymerization – Polymerization is occasionally used to convert propene and butene to high octane gasoline blending components. The process is similar to alkylation in its feed and products, but is often used as a less expensive alternative to alkylation. The reactions typically take place under high pressure in the presence of a phosphoric acid catalyst. The feed must be free of sulfur, which poisons the catalyst; basic materials, which neutralize the catalyst; and oxygen, which affects the reactions. The propene & butene feed is washed first with caustic to remove mercaptans (molecules containing sulfur), then with an amine solution to remove hydrogen sulfide, then with water to remove caustics & amines, and finally dried by passing through a silica gel or molecular sieve dryer. Air emissions of sulfur dioxide may arise during the caustic washing operation.

Catalytic Reforming – Catalytic reforming uses catalytic reactions to process primarily low octane heavy straight run (from the crude distillation unit) gasolines and naphthas into high octane aromatics (including benzene). There are four major types of reactions which occur during reforming processes: (1) dehydrogenation of naphthenes to aromatics; (2) dehydrocyclization of paraffins to aromatics; (3) isomerization; and (4) hydrocracking. The dehydrogenation reactions are very endothermic, requiring that the hydrocarbon stream be heated between each catalyst bed. All but the hydrocracking reaction release hydrogen which can be used in the hydrotreating or hydrocracking processes. Fixed-bed or moving bed processes are utilized in a series of three to six reactors. Feedstocks to catalytic reforming processes are usually hydrotreated first to remove sulfur, nitrogen and metallic contaminants. In continuous reforming processes, catalysts can be regenerated one reactor at a time, once or twice per day, without disrupting the operation of the unit. In semi-regenerative units, the generation of all reactors can be carried out simultaneously after 3 to 24 months of operation by first shutting down the process. Because the recent reformulated gasoline rules have limited the allowable amount of benzene in gasoline, catalytic reforming is being used less as an octane enhancer than in past years. Air emissions from catalytic reforming arise from the process heater gas and fugitive emissions. The catalysts used in catalytic reforming processes are usually very expensive and extra precautions are taken to ensure that catalyst is not lost. When the catalyst has lost its activity and can no longer be regenerated, the catalyst is usually sent off-site for recovery of the metals. Subsequent air emissions from catalyst regeneration are, therefore, relatively low. Relatively small volumes of wastewater containing sulfides, ammonia, & mercaptans may be generated from the stripping tower used to remove light ends from the reactor effluent.

Solvent Extraction - Solvent extraction uses solvents to dissolve & remove aromatics from lube oil feed stocks, improving viscosity, oxidation resistance color, and gum formation. A number of different solvents are used, with the two most common being furfural and phenol. Typically, feed lube stocks are contacted with the solvent in a packed tower or rotating disc contactor. Each solvent has a different solvent-to-oil ratio and recycle ratio within the tower.

Solvents are recovered from the oil stream through distillation & steam stripping in a fractionator. The stream extracted from the solvent contains high concentrations of hydrogen sulfide, aromatics, naphthenes and other hydrocarbons, and is often fed to the hydrocracking unit.

Chemical Treating – Chemical treating is used to remove or change the undesirable properties associated with sulfur, nitrogen, or oxygen compound contaminants in petroleum products. Chemical treating is accomplished by either extraction or oxidation (also known as sweetening), depending upon the product. Extraction is used to remove sulfur from the very light petroleum fractions, such as propane/propylene (PP) and butane /butylene (BB). Sweetening, though, is more effective on gasoline and middle distillate products.

A typical extraction process is "Merox" extraction. Merox extraction is used to remove mercaptans (organic sulfur compounds) from PP and BB streams. PP streams may undergo amine treating before the Merox extraction to remove excess H_2S which tends to fractionate with PP and interferes with the Merox process. A caustic prewash of the PP and BB removes any remaining trace H_2S prior to Merox extraction. The PP and BB streams are passed up through the trays of an extraction tower. Caustic solution flowing down the extraction tower absorbs mercaptan from the PP and BB streams. The rich caustic is then regenerated by oxidizing the mercaptans to disulfide in the presence of aqueous Merox catalyst and the lean caustic recirculated to the extraction tower. The disulfide is insoluble in the caustic and can be separated.

Oxidation or "sweetening" – is used on gasoline and distillate fractions. A common oxidation process is also a Merox process that uses a solid catalyst bed. Air and a minimum amount of alkaline caustic ("mini-alkyl' operation) is injected into the hydrocarbon stream. As the hydrocarbon passes through the Merox catalyst bed, sulfur mercaptans are oxidized to disulfide. In the sweetening Merox process, the caustic is not regenerated. The disulfide can remain with the gasoline product, since it does not possess the objectionable odor properties of mercaptans; hence, the product has been sweetened.

In the extraction process, a waste oily disulfide stream leaves the separator. Air emissions arise from fugitive hydrocarbons and the process vents.

Appendix 11.24: Occupational health and safety (OHS) consideration in various petroleum refinery processes

Occupational health & safety(OHS) consideration s and precautions in various units of a petroleum refinery is discussed below unitwise:

1. Crude Oil Pretreatment (Desalting)

Crude oil contains water, inorganic salts, suspended solids, and water soluble trace metals. To reduce corrosion, plugging & fouling of equipments and to prevent poisoning of the catalysts in processing units, these contaminants are removed by desalting (dehydration).Crude-oil desalting by chemical and electrostatic separation, use hot water as the extraction agent. In chemical desalting, water and chemical surfactant (demulsifiers) are added to the crude, heated so that salts & other impurities dissolve into the water and then held in a tank where they settle out. Electrical desalting employs high-voltage electrostatic charges to concentrate suspended water globules in the bottom of the settling tank. Another less common process involves filtering heated crude using diatomaceous earth.Crude oil is heated to 65-180^{0}C to reduce its viscosity & surface tension for easier mixing and separation of water. Ammonia is used to reduce corrosion. Caustic or acid may be added to adjust the pH of the water wash. Wastewater and contaminants are discharged from the bottom of the settling tank to the wastewater treatment facility. The desalted crude is continuously drawn from the top of the settling tanks and sent to the crude distillation (fractionating) tower.

OHS Considerations

Because this is a closed process, there is little potential for exposure to crude oil unless a leak occurs. During desalting sour crudes, hydrogen sulfide will be present. There may be exposure to ammonia, dry chemical demulsifiers, caustics, and/or acids during this operation. Safe work practices and use of personal protective equipment (PPE) may be needed for avoiding exposures to chemicals & other hazards such as heat, and during process sampling, inspection, maintenance etc. The wastewater may contain chlorides, sulfides, bicarbonates, ammonia, hydrocarbons, phenol, and suspended solids. If diatomaceous earth is used in filtration, exposures should be minimized because it contains silica which is a respiratory hazard.

The potential exists for a fire due to a leak of crude oil from heaters in the desalting unit. Inadequate desalting can cause fouling of heater tubes & heat exchangers throughout the refinery. Fouling restricts product flow & heat transfer and leads to failures due to increased pressures & temperatures. Corrosion, due to the presence of hydrogen sulfide, hydrogen chloride, naphthenic (organic) acids, and other contaminants in the crude oil, also causes equipment failure. Neutralized salts (ammonium chlorides & sulfides), when moistened by condensed water, can cause corrosion. Over pressurising the unit is another potential hazard that causes failures.

2. Crude Oil Distillation

During refining of crude oil by distillation in atmospheric & vacuum distillation towers, the main fractions obtained have specific boiling-point ranges .At the refinery, the desalted crude feedstock is preheated using recovered process heat. The feedstock then flows to a direct-fired crude charge heater where it is fed into the vertical distillation column . All but the heaviest fractions flash into vapor. As the hot vapor rises in the tower, its temperature is reduced. Heavy fuel oil or asphalt residue is taken from the bottom. At successively higher points on the tower, the various major products including lubricating oil, heating oil, kerosene, gasoline, and uncondensed gases are drawn off. . Products ranging from uncondensed fixed gases at the top to heavy fuel oils at the bottom are taken out continuously from atmospheric distillation tower. Steam is often used in towers to lower the vapor pressure and create a partial vacuum. The distillation process separates the major constituents of crude oil into so-called straight-run products. Sometimes crude oil is "topped" by distilling off only the lighter fractions, leaving a heavy residue that is often distilled further under high vacuum.

In order to further distill the residuum or topped crude from the atmospheric tower at higher temperatures, reduced pressure is required to prevent thermal cracking. The process takes place in vacuum distillation towers. A typical first-phase vacuum tower may produce gas oils, lubricating-oil base stocks, and heavy residual for propane deasphalting. A second-phase tower operating at lower vacuum may distill surplus residuum from the atmospheric tower, which is not used for lube-stock processing, and surplus residuum from the first vacuum tower not used for deasphalting. Vacuum towers are typically used to separate catalytic cracking feedstock from surplus residuum. Within refineries there are numerous other, smaller distillation towers called columns, designed to separate specific and unique products.

OHS Considerations

Atmospheric & vacuum distillation are closed processes and exposures are expected to be minimal. When sour crudes are processed, there is potential for exposure to hydrogen sulfide in the preheat exchanger & furnace, tower flash zone & overhead system, vacuum furnace & tower, and bottoms exchanger. Hydrogen chloride may be present in the preheat exchanger, tower top zones, and overheads. Wastewater may contain water soluble sulfides in high concentrations & other water soluble compounds such as ammonia, chlorides, phenol, mercaptans, etc., Safe work practices and/or the use of appropriate personal protective equipment may be needed for avoiding exposures to chemicals, heat & noise, and during sampling, inspection, maintenance etc.

Heaters and heat exchangers in the atmospheric& vacuum distillation units could provide a source of ignition, and the potential for a fire exists in case leak occurs. An excursion in pressure, temperature, or liquid levels may occur if automatic control devices fail. Control of temperature, pressure, and reflux within operating parameters is needed to prevent thermal cracking within the distillation towers. Relief systems are provided for overpressure and operations monitored to prevent crude from entering the reformer charge.

Corrosion may occur in preheat exchanger (by HCl & H_2S), preheat furnace & bottoms exchanger (by H_2S & sulfur compounds), atmospheric tower & vacuum furnace (by H_2S, sulfur compounds,& organic acids), vacuum tower (by H_2S & organic acids), and overhead equipments (by H_2S, HCl, & water). Where sour crudes are processed, severe corrosion can occur in furnace tubing and in both atmospheric & vacuum towers where metal temperatures exceed 230°C. Wet H_2S also causes cracks in steel. When processing high-nitrogen crudes, nitrogen oxides can form in the flue gases of furnaces. Nitrogen oxides are corrosive to steel when cooled to low temperatures in the presence of water.

Chemicals are used to control corrosion by hydrochloric acid produced in distillation units. Ammonia may be injected into the overhead stream prior to initial condensation and/or an alkaline solution may be injected into the hot crude-oil feed. If sufficient wash-water is not injected, deposits of ammonium chloride can cause serious corrosion. Crude feedstock may contain appreciable amounts of water in suspension which can separate during startup and, along with water remaining in the tower from steam purging, settle in the bottom of the tower. This water can be heated to the boiling point and create an instantaneous vaporization explosion upon contact with the oil in the unit.

3. Solvent Extraction & Dewaxing

Solvent treating is used for refining lubricating oils as well as a host of other refinery stocks. Since distillation (fractionation) separates petroleum products into groups only by their boiling-point ranges, impurities may remain. These include organic compounds containing sulfur, nitrogen, & oxygen; inorganic salts & dissolved metals; and soluble salts that were present in the crude feedstock. In addition, kerosene and distillates may have trace amounts of aromatics & naphthenes, and lubricating oil base-stocks may contain wax. Solvent refining processes including solvent extraction and solvent dewaxing remove these undesirables at intermediate refining stages or just before sending the product to storage.

Solvent Extraction

The purpose of solvent extraction is to prevent corrosion, protect catalyst in subsequent processes, and improve finished products by removing unsaturated, aromatic hydrocarbons from lubricant & grease stocks. The solvent extraction process separates aromatics, naphthenes, & impurities from the product stream by dissolving or precipitation. The feedstock is first dried and then treated using a continuous countercurrent solvent treatment operation. In one type of process, the feedstock is washed with a liquid in which the substances to be removed are more soluble than in the desired resultant product. In another process, selected solvents are added to cause impurities to precipitate out of the product. In the adsorption process, highly porous solid materials collect liquid molecules on their surfaces.The solvent is separated from the product stream by heating, evaporation, or fractionation, and residual trace amounts are subsequently removed from the raffinate by steam stripping or vacuum flashing. Electric precipitation may be used for separation of inorganic compounds. The solvent is then regenerated to be used again in the process.The most widely used extraction solvents are phenol, furfural, and cresylic acid. Other solvents less frequently used are liquid sulfur dioxide, nitrobenzene, and 2,2-dichloroethyl ether.

Solvent Dewaxing

Solvent dewaxing is used to remove wax from either distillate or residual base stocks at any stage in the refining process. General steps in solvent dewaxing are : (1) mixing the feedstock with a solvent, (2) precipitating the wax from the mixture by chilling, and (3) recovering the solvent from the wax & dewaxed oil for recycling by distillation and steam stripping. Usually two solvents are used: toluene, which dissolves the oil and maintains fluidity at low temperatures, and methyl ethyl ketone (MEK), which dissolves little wax at low temperatures and acts as a wax precipitating agent. Other solvents that are sometimes used include benzene, methyl isobutyl ketone, propane, petroleum naphtha, ethylene dichloride, methylene chloride, and sulfur dioxide. In addition, there is a catalytic process used as an alternate to solvent dewaxing.

OHS Considerations

Because solvent extraction is a closed process, exposures are expected to be minimal under normal operating conditions. However, there is a potential for exposure to extraction solvents such as phenol, furfural, glycols, methyl ethyl ketone, amines, and other process chemicals. Safe work practices and/or the use of appropriate personal protective equipment may be needed for avoiding exposures to chemicals, noise & heat, and during repair, inspection, maintenance etc. Solvent treatment is essentially a closed process and, although operating pressures are relatively low, the potential exists for fire from a leak or spill contacting a source of ignition such as the drier or extraction heater. In solvent dewaxing, disruption of the vacuum will create a potential fire hazard by allowing air to enter the unit.

4. Thermal Cracking

Because the simple distillation of crude oil produces amounts and types of products that are not consistent with those required by the marketplace, subsequent refinery processes change the product mix by altering the molecular structure of the hydrocarbons. One of the ways of accomplishing this change is through "cracking," a process that breaks or cracks the heavier, higher boiling-point petroleum fractions into more valuable products such as gasoline, fuel oil, and gas oils. The two basic types of cracking are thermal cracking, using heat and pressure, and catalytic cracking. In the first thermal cracking process developed around 1913 distillate fuels & heavy oils were heated under pressure in large drums until they cracked into smaller molecules with better antiknock characteristics. However, this method produced large amounts of solid, unwanted coke. This early process has evolved into the following applications of thermal cracking: visbreaking, steam cracking, and coking.

Visbreaking Process. Visbreaking, a mild form of thermal cracking, significantly lowers the viscosity of heavy crude oil residue without affecting the boiling point range. Residual from the atmospheric distillation tower is heated to 425-510 °C at atmospheric pressure and mildly cracked in a heater. It is then quenched with cool gas oil to control overcracking, and flashed in a distillation tower. Visbreaking is used to reduce the pour point of waxy residues and reduce the viscosity of residues used for blending with lighter fuel oils. Middle distillates may also be produced, depending on product demand. The thermally cracked residue tar, which accumulates in the bottom of the fractionation tower, is vacuum flashed in a stripper and the distillate recycled.

Steam Cracking Process. Steam cracking is a sometimes used in refineries to produce olefinic raw materials (e.g., ethylene) from various feedstock for petrochemicals manufacture. The feedstock range from ethane to vacuum gas oil, with heavier feeds giving higher yields of by-products such as naphtha. The most common feeds are ethane, butane, and naphtha. Steam cracking is carried out at temperatures of 815-870^{0}C and at pressures slightly above atmospheric. Naphtha produced from steam cracking contains benzene, which is extracted prior to hydrotreating. Residual from steam cracking is sometimes blended into heavy fuels.

Coking Processes. Coking is a severe method of thermal cracking used to upgrade heavy residuals into lighter products or distillates. Coking produces straight-run gasoline (coker naphtha) and various middle-distillate fractions used as catalytic cracking feedstock. The process so completely reduces hydrogen that the residue is a form of carbon called "coke." The two most common processes are delayed coking and continuous (contact or fluid) coking process . Three typical types of coke are obtained (sponge coke, honeycomb coke, and needle coke) depending upon the reaction mechanism, time, temperature, & the crude feedstock.

Delayed Coking. In delayed coking the heated charge (typically residuum from atmospheric distillation towers) is transferred to large coke drums which provide the long residence time needed to allow the cracking reactions to proceed to completion. Initially the heavy feedstock is fed to a furnace which heats the residuum to high temperatures of 480-510^{0}C at low pressures of

1.5-2.0 kg/cm^2 and is designed & controlled to prevent premature coking in the heater tubes. The mixture is passed from the heater to one or more coker drums where the hot material is held approximately 24 hours (delayed) at pressures of 1.5-5.0 kg/cm^2, until it cracks into lighter products. Vapors from the drums are returned to a fractionator where gas, naphtha, & gas oils are separated out. The heavier hydrocarbons produced in the fractionator are recycled through the furnace.

After the coke reaches a predetermined level in one drum, the flow is diverted to another drum to maintain continuous operation. The full drum is steamed to strip out uncracked hydrocarbons, cooled by water injection, and decoked by mechanical or hydraulic methods. The coke is mechanically removed by an auger rising from the bottom of the drum. Hydraulic decoking consists of fracturing the coke bed with high-pressure water ejected from a rotating cutter.

Continuous Coking. Continuous (contact or fluid) coking is a moving-bed process that operates at temperatures higher than delayed coking. In continuous coking, thermal cracking occurs by using heat transferred from hot, recycled coke particles to feedstock in a radial mixer, called a reactor, at a pressure of 3.5 kg/cm^2. Gases and vapors are taken from the reactor, quenched to stop any further reaction, and fractionated. The reacted coke enters a surge drum and is lifted to a feeder and classifier where the larger coke particles are removed as product. The remaining coke is dropped into the preheater for recycling with feedstock. Coking occurs both in the reactor and in the surge drum. The process is automatic in that there is a continuous flow of coke and feedstock.

OHS Considerations

The potential exists for exposure to hazardous gases like hydrogen sulfide, carbon monoxide,& trace polynuclear aromatics hydrocarbons (PAHs) associated with coking operations. When coke is moved as a slurry, oxygen depletion may occur within confined spaces such as storage silos, since wet carbon will adsorb oxygen. Wastewater may be highly alkaline and contain oil, sulfides, ammonia, and/or phenol. The potential exists in the coking process for exposure to burns when handling hot coke or in the event of a steam line leak, or from steam, hot water, hot coke, or hot slurry that may be expelled when opening cokers. Safe work practices and use of appropriate personal protective equipment is needed for avoiding exposures to chemicals, heat & noise, and during process sampling, inspection, maintenance, and turnaround activities.

Because thermal cracking is a closed process, the primary potential for fire is from leaks or releases of liquids, gases, or vapors reaching an ignition source such as a heater. The potential for fire is present in coking operations due to vapor or product leaks. Should coking temperatures get out of control, an exothermic reaction could occur within the coker. In thermal cracking when sour crudes are processed, corrosion can occur where metal temperatures are between 230-480^{0}C . Above 480^{0}C, coke forms a protective layer on the metal. The furnace, soaking drums, lower part of the tower, and high-temperature exchangers are usually subject to corrosion. Hydrogen sulfide corrosion in coking can also occur when temperatures are not properly controlled above 480^{0}C. Continuous thermal changes can lead to bulging and cracking of coke drum shells. In coking, temperature control must often be held within a close range, as high temperatures will produce coke that is too hard to cut out of the drum. Conversely, temperatures that are too low will result in a high asphaltic content slurry. Water or steam injection may be used to prevent buildup of coke in delayed coker furnace tubes. Water must be completely drained from the coker, so as not to cause an explosion upon recharging with hot coke.

5. Catalytic Cracking

Catalytic cracking breaks complex hydrocarbons into simpler molecules in order to increase the quality & quantity of lighter, more desirable products and decrease the amount of residuals. This process rearranges the molecular structure of hydrocarbon compounds to convert heavy hydrocarbon feedstock into lighter fractions such as kerosene, gasoline, LPG, heating oil, and petrochemical feedstock.Catalytic cracking is similar to thermal cracking except that catalysts facilitate the conversion of the heavier molecules into lighter products. Use of a catalyst in the cracking reaction increases the yield of improved quality products under much less severe operating conditions than in thermal cracking. Typical temperatures are from 450-510^{0}C at much lower pressures of 0.7-1.5 kg/cm^2. The catalysts used in refinery cracking units are typically solid materials (zeolite, aluminum hydrosilicate, treated bentonite clay, fuller's earth, bauxite, and silica-alumina) that come in the form of powders, beads, pellets or shaped materials called extrudites.

There are three basic functions in the catalytic cracking process:

- **Reaction:** Feedstock reacts with catalyst and cracks into different hydrocarbons;
- **Regeneration:** Catalyst is reactivated by burning off coke; and
- **Fractionation:** Cracked hydrocarbon stream is separated into various products.

The three types of catalytic cracking processes are fluid catalytic cracking (FCC), moving-bed catalytic cracking, and Thermofor catalytic cracking (TCC) process . The catalytic cracking process is very flexible, and operating parameters can be adjusted to meet changing product demand. In addition to cracking, catalytic activities include dehydrogenation, hydrogenation, and isomerization.

Fluid Catalytic Cracking(FCC) . The most common process is FCC, in which the oil is cracked in the presence of a finely divided catalyst which is maintained in an aerated or fluidized state by the oil vapors. The fluid cracker consists of a catalyst section and a fractionating section that operate together as an integrated processing unit. The catalyst section contains the reactor and regenerator, which, with the standpipe and riser, forms the catalyst circulation unit. The fluid catalyst is continuously circulated between the reactor and the regenerator using air, oil vapors, and steam as the conveying media.

A FCC process involves mixing a preheated hydrocarbon charge with hot, regenerated catalyst as it enters the riser leading to the reactor. The charge is combined with a recycle stream within the riser, vaporized, and raised to reactor temperature 480-540^{0}C by the hot catalyst. As the mixture travels up the riser, the charge is cracked at 0.7-2.0 kg/cm ^{2}. In the modern FCC units, all cracking takes place in the riser. The "reactor" no longer functions as a reactor; it merely serves as a holding vessel for the cyclones. This cracking continues until the oil vapors are separated from the catalyst in the reactor cyclones. The resultant product stream (cracked product) is then charged to a fractionating column where it is separated into fractions, and some of the heavy oil is recycled to the riser. Spent catalyst is regenerated to get rid of coke that collects on the catalyst during the process. Spent catalyst flows through catalyst stripper to the regenerator, where most of the coke deposits burn off at the bottom where preheated air and spent catalyst are mixed. Fresh catalyst is added and worn out catalyst is removed to optimize the cracking process.

Moving Bed Catalytic Cracking. The moving bed catalytic cracking process is similar to the FCC process. The catalyst is in the form of pellets that are moved continuously to the top of the unit by conveyor or pneumatic lift tubes to a storage hopper, then flow downward by gravity through the reactor, and finally to a regenerator. The regenerator and hopper are isolated from the reactor by steam seals. The cracked product is separated into recycle gas, oil, clarified oil, distillate, naphtha, and wet gas.

Thermofor Catalytic Cracking (TCC) . In a typical thermofor catalytic cracking unit, the preheated feedstock flows by gravity through the catalytic reactor bed. The vapors are separated from the catalyst and sent to a fractionating tower. The spent catalyst is regenerated, cooled, and recycled. The flue gas from regeneration is sent to a carbon-monoxide boiler for heat recovery.

OHS Considerations

Because the catalytic cracker is a closed system, there is normally little opportunity for exposure to hazardous substances during normal operations. The possibility exists of exposure to extremely hot (370^{0}C) hydrocarbon liquids or vapors during process sampling or if a leak or release occurs. In addition, exposure to hydrogen sulfide and/or carbon monoxide gas may occur during a release of product or vapor.Catalyst regeneration involves steam stripping & decoking, and produces fluid waste streams that may contain hydrocarbon, phenol, ammonia, hydrogen sulfide, mercaptan, and other materials depending upon the feedstock, crudes, and processes. Formation of nickel carbonyl occurs in cracking processes using nickel catalysts, with resultant potential for hazardous exposures. Safe work practices and/or the use of appropriate personal protective equipment is needed for avoiding exposures to chemicals, noise & heat; during process sampling, inspection, maintenance and when handling spent catalyst, recharging catalyst, or if leaks or releases occur.

Liquid hydrocarbons in the catalyst or entering the heated combustion air stream should be controlled to avoid exothermic reactions. Because of the presence of heaters in catalytic cracking units, the possibility exists for fire due to a leak or vapor release. Fire protection including concrete or other insulation on columns and supports, or fixed water spray or fog systems where insulation is not feasible and in areas where firewater hose streams cannot reach, should be considered.

In some processes, caution must be taken to prevent explosive concentrations of catalyst dust during recharge or disposal. When unloading any coked catalyst, the possibility exists for iron sulfide fires. Iron sulfide will ignite spontaneously when exposed to air and therefore must be wetted with water to prevent it from igniting vapors. Coked catalyst may be either cooled below 50^{0}C before it is dumped from the reactor, or dumped into containers that have been purged & inerted with nitrogen and then cooled before further handling.

Regular sampling & testing of the feedstock, product, and recycle streams should be performed to assure that the cracking process is working as intended and that no contaminants have entered the process stream. Corrosives or deposits in the feedstock can foul gas compressors. Inspections of critical equipment including pumps, compressors, furnaces, and heat exchangers should be conducted as needed. When processing sour crude, corrosion may be expected where temperatures are below 480^{0}C. Corrosion takes place where both liquid & vapor phases exist, and at areas subject to local cooling such as nozzles & platform supports.

When processing high-nitrogen feedstock, exposure to ammonia and cyanide may occur, subjecting carbon steel equipment in the FCC overhead system to corrosion, cracking, or hydrogen embrittlement. These effects may be minimized by water wash or corrosion inhibitors. Water wash may also be used to protect overhead condensers in the main column subjected to fouling from ammonium hydrosulfide. Inspections should include checking for leaks due to erosion or other malfunctions such as catalyst buildup on the expanders, coking in the overhead feeder lines from feedstock residues, and other unusual operating conditions.

6. Hydrocracking

Hydrocracking is a two-stage process combining catalytic cracking & hydrogenation, where heavier feedstocks are cracked in the presence of hydrogen to produce more desirable products . Heavy aromatic feedstock is converted into lighter products under high pressures 70-140 kg/cm^2) & high temperatures (410-815^{0}C), in the presence of hydrogen and special catalysts. When the feedstock has a high paraffinic content, the primary function of hydrogen is to prevent the formation of polycyclic aromatic compounds. Another role of hydrogen is to reduce tar formation and prevent buildup of coke on the catalyst. Hydrogenation also serves to converts sulfur & nitrogen compounds present in the feedstock to hydrogen sulfide & ammonia. Hydrocracking produces large amounts of isobutane for alkylation feedstock. Hydrocracking also performs isomerization for pour-point control and smoke-point control, both of which are important in high-quality jet fuel.

OHS Considerations

Because this is a closed process, exposures are expected to be minimal under normal operating conditions. There is a potential for exposure to hydrocarbon gas & vapor emissions, hydrogen and hydrogen sulfide gas due to high-pressure leaks. Large quantities of carbon monoxide may be released during catalyst regeneration & changeover. Catalyst steam stripping & regeneration create waste streams containing sour water and ammonia. Safe work practices and use of appropriate personal protective equipment is needed for avoiding exposure to chemicals, noise & heat, during process sampling, inspection, maintenance etc. and when handling spent catalyst.

Because this unit operates at very high pressures & temperatures, control of both hydrocarbon & hydrogen leakage is important to prevent fires . Care is needed to ensure that explosive concentrations of catalytic dust do not form during recharging. Inspection and testing of safety relief devices are important due to the very high pressures in this unit. Proper process control is needed to protect against plugging reactor beds. Unloading coked catalyst requires precautions to prevent iron sulfide induced fires. The coked catalyst should either be cooled to below 50^{0}C before dumping, or be placed in nitrogen inerted containers until cooled.Because of the operating temperatures & presence of hydrogen, the hydrogen-sulfide content of the feedstock must be strictly controlled to a minimum to reduce the possibility of severe corrosion.

Corrosion by wet carbon dioxide in areas of condensation is also important. When processing high-nitrogen feedstock, the ammonia and hydrogen sulfide form ammonium hydrosulfide, which causes serious corrosion at temperatures below the water dew point. Ammonium hydrosulfide is also present in sour water stripping.

7. Catalytic Reforming

Catalytic reforming converts low-octane naphthas into high-octane gasoline blending components called reformates. Reforming represents the total effect of numerous reactions such as cracking, polymerization, dehydrogenation, & isomerization taking place simultaneously. Depending on the properties of the naphtha feedstock (as measured by the paraffin, olefin, naphthene, & aromatic content)& catalysts used, reformates can be produced with very high concentrations of toluene, benzene, xylene, & other aromatics useful in gasoline blending and petrochemical processing. Hydrogen, a by-product, is separated from the reformate for recycling & use in other processes.

There are many different commercial catalytic reforming processes including platforming, powerforming, ultraforming, & Thermofor catalytic reforming. In the platforming process, the first step is preparation of the naphtha feed to remove impurities from the naphtha & reduce catalyst degradation. The naphtha feedstock is then mixed with hydrogen, vaporized, and passed through a series of alternating furnace & fixed-bed reactors containing a platinum catalyst. The effluent from the last reactor is cooled and sent to a separator to permit removal of the hydrogen rich gas stream from the top of the separator for recycling. The liquid product from the bottom of the separator is sent to a fractionator called a stabilizer (butanizer). It makes a bottom product called reformate; butanes & lighter hydrocarbons go overhead and are sent to the saturated gas plant.Some catalytic reformers operate at low pressure (3.5-14 kg/cm^2), and others operate at high pressures (up to 70 kg/cm^2). Some catalytic reforming systems continuously regenerate the catalyst while in other systems one reactor at a time is taken off stream for catalyst regeneration, and some facilities regenerate all of the reactors during turnarounds.

OHS Considerations

Because this is a closed process, exposures are expected to be minimal under normal operating conditions. There is potential for exposure to hydrogen sulfide & benzene if leakage occurs. Small emissions of carbon monoxide & hydrogen sulfide may occur during regeneration of catalyst. Safe work practices and appropriate personal protective equipment is needed for avoiding exposures to chemicals, noise & heat during testing, inspecting, maintenance etc. and when handling regenerated or spent catalyst.

This is a closed system; however, the potential for fire exists if leakage of reformate gas or hydrogen occur. Operating procedures should ensure control of hot spots during start-up. Safe catalyst handling is very important. Care must be taken not to break or crush the catalyst when loading the beds, as the small fines will plug up the reformer screens. Precautions against dust when regenerating or replacing catalyst is also important. Also, water wash should be considered where stabilizer fouling has occurred due to the formation of ammonium chloride & iron salts. Ammonium chloride may form in pretreater exchangers and cause corrosion & fouling. Hydrogen chloride from the hydrogenation of chlorine compounds may form acid or ammonium chloride salt.

8. Catalytic Hydrotreating

Catalytic hydrotreating is a hydrogenation process used to remove about 90% of contaminants such as nitrogen, sulfur, oxygen, & metals from liquid petroleum fractions. These contaminants, if not removed from the petroleum fractions as they travel through the refinery processing units, can have detrimental effects on the equipment, the catalysts, and the quality of the finished product. Typically, hydrotreating is done prior to processes such as catalytic reforming so that the catalyst is not contaminated by untreated feedstock. Hydrotreating is also used prior to catalytic cracking to reduce sulfur & improve product yields, and to upgrade middle distillate petroleum fractions into finished kerosene, diesel fuel, & heating fuel oils. In addition, hydrotreating converts olefins & aromatics to saturated compounds.

Catalytic Hydrodesulfurization Process. Hydrotreating for sulfur removal is called hydrodesulfurization in which the feedstock is deaerated & mixed with hydrogen, preheated in a fired heater (315-425^{0}C) and then charged under pressure (up to 70 kg/cm^2) through a fixed-bed catalytic reactor. In the reactor, the sulfur & nitrogen compounds in the feedstock are converted into H_2S & NH_3. The reaction products leave the reactor and after cooling to a low temperature enter a liquid/gas separator. The hydrogen-rich gas from the high-pressure separation is recycled to combine with the feedstock, and the low-pressure gas stream rich in H_2S is sent to a gas treating unit where H_2S is removed. The clean gas is then suitable as fuel for the refinery furnaces. The liquid stream is the product from hydrotreating and is normally sent to a stripping column for removal of H_2S & other undesirable components. In cases where steam is used for stripping, the product is sent to a vacuum drier for removal of water. Hydrodesulfurized products are blended or used as catalytic reforming feedstock.

Other Hydrotreating Processes. Hydrotreating processes differ depending upon the feedstock available & catalysts used. Hydrotreating can be used to improve the burning characteristics of kerosene by converting aromatics into naphthenes, which are cleaner-burning compounds.Lube-oil hydrotreating uses catalytic treatment of the oil with hydrogen to improve product quality by saturation of olefins and improvements in color, odor, & acid nature of the oil. Mild lube hydrotreating also may be used following solvent processing. Operating temperatures are usually below 315 ^{0}C and operating pressures below 55 kg/cm^2. Severe lube hydrotreating, at temperatures in the 315-410^{0}C range & hydrogen pressures up to 210 kg/cm^2, is capable of saturating aromatic rings, along with sulfur & nitrogen removal, to impart specific properties not achieved at mild conditions. Hydrotreating also can be employed to improve the quality of pyrolysis gasoline (pygas), a by-product from the manufacture of ethylene. Traditionally, the outlet for pygas has been motor gasoline blending, a suitable route in view of its high octane number. However, only small portions can be blended untreated owing to the unacceptable odor, color, & gum-forming tendencies of this material. The quality of pygas, which is high in diolefin content, can be satisfactorily improved by hydrotreating, whereby conversion of diolefins into mono olefins provides an acceptable product for motor gas blending.

OHS Considerations

There is a potential for exposure to hydrogen sulfide or hydrogen gas in the event of leakage, or to ammonia if sour water leak or spill occurs . Phenol also may be present if high boiling-point feedstocks are processed. Safe work practices and personal protective equipment is needed for avoiding exposures to chemicals noise & heat; during process sampling, inspection, maintenance etc. and when handling amine or exposed to catalyst.

The potential exists for fire in the event of a leak or release of product or hydrogen gas. Many processes require hydrogen generation to provide for a continuous supply. Because of the operating temperatures and presence of hydrogen, the hydrogen sulfide content of the feedstock must be strictly controlled to a minimum to reduce corrosion. Hydrogen chloride may form and condense as hydrochloric acid in the lower-temperature parts of the unit. Ammonium hydrosulfide may form in high-temperature, high-pressure units. Excessive contact time and/or temperature will create coking. Precautions need to be taken when unloading coked catalyst from the unit to prevent iron sulfide fires. The coked catalyst should be cooled to below 50^{0}C before removal, or dumped into nitrogen inerted bins where it can be cooled before further handling. Special antifoam additives may be used to prevent catalyst poisoning from silicone carryover in the coker feedstock.

9. Polymerization

Polymerization in the petroleum industry is the process of converting light olefin gases including ethylene, propylene, & butylene into hydrocarbons of higher molecular weight & higher octane number that can be used as gasoline blending stocks. Polymerization combines two or more identical olefin molecules to form a single molecule with the same elements in the same proportions as the original molecules. Polymerization may be accomplished thermally or in the presence of a catalyst at lower temperatures. The olefin feedstock is pretreated to remove sulfur and other undesirable compounds. In the catalytic process the feedstock is either passed over a solid phosphoric acid catalyst or comes in contact with liquid phosphoric acid, where an exothermic polymeric reaction occurs. This reaction requires cooling water and the injection of cold feedstock into the reactor to control temperatures between 150 to 230^{0}c at pressures from 14-84 kg/cm^2. The reaction products leaving the reactor are sent to stabilization and/or fractionator systems to separate saturated & unreacted gases from the polymer gasoline product.

In the petroleum industry, polymerization is used to indicate the production of gasoline components, hence the term "polymer" gasoline. It is not essential that only one type of monomer be involved. If unlike olefin molecules are combined, the process is referred to as "copolymerization." Polymerization in the true sense of the word is normally prevented, & attempts are made to terminate the reaction at the dimer or trimer (two or three monomers joined together) stage. However, in the petrochemical section of a refinery, polymerization, which results in the production of, for instance, polyethylene, is allowed to proceed until materials of the required high molecular weight are produced.

OHS Considerations

Because this is a closed system, exposures are minimal under normal operating conditions. There is a potential for exposure to caustic wash (sodium hydroxide), to phosphoric acid used in the process or washed out during turnarounds, and to catalyst dust. Safe work practices and personal protective equipment is needed for avoiding exposures to chemicals, noise & heat, and during process sampling, inspection, maintenance, etc.

Polymerization is a closed process where the potential for a fire exists due to leaks or releases reaching a source of ignition.The potential for an uncontrolled exothermic reaction exists if loss of cooling water occur. Severe corrosion leading to equipment failure will occur if water makes contact with the phosphoric acid, such as during water washing at shutdowns. Corrosion may also occur in piping manifolds, reboilers, exchangers, and other locations where acid may settle out.

10. Alkylation

Alkylation combines low-molecular-weight olefins (primarily a mixture of propylene & butylene) with isobutene in the presence of a catalyst, either sulfuric acid or hydrofluoric acid. The product is called alkylate and is composed of a mixture of high-octane, branched-chain paraffinic hydrocarbons. Alkylate is a premium blending stock for gasoline because it has high antiknock properties and is clean burning. The octane number of the alkylate depends mainly upon the kind of olefins used and upon operating conditions.

Sulfuric Acid Alkylation Process. In cascade type sulfuric acid (H_2SO_4) alkylation units, the feedstock (propylene, butylene, amylene, and fresh isobutane) enters the reactor and contacts the concentrated sulfuric acid catalyst (in concentrations of 85 to 95% for good operation and to minimize corrosion). The reactor is divided into zones, with olefins fed through distributors to each zone, and the sulfuric acid and isobutanes flowing over baffles from zone to zone.The reactor effluent is separated into hydrocarbon and acid phases in a settler, and the acid is returned to the reactor. The hydrocarbon phase is hot-water washed with caustic for pH control before being successively depropanized, deisobutanized, and debutanized. The alkylate obtained from the deisobutanizer can then go directly to motor-fuel blending or be rerun to produce aviation-grade blending stock. The isobutane is recycled to the feed.

Hydrofluoric Acid Alkylation Process. Phillips and UOP are the two common types of hydrofluoric acid alkylation processes in use. In the Phillips process, olefin and isobutane feedstock are dried and fed to a combination reactor/settler system. Upon leaving the reaction zone, the reactor effluent flows to a settler (separating vessel) where the acid separates from the hydrocarbons. The acid layer at the bottom of the separating vessel is recycled. The top layer of hydrocarbons (hydrocarbon phase), consisting of propane, normal butane, alkylate, and excess (recycle) isobutanc, is charged to the main fractionator, the bottom product of which is motor alkylate. The main fractionator overhead, consisting mainly of propane, isobutane, and HF, goes to a depropanizer. Propane with trace amount of HF goes to an HF stripper for HF removal and is then catalytically defluorinated, treated, and sent to storage. Isobutane is withdrawn from the main fractionator and recycled to the reactor/settler, and alkylate from the bottom of the main fractionator is sent to product blending.

The UOP process uses two reactors with separate settlers. Half of the dried feedstock is charged to the first reactor, along with recycle and makeup isobutane. The reactor effluent then goes to its settler, where the acid is recycled and the hydrocarbon charged to the second reactor. The other half of the feedstock also goes to the second reactor, with the settler acid being recycled and the hydrocarbons charged to the main fractionator. Subsequent processing is similar to the Phillips process. Overhead from the main fractionator goes to a depropanizer. Isobutane is recycled to the reaction zone and alkylate is sent to product blending.

OHS Considerations

Because alkylation is a closed process, exposures are minimal during normal operations. There is a potential for exposure if leaks, spills, or releases occur. Sulfuric acid and particularly hydrofluoric acid are hazardous chemicals. Special precautionary emergency preparedness measures and protection appropriate to the potential hazard and areas possibly affected needs to be provided. Safe work practices and appropriate skin & respiratory personal protective equipment are needed for avoiding exposures to hydrofluoric & sulfuric acids during normal operations such as reading gauges, inspecting & process sampling, as well as during emergency response, maintenance, and turnaround activities. Procedures should ensure protective equipment & clothing worn in hydrofluoric acid activities are decontaminated and inspected before reissue. Appropriate personal protection for exposure to heat and noise is also required.

Alkylation units are closed processes; however, the potential exists for fire if a leak occurs that allows product or vapor to reach a source of ignition. Loss of coolant water, which is needed to maintain process temperatures, could result in an upset. Precautions are necessary to ensure that equipment and materials that have been in contact with acid are handled carefully and are thoroughly cleaned before they leave the process area or refinery. Immersion wash vats are often provided for neutralization of equipment that has come into contact with hydrofluoric acid. Hydrofluoric acid units should be thoroughly drained and chemically cleaned prior to turnarounds and entry to remove all traces of iron fluoride & hydrofluoric acid. Following shutdown, where water has been used the unit should be thoroughly dried before hydrofluoric acid is introduced.

Leaks, spills, or releases involving hydrofluoric acid or hydrocarbons containing hydrofluoric acid can be extremely hazardous. Care during delivery and unloading of acid is needed. Process unit containment by curbs, drainage, and isolation so that effluent can be neutralized before release to the sewer system is considered. Vents can be routed to soda-ash scrubbers to neutralize hydrogen fluoride gas or hydrofluoric acid vapors before release. Pressure on the cooling water and steam side of exchangers should be kept below the minimum pressure on the acid service side to prevent water contamination. Some corrosion and fouling in sulfuric acid units may occur from the breakdown of sulfuric acid esters or where caustic is added for neutralization. These esters can be removed by fresh acid treating and hot-water washing. To prevent corrosion from hydrofluoric acid, the acid concentration inside the process unit should be maintained above 65% and moisture below 4%.

11. Isomerization

Isomerization converts n-butane, n-pentane & n-hexane into their respective isoparaffins of substantially higher octane number. The straight-chain paraffins are converted to their branched-chain counterparts whose component atoms are the same but are arranged in a different geometric structure. Isomerization is important for the conversion of n-butane into isobutane, to provide additional feedstock for alkylation units, and the conversion of normal pentanes & hexanes into higher branched isomers for gasoline blending. Isomerization is similar to catalytic reforming in that the hydrocarbon molecules are rearranged, but unlike catalytic reforming, isomerization just converts normal paraffins to isoparaffins.

There are two isomerization processes, butane (C_4) and pentane/hexane (C_5/C_6). Butane isomerization produces feedstock for alkylation. Aluminum chloride catalyst plus hydrogen chloride are universally used for the low-temperature isomerisation processes. Platinum or another metal catalyst is used for the higher-temperature processes. In a typical low-temperature process, the feed to the isomerization plant is n-butane or mixed butanes mixed with hydrogen (to inhibit olefin formation) and passed to the reactor at 110-170^{0}C and 14-21 kg/cm^2 . Hydrogen is flashed off in a high-pressure separator and the hydrogen chloride is removed in a stripper column. The resultant butane mixture is sent to a fractionator (deisobutanizer) to separate n-butane from the isobutane product.

Pentane/hexane isomerization increases the octane number of the light gasoline components n-pentane & n-hexane, which are in abundance in straight-run gasoline. In a typical C_5/C_6 isomerization process, dried and desulfurized feedstock is mixed with a small amount of organic chloride & recycled hydrogen, and then heated to reactor temperature. It is then passed over supported-metal catalyst in the first reactor where benzene & olefins are hydrogenated. The feed next goes to the isomerization reactor where the paraffins are catalytically isomerized to isoparaffins. The reactor effluent is then cooled & subsequently separated in the product separator into two streams: a liquid product (isomerate) and a recycle hydrogen-gas stream. The isomerate is washed (with caustic & water), acid stripped, and stabilized before going to storage.

OHS Considerations

Because this is a closed process, exposures are minimal during normal operating conditions. There is a potential for exposure to hydrogen gas, hydrochloric acid, hydrogen chloride and to dust when solid catalyst is used. Safe work practices and personal protective equipment is needed for avoiding exposures to chemicals, heat & noise, and during process sampling, inspection, maintenance, and turnaround activities. Although this is a closed process, the potential for a fire exists should a release or leak contact a source of ignition such as the heater.

If the feedstock is not completely dried & desulfurized, the potential exists for acid formation leading to catalyst poisoning and metal corrosion. Water or steam must not be allowed to enter areas where hydrogen chloride is present. Precautions are needed to prevent HCl from entering sewers and drains.

12. Sweetening & Treating Processes

Treating is a means by which contaminants such as organic compounds containing sulfur, nitrogen,& oxygen; dissolved metals &inorganic salts; and soluble salts dissolved in emulsified water are removed from petroleum fractions / streams. A variety of intermediate & finished products, including middle distillates, gasoline, kerosene, jet fuel, and sour gases are dried & sweetened. Sweetening, of gasoline, treats sulfur compounds (hydrogen sulfide, thiophene & mercaptan) to improve color, odour & oxidation stability. Sweetening also reduces concentrations of carbon dioxide.Treating is done at an intermediate stage in the refining process, or just before sending the finished product to storage. Choices of a treating method depend on the nature of the petroleum fractions, amount & type of impurities, the extent to which the process removes the impurities, and end-product specifications. Treating materials include acids, solvents, alkalis and oxidizing, & adsorption agents.

- **Acid, Caustic, or Clay Treating.** Sulfuric acid is the most commonly used acid treating process. Sulfuric acid treating results in partial or complete removal of unsaturated hydrocarbons, sulfur, nitrogen, & oxygen compounds, and resinous & asphaltic compounds. It is used to improve the odor, color, stability, carbon residue, and other properties of the oil. Clay/lime treatment of acid-refined oil removes traces of asphaltic materials and other compounds improving product color, odor, & stability. Caustic treating with sodium (or potassium) hydroxide is used to improve odor & color by removing organic acids (naphthenic acids, phenols) and sulfur compounds (mercaptans, H_2S) by a caustic wash. By combining caustic soda solution with various solubility promoters (e.g., methyl alcohol and cresols), up to 99%

of all mercaptans as well as oxygen & nitrogen compounds can be dissolved from petroleum fractions.

- **Drying and Sweetening.** Feedstocks from various refinery units are sent to gas treating plants where butanes and butenes are removed for use as alkylation feedstock, heavier components are sent to gasoline blending, propane is recovered for LPG, and propylene is removed for use in petrochemicals. Some mercaptans are removed by water-soluble chemicals that react with the mercaptans. Caustic liquid (sodium hydroxide), amine compounds (diethanolamine) or fixed-bed catalyst sweetening also are used. Drying is accomplished by the use of water absorption or adsorption agents to remove water from the products. Some processes simultaneously dry & sweeten by adsorption on molecular sieves.
- **Sulfur Recovery.** Sulfur recovery converts hydrogen sulfide in sour gases & hydrocarbon streams to elemental sulfur. The most widely used recovery system is the Claus process, which uses both thermal & catalytic-conversion reactions. A typical process produces elemental sulfur by burning hydrogen sulfide under controlled conditions. Knockout pots are used to remove water & hydrocarbons from feed gas streams. The gases are then exposed to a catalyst to recover additional sulfur. Sulfur vapor from burning and conversion is condensed & recovered.
- **Hydrogen Sulfide Scrubbing.** In hydrogen sulfide scrubbing process, the hydrocarbon feedstock is first scrubbed to prevent catalyst poisoning. Depending on the feedstock & the nature of contaminants, desulfurization methods vary from ambient temperature activated charcoal adsorption to high temperature catalytic hydrogenation followed by zinc oxide treating.

OHS Consideration

Because these are closed processes, exposures are minimal under normal operating conditions. There is a potential for exposure to hydrogen sulfide, sodium hydroxide, spent caustic, spent catalyst (Merox process), catalyst dust and sweetening agents (sodium carbonate & sodium bicarbonate). Safe work practices and personal protective equipment is needed for avoiding exposures to chemicals, noise & heat, and during process sampling, inspection, maintenance, and turnaround activities .The potential exists for fire from a leak or release of feedstock or product. Sweetening processes use air or oxygen. If excess oxygen enters these processes, it is possible for a fire to occur in the settler due to the generation of static electricity, which acts as the ignition source..

13. Lubricant, Wax, & Grease Manufacturing Processes

Lubricating oils and waxes are refined from the residual fractions of atmospheric & vacuum distillation. The primary objective of the various lubricating oil refinery processes is to remove asphalts, sulfonated aromatics, and paraffinic & isoparaffinic waxes from residual fractions. Reduced crude from the vacuum unit is deasphalted & combined with straight-run lubricating oil feedstock, preheated, and solvent-extracted (with phenol or furfural) to produce raffinate.

- **Wax Manufacturing Process.** Raffinate from the extraction unit contains wax that must be removed by solvent extraction and crystallization. The raffinate is mixed with a solvent (propane) and precooled in heat exchangers. The crystallization temperature is attained by the evaporation of propane in the chiller and filter feed tanks. The wax is continuously removed by filters and cold solvent-washed to recover retained oil. The solvent is recovered from the oil by flashing & steam stripping. The wax is then heated with hot solvent, chilled, filtered & given a final wash to remove all oil.
- **Lubricating Oil Manufacturing Process.** The dewaxed raffinate is blended with other distillate fractions and further treated for viscosity index, color, stability, carbon residue, sulfur, additive response, and oxidation stability in extremely selective extraction processes using solvents (furfural, phenol, etc.). In a typical phenol unit, the raffinate is mixed with phenol in the treating section at temperatures below 205^{0}C. Phenol is then separated from the treated oil and recycled. The treated lube-oil base stocks are then mixed and/or compounded with additives to meet the required physical & chemical characteristics of motor oils, industrial lubricants, and metal working oils.
- **Grease Manufacturing.** Grease is made by blending metallic soaps (salts of long-chained fatty acids) and additives into a lubricating oil medium at temperatures of 205-315^{0}C. Grease may be either batch-produced or continuously compounded. The characteristics of the grease depend to a great extent on the metallic element (calcium, sodium, aluminum, lithium, etc.) in the soap and the additives used.

OHS Considerations

When blending, sampling, and compounding, personal protection from steam, dusts, mists, vapors, metallic salts, & other additives is required. Skin contact with any formulated grease or lubricant should be avoided. Safe work practices and/or appropriate personal protection is needed for avoiding exposures to chemicals, noise & heat; during inspection, maintenance, & turnaround activities; and while sampling & handling hydrocarbons & chemicals during the production of lubricating oil and wax.

The potential for fire exists if a product or vapor leak or release in the lube blending and wax processing areas reaches a source of ignition. Storage of finished products should be in accordance with recognized safe practices. While the potential for fire is reduced in lube oil blending, care must be taken when making metal-working oils and compounding greases due to the use of higher blending & compounding temperatures & lower flash point products. Control of treater temperature is important as phenol can cause corrosion above 205^{0}C . Spills should be cleaned and leaks repaired to avoid slips & falls. Additives in drums and bags need to be handled properly to avoid strain. Wax can clog sewer or oil drainage systems and interfere with wastewater treatment.

14. Asphalt Production

Asphalt is a portion of the residual fraction that remains after primary distillation operations. It is further processed to impart characteristics required by its final use. In vacuum distillation, generally used to produce road-tar asphalt, the residual is heated to about 400 °C and charged to a column where vacuum is applied to prevent cracking. Asphalt for roofing materials is produced by air blowing. Residual is heated in a pipe still almost to its flash point and charged to a blowing tower where hot air is injected for a predetermined time. The dehydrogenization of the asphalt forms hydrogen sulfide, and the oxidation creates sulfur dioxide. Steam, used to blanket the top of the tower to entrain the various contaminants, is then passed through a scrubber to condense the hydrocarbons.

Another process used to produce asphalt is solvent deasphalting. In this extraction process, which uses propane (or hexane) as a solvent, heavy oil fractions are separated to produce heavy lubricating oil, catalytic cracking feedstock & asphalt. Feedstock & liquid propane are pumped to an extraction tower at precisely controlled mixtures, temperatures (65-120 °C), & pressures of 25-40 kg/cm^2. Separation occurs in a rotating disc contactor, based on differences in solubility. The products are then evaporated & steam stripped to recover the propane, which is recycled. Deasphalting also removes some sulfur & nitrogen compounds, metals, carbon residues, and paraffins from the feedstock.

OHS Considerations

Because these are closed processes, exposures are minimal during normal operations. If a spill occurs, there is a potential for exposure to residuals & asphalt. Air blowing can create some polynuclear aromatic hydrocarbons. Condensed steam from the air-blowing asphalt process may also contain contaminants. The potential for exposure to hydrogen sulfide & sulfur dioxide exists in the production of asphalt. Safe work practices and personal protective equipment is needed for avoiding exposures to chemicals, noise and heat, and during process sampling, inspection, maintenance, and turnaround activities.

The potential for a fire exists if a product leak or release contacts a source of ignition such as the process heater. Condensed steam from the various asphalt & deasphalting processes will contain trace amounts of hydrocarbons. Any disruption of the vacuum can result in the entry of atmospheric air and subsequent fire. In addition, raising the temperature of the vacuum tower bottom to improve efficiency can generate methane by thermal cracking. This can create vapors in asphalt storage tanks that are not detectable by flash testing but are high enough to be flammable. Deasphalting requires exact temperature & pressure control. In addition, moisture, excess solvent, or a drop in operating temperature may cause foaming, which affects the product temperature control and may create an upset.

15. Hydrogen Production

High-purity hydrogen (95%-99%) is required in refineries for hydrodesulfurization, hydrogenation, hydrocracking, and petrochemical processes. Hydrogen, produced as a by-product of refinery processes (principally hydrogen recovery from catalytic reformer product gases), is not enough to meet the total refinery requirements, necessitating the manufacturing of additional hydrogen .

In steam-methane reforming, desulfurized gases are mixed with superheated steam (at 590-870^{0}C) and reformed in tubes containing a nickel base catalyst. The reformed gas, which consists of steam, hydrogen, carbon monoxide & carbon dioxide, is cooled and passed through converters containing an iron catalyst where the carbon monoxide reacts with steam to form carbon dioxide and more hydrogen. The carbon dioxide is removed by amine washing. Any remaining carbon monoxide in the product stream is converted to methane. Steam-naphtha reforming is a continuous process for the production of hydrogen from liquid hydrocarbons and is similar to steam-methane reforming. A variety of naphthas in the gasoline boiling range may be employed, including fuel containing up to 35% aromatics. Following pretreatment to remove sulfur compounds, the feedstock is mixed with steam and taken to the reforming furnace (at 675-815^{0}C) where hydrogen is produced.

OHS Considerations

Because these are closed processes, exposures are minimal during normal operating conditions. There is a potential for exposure to excess hydrogen, carbon monoxide, and/or carbon dioxide. Condensate can be contaminated by process materials such as caustics & amine compounds, with resultant exposures. Depending on the specific process used, safe work practices and appropriate personal protective equipment may be needed for avoiding exposures to chemicals, noise & heat, and during process sampling, inspection, maintenance, & turnaround activities.

The possibility of fire exists if a leakage occurs and reaches an ignition source. The potential exists for burns from hot gases & superheated steam, if leakage occurs. Inspections and testing should be done where the possibility exists for valve failure due to contaminants in the hydrogen. Carryover from caustic scrubbers should be controlled to prevent corrosion in preheaters. Chlorides from the feedstock or steam system should be prevented from entering reformer tubes and contaminating the catalyst.

16. Amine Plants

Amine plants remove acid contaminants from sour gas & hydrocarbon streams. In amine plants, gas & liquid hydrocarbon streams containing carbon dioxide and/or hydrogen sulfide are charged to a gas absorption tower or liquid contactor where the acid contaminants are absorbed by counterflowing amine solutions (i.e., MEA, DEA, MDEA). The stripped gas or liquid is removed overhead, and the amine is sent to a regenerator. In the regenerator, the acidic components are stripped by heat & reboiling action and disposed of, and the amine is recycled.

OHS Considerations

Because this is a closed process, exposures are minimal during normal operations. There is potential for exposure to amine compounds (i.e. monoethanolamine, diethanolamine, methyldiethanolamine), hydrogen sulfide and carbon dioxide. Safe work practices and personal protective equipment is needed for exposures to chemicals, noise & heat, and during process sampling, inspection, maintenance etc. The potential for fire exists where a spill or leak could reach a source of ignition. To minimize corrosion, proper operating practices should be established and regenerator bottom & reboiler temperatures controlled. Oxygen should be kept out of the system to prevent amine oxidation.

17. Unsaturated Gas Plants

Unsaturated (unsat) gas plants recover light hydrocarbons (C_3 & C_4 olefins) from wet gas streams from the FCC, TCC, and delayed coker overhead accumulators or fractionation receivers. In a unsat gas plant, the gases are compressed and treated with amine to remove hydrogen sulfide either before or after they are sent to a fractionating absorber where they are mixed into a concurrent flow of debutanized gasoline. The light fractions are separated by heat in a reboiler, the offgas is sent to a sponge absorber, and the bottoms are sent to a debutanizer. A portion of the debutanized hydrocarbon is recycled, with the balance sent to the splitter for separation. The overhead gases go to a depropanizer for use as alkylation unit feedstock.

OHS Considerations

Because these are closed processes, exposures are minimal under normal operating conditions. There is a potential for exposures to amine compounds such as monoethanolamine (MEA), diethanolamine (DEA), methyldiethanolamine (MDEA) and hydrocarbons. Safe work practices and personal protective equipment is needed for avoiding exposures to chemicals, noise & heat, and during process sampling, inspection, maintenance, etc.

The potential of a fire exists in case spills, releases, or vapors reach a source of ignition. In unsat gas plants handling FCC feedstock, the potential exists for corrosion from moist hydrogen sulfide and cyanides. When feedstocks are from the delayed coker or the TCC, corrosion from hydrogen sulfide and deposits in the high pressure sections of gas compressors from ammonium compounds is possible.

18. Saturate Gas Plant

Saturate (sat) gas plants separate refinery gas components including butanes for alkylation, pentanes for gasoline blending, LPG›s for fuel, and ethane for petrochemicals. Because sat gas processes depend on the feedstock and product demand, each refinery uses different systems, usually absorption-fractionation or straight fractionation. In absorption-fractionation, gases & liquids from various refinery units are fed to an absorber-deethanizer where C_2 and lighter fractions are separated from heavier fractions by lean oil absorption and removed for use as fuel gas or petrochemical feed. The heavier fractions are stripped & sent to a debutanizer, and the lean oil is recycled back to the absorber-deethanizer. C_3/C_4 is separated from pentanes in the debutanizer, scrubbed to remove hydrogen sulfide, and fed to a splitter where propane & butane are separated. In fractionation sat gas plants, the absorption stage is eliminated.

OHS Considerations

Because this is a closed process, exposures are minimal during normal operations. There is potential for exposure to hydrogen sulfide, carbon dioxide, diethanolamine or sodium hydroxide carried over from prior treating. Safe work practices and personal protective equipment is needed for avoiding exposures to chemicals, noise & heat, and during process sampling, inspection, maintenance, and turnaround activities.

There is potential for fire if a leak or release reaches a source of ignition such as the unit reboiler. Corrosion could occur from the presence of hydrogen sulfide, carbon dioxide, and other compounds as a result of prior treating. Streams containing ammonia should be dried before processing. Antifouling additives may be used in absorption oil to protect heat exchangers. Corrosion inhibitors may be used to control corrosion in overhead systems.

19. Blending

Blending is the physical mixture of a number of different liquid hydrocarbons to produce a finished product with certain desired characteristics. Products can be blended in-line through a manifold system, or batch blended in tanks & vessels. In-line blending of gasoline, distillates, jet fuel, & kerosene is accomplished by injecting proportionate amounts of each component into the main stream where turbulence promotes thorough mixing. Additives including octane enhancers, metal deactivators, anti-oxidants, anti-knock agents, gum & rust inhibitors, detergents, etc. are added during and/or after blending to provide specific properties not inherent in hydrocarbons.

OHS Considerations

Ignition sources in the area need to be controlled in the event of a leakage. Safe work practices and/or appropriate personal protective equipment may be needed for avoiding exposures to chemicals & hazards such as noise & heat; when handling additives; and during inspection, maintenance etc.

20. Steam Generation

Heater and Boiler Operations. Steam is generated in main generation plants, and/or at various process units using heat from flue gas or other sources. Heaters (furnaces) include burners and a combustion air system, the boiler enclosure in which heat transfer takes place, a draft or pressure system to remove flue gas from the furnace, soot blowers, and compressed-air systems that seal openings to prevent the escape of flue gas. Boilers consist of a number of tubes that carry the water-steam mixture through the furnace for maximum heat transfer. These tubes run between steam-distribution drums at the top of the boiler and water-collecting drums at the bottom of the boiler. Steam flows from the steam drum to the superheater before entering the steam distribution system.

Heater Fuel. Heaters may use any one or combination of fuels including refinery gas, natural gas, fuel oil, and pulverised coal. Refinery off-gas is collected from process units and combined with natural gas & LPG in a fuel-gas balance drum. The balance drum provides constant system pressure, stable heat content fuel, & automatic separation of suspended liquids in gas vapors, and it prevents carryover of large slugs of condensate into the distribution system. Fuel oil is typically a mix of refinery crude oil with straight-run & cracked residues and other products. The fuel-oil system delivers fuel to process-unit heaters and steam generators at required temperatures & pressures. The fuel oil is heated to pumping temperature, sucked through a coarse suction strainer, pumped to a temperature-control heater, and then pumped through a fine-mesh strainer before being burned.In one example of process-unit heat generation, carbon monoxide boilers recover heat in catalytic cracking units as carbon monoxide in flue gas is burned to complete combustion. In other processes, waste-heat recovery units use heat from the flue gas to make steam.

Steam Distribution. The steam distribution system consists of valves, fittings, piping, and connections suitable for the pressure of the steam transported. Steam leaves the boilers at the highest pressure required by the process units or electrical power generation. The steam pressure is then reduced in turbines that drive process pumps and compressors. Most steam used in the refinery is condensed to water in various types of heat exchangers. The condensate is reused as boiler feedwater or discharged to wastewater treatment. When refinery steam is also used to drive steam turbine generators to produce electricity, the steam must be produced at much higher pressure than required for process steam. Steam typically is generated by heaters (furnaces) and boilers combined in one unit.

Feedwater. Feedwater supply is an important part of steam generation. Water used in steam generation must be free of contaminants including minerals & dissolved impurities that can damage the system or affect its operation. Suspended materials such as silt, sewage, and oil, which form scale & sludge, must be coagulated or filtered out of the water. Dissolved gases, particularly carbon dioxide & oxygen, cause boiler corrosion and are removed by deaeration & treatment. Dissolved minerals including metallic salts, calcium, carbonates, etc., that cause scale, corrosion, and turbine blade deposits are treated with lime or soda ash to precipitate them from the water. Recirculated cooling water must also be treated for hydrocarbons and other contaminants

OHS Considerations

The most potentially hazardous operation in steam generation is heater startup. A flammable mixture of gas and air can build up as a result of loss of flame at one or more burners during light-off. Each type of unit requires specific startup & emergency procedures for purging before light off and in the event of misfire or loss of burner flame. If feedwater runs low and boilers are dry, the tubes will overheat & fail. Conversely, excess water will be carried over into the steam distribution system and damage the turbines. Feedwater must be free of contaminants that could affect operations. Boilers should have continuous or intermittent blowdown systems to remove water from steam drums and limit buildup of scale on turbine blades & superheater tubes. Care must be taken not to overheat the superheater during startup and shut-down.

Alternate fuel sources should be provided in the event of loss of gas due to refinery unit shutdown or emergency. Knockout pots provided at process units remove liquids from fuel gas before burning.

Safe work practices and/or appropriate personal protective equipment may be needed for avoiding exposures to feedwater chemicals, steam, hot water, radiant heat, & noise, and during process sampling, inspection, maintenance, and turnaround activities.

21. Wastewater Treatment

Wastewater treatment is used for process, runoff, and sewerage water prior to discharge or recycling. Wastewater typically contains hydrocarbons, dissolved materials, suspended solids, phenols, ammonia, sulfides, & other compounds. Wastewater includes condensed steam, stripping water, spent caustic solutions, cooling tower & boiler blowdown, wash water, alkaline & acid waste neutralization water, and other process water.

Pretreatment of waste water. Pretreatment is the separation of hydrocarbons and solids from wastewater. API separators, interceptor plates, and settling ponds remove suspended hydrocarbons, oily sludge, & solids by gravity separation, skimming, and filtration. Some oil-in-water emulsions must be heated to assist in separating the oil and water. Gravity separation depends on the specific gravity differences between water and immiscible oil globules and allows free oil to be skimmed off the surface of the wastewater. Acidic wastewater is neutralized using ammonia, lime, or soda ash. Alkaline wastewater is treated with sulfuric acid, hydrochloric acid, carbon dioxide rich flue gas, or sulfur.

Secondary Treatment. After pretreatment, suspended solids are removed by sedimentation or air flotation. Wastewater with low levels of solids may be screened or filtered. Flocculation agents are sometimes added to help separation. Secondary treatment processes biologically degrade & oxidize soluble organic matter by the use of activated sludge, unaerated or aerated lagoons, trickling filter methods, or anaerobic treatments. Materials with high adsorption characteristics are used in fixed-bed filters or added to the wastewater to form a slurry which is removed by sedimentation or filtration. Additional treatment methods are used to remove oils and chemicals from wastewater. Stripping is used on wastewater containing sulfides and/or ammonia, and solvent extraction is used to remove phenols.

Tertiary Treatment. Tertiary treatments remove specific pollutants to meet environmental regulatory discharge requirements. These treatments include chlorination, ozonation, ion exchange, reverse osmosis, activated carbon adsorption, etc. Compressed oxygen is diffused into wastewater streams to oxidize certain chemicals or to satisfy regulatory oxygen content requirements. Wastewater that is to be recycled may require cooling to remove heat and/or oxidation by spraying or air stripping to remove any remaining phenols, nitrates & ammonia.

OHS Considerations

The potential for fire exists if vapors from wastewater containing hydrocarbons reach a source of ignition during treatment. Safe work practices and/or appropriate personal protective equipment may be needed for avoiding exposures to chemicals & waste products during process sampling, inspection, maintenance etc. as well as to noise, gases, and heat.

22. Marine, Tank Car, and Tank Truck Loading & Unloading

Facilities for loading liquid hydrocarbons into tank cars, tank trucks, and marine vessels & barges are usually part of the refinery operations. Product characteristics, distribution needs, shipping requirements, and operating criteria are important when designing loading facilities. Tank trucks and rail tank cars are either top or bottom loaded, and vapor-recovery systems may be provided where required. Loading & unloading liquefied petroleum gas (LPG) require special considerations in addition to those for liquid hydrocarbons.

OHS Considerations

The nature of the health hazards at loading & unloading facilities depends upon the products being loaded and the products previously transported in the tank cars, tank trucks, or marine vessels. Safe work practices and/ or appropriate personal protective equipment is needed to protect against hazardous exposures when loading or unloading, cleaning up spills or leaks, or when gauging, inspecting, sampling, or performing maintenance activities on loading facilities or vapor-recovery systems.

The potential for fire exists where flammable vapors from spills or releases can reach a source of ignition. Where switch loading is permitted, safe practices need to be established and followed. Bonding is used to equalize the electrical charge between the loading rack and the tank truck or tank car. Grounding is used at truck and rail loading facilities to prevent flow of stray currents. Insulating flanges are used on marine dock piping connections to prevent static electricity buildup and discharge. Flame arrestors should be installed in loading rack and marine vapor-recovery lines to prevent flashback. Automatic or manual shutoff systems at supply headers are needed for top & bottom loading in the event of leaks or overfills. Fall protection such as railings are needed for top-loading racks where employees are exposed to falls. Drainage and recovery systems may be provided for storm drainage and to handle spills & leaks. Precautions must be taken at LPG loading facilities not to overload or over pressurize tank cars and trucks.

23. Tank Storage

Atmospheric storage tanks and pressure storage tanks are used throughout the refinery for storage of crudes, intermediate hydrocarbons (during the process), & finished products. Tanks are also provided for fire water, process & treatment water, acids, additives, and other chemicals. The type, construction, capacity & location of tanks depends on their use and materials stored .

OHS Considerations

Safe work practices and personal protective equipment is needed for avoiding exposure to hazards related to product sampling, manual gauging, inspection & maintenance activities including confined space entry .

The potential for fire exists should hydrocarbon storage tanks be overfilled or develop leaks that allow vapors to escape & reach sources of ignition. Remote sensors, control valves, isolation valves, and fire valves may be provided at tanks for pump out or closure in the event of a fire in the tank, or in the tank dike or storage area. Tanks may be provided with automatic overflow control & alarm systems, or manual gauging and checking procedures may be established to control overfills.

24. Heat Exchangers, Coolers, & Process Heaters

Heating Operations. Process heaters and heat exchangers preheat feedstock in distillation towers and in refinery processes to reaction temperatures. Heat exchangers use either steam or hot hydrocarbon transferred from some other section of the process for heat input. The heaters are usually designed for specific process operations, and most are of cylindrical vertical or box-type designs. The major portion of heat provided to process units comes from fired heaters fueled by refinery or natural gas, distillate, and residual oils. Fired heaters are used on crude & reformer preheaters, coker heaters, and large-column reboilers.

Cooling Operations. Heat also may be removed from some processes by air & water exchangers, fin fans, gas & liquid coolers, and overhead condensers, or by transferring heat to other systems. The basic mechanical vapor-compression refrigeration system, which may serve one or more process units, includes an evaporator, compressor, condenser, controls, and piping. Common coolants are water, alcohol/water mixtures, or various glycol solutions.

OHS Considerations

Because these are closed systems, exposures under normal operating conditions are minimal. Depending on the fuel, process operation, and unit design, there is a potential for exposure to hydrogen sulfide, carbon monoxide, hydrocarbons, steam boiler feed-water sludge, & water-treatment chemicals. Skin contact should be avoided with boiler blowdown, which may contain phenolic compounds. Safe work practices and/or appropriate personal protective equipment against hazards may be needed during process maintenance, inspection, & turnaround activities and for protection from radiant heat, superheated steam, hot hydrocarbon, & noise exposures.

A means of providing adequate draft or steam purging is required to reduce the chance of explosions when lighting fires in heater furnaces. Specific start-up and emergency procedures are required for each type of unit. If fire impinges on fin fans, failure could occur due to overheating. If flammable product escapes from a heat exchanger or cooler due to a leak, fire could occur. Care must be taken to ensure that all pressure is removed from heater tubes before removing header or fitting plugs. Provision for pressure relief in heat-exchanger piping systems in the event they are blocked off while full of liquid should be made . If controls fail, variations of temperature and pressure could occur on either side of the heat exchanger. If heat exchanger tubes fail and process pressure is greater than heater pressure, product could enter the heater with downstream consequences. If the process pressure is less than heater pressure, the heater stream could enter into the process fluid. If loss of circulation occurs in liquid or gas coolers, increased product temperature could affect downstream operations and require pressure relief.

25. Cooling Towers

Cooling towers remove heat from process water by evaporation and latent heat transfer between hot water & air. The two types of towers are crossflow and counterflow. Crossflow towers introduce the airflow at right angles to the water flow throughout the structure. In counterflow cooling towers, hot process water is pumped to the uppermost plenum and allowed to fall through the tower. Spray nozzles located throughout the length of the tower disperse the water and help in cooling. Air enters at the tower bottom and flows upward against the water. Recirculated cooling water must be treated to remove impurities and dissolved hydrocarbons. Because the water is saturated with oxygen from being cooled with air, the chances for corrosion are increased. One means of corrosion prevention is the addition of a material to the cooling water that forms a protective film on pipes & other metal surfaces.

OHS Considerations

When cooling water is contaminated by hydrocarbons, flammable vapors can be evaporated into the discharge air. If a source of ignition is present, or if lightning occurs, a fire may result. A potential fire hazard also exists where there are relatively dry areas in induced-draft cooling towers of combustible construction.Loss of power to cooling tower fans or water pumps could have serious consequences in the operation of the refinery. Impurities in cooling water can corrode & foul pipes & heat exchangers, scale from dissolved salts can deposit on pipes, and wooden cooling towers can be damaged by microorganisms.

Cooling-tower water can be contaminated by process materials and by-products including sulfur dioxide, hydrogen sulfide & carbon dioxide, with resultant exposures. Safe work practices and personal protective equipment is needed during process sampling, inspection, maintenance etc. and for exposure to hazards such as those related to noise, water-treatment chemicals, & hydrogen sulfide when wastewater is treated in conjunction with cooling towers.

26. Pressure-Relief and Flare Systems

Pressure-Relief Systems. Pressure-relief systems control vapors and liquids that are released by pressure-relieving devices and blow-downs. Pressure relief is an automatic, planned release when operating pressure reaches a predetermined level. Blowdown normally refers to the intentional release of material, such as blowdowns from process unit startups, furnace blowdowns, shutdowns, & emergencies. Vapor depressuring is the rapid removal of vapors from pressure vessels in case of fire. This may be accomplished by the use of a rupture disc, usually set at a higher pressure than the relief valve. Safety relief valves, used for air, steam, & gas as well as for vapor and liquid, allow the valve to open in proportion to the increase in pressure over the normal operating pressure. Safety valves designed primarily to release high volumes of steam usually pop open to full capacity. The over pressure needed to open liquid relief valves where large-volume discharge is not required increases as the valve lifts due to increased spring resistance. Pilot operated safety relief valves, with up to six times the capacity of normal relief valves, are used where tighter sealing and larger volume discharges are required. Nonvolatile liquids are usually pumped to oil-water separation & recovery systems, and volatile liquids are sent to units operating at a lower pressure.

Flare Systems / bleeder. A typical closed pressure release & flare system includes relief valves and lines from process units for collection of discharges, knockout drums to separate vapors & liquids, seals, and/or purge gas for flashback protection, and a flare & igniter system which combusts vapors when discharging directly to the atmosphere is not permitted. Steam may be injected into the flare tip to reduce visible smoke.

OHS Considerations

Safe work practices and appropriate personal protective equipment may be needed to protect against hazards during inspection, maintenance, and turnaround activities.

Vapors and gases must not discharge where sources of ignition could be present. Liquids should not be discharged directly to a vapor disposal system. Flare knockout drums & flares should be large enough to handle emergency blowdowns. Drums should be provided with relief in the event of over pressure.

Pressure relief valves must be provided where the potential exists for overpressure in refinery processes due to the following causes:

- Loss of cooling water, which may greatly reduce pressure in condensers & increase the pressure in the process unit.
- Loss of reflux volume, which may cause a pressure drop in condensers & a pressure rise in distillation towers because the quantity of reflux affects the volume of vapors leaving the distillation tower.
- Rapid vaporization & pressure increase from injection of a lower boiling-point liquid including water into a process vessel operating at higher temperatures.
- Expansion of vapor & resultant over-pressure due to overheated process steam, malfunctioning heaters, or fire.
- Failure of automatic controls, closed outlets, heat exchanger failure, etc.
- Internal explosion, chemical reaction, thermal expansion, or accumulated gases.

Maintenance is important because valves are required to function properly. The most common operating problems include:

- Failure to open at set pressure, because of plugging of the valve inlet or outlet, or because corrosion prevents proper operation of the disc holder & guides.
- Failure to reseat after popping open due to fouling, corrosion, or deposits on the seat or moving parts, or because solids in the gas stream have cut the valve disc.
- Chattering and premature opening, because operating pressure is too close to the set point.

27. Turbines

Turbines are usually gas or steam powered and are typically used to drive pumps, compressors, blowers, and other refinery process equipments. Steam enters turbines at high temperatures & pressures, expands across and drives rotating blades while directed by fixed blades.

OHS Considerations

Safe work practices and/or appropriate personal protective equipment may be needed for avoiding noise, steam & heat exposures, and during inspection & maintenance activities.

Steam turbines used for exhaust operating under vacuum should have safety relief valves on the discharge side, both for protection and to maintain steam in the event of vacuum failure. Where maximum operating pressure could be greater than design pressure, steam turbines should be provided with relief devices. Consideration should be given to providing governors and overspeed control devices on turbines.

28. Gas & Air Compressors

Both reciprocating & centrifugal compressors are used throughout the refinery for handling gas and compressed air. Air compressor systems include compressors, coolers, air receivers, air dryers, controls, and distribution piping. Blowers are used to provide air to certain processes. Plant air is provided for the operation of pneumatic tools, catalyst regeneration, process heaters, steam air decoking, sour water oxidation, gasoline sweetening, asphalt blowing, and other uses. Instrument air is provided for use in pneumatic instruments & controls, air motors and purge connections.

OHS Considerations

Air compressors should be located so that the suction does not take in flammable vapors or corrosive gases. There is a potential for fire if a leak occurs in gas compressors. Knockout drums are needed to prevent liquid surges from entering gas compressors. If gases are contaminated with solid materials, strainers are needed. Failure of automatic compressor controls will affect processes. If maximum pressure could be greater than compressor or process-equipment design pressure, pressure relief should be provided. Guarding is needed for exposed moving parts on compressors. Compressor buildings should be properly electrically classified, and provisions should be made for proper ventilation. Where plant air is used to back up instrument air, interconnections must be upstream of the instrument air drying system to prevent contamination of instruments with moisture. Alternate sources of instrument air supply, such as use of nitrogen, may be needed in the event of power compressor failure.

Safe work practices and/or appropriate personal protective equipment may be needed for avoiding exposure to hazards such as noise and during inspection & maintenance activities. The use of appropriate safeguards must be considered so that plant & instrument air is not used for breathing or pressurising potable water systems.

29. Pumps, Piping & Valves

Centrifugal & reciprocating pumps are used to move hydrocarbons, process water, fire water, and wastewater through piping within the refinery. Pumps are driven by electric motors, steam turbines, or internal combustion engines. The pump type, capacity, and construction materials depend on the service for which it is used. Process & utility piping distribute hydrocarbons, steam, water, and other products throughout the facility. Their size & construction depend on the type of service, pressure, temperature, and nature of the products. Vent, drain, and sample connections are provided on piping, as well as provisions for blanking. Different types of valves are used depending on their operating purpose. These include gate valves, bypass valves, globe & ball valves, plug valves, block & bleed valves, and check valves. Valves can be operated manually or automatically.

OHS Considerations

The potential for fire exists, if hydrocarbon pumps, valves, or lines develop leaks that could allow vapors to reach sources of ignition. Remote sensors, control valves, fire valves, and isolation valves should be used to limit the release of hydrocarbons at pump suction lines in the event of leakage and/ or fire. Depending on the product and service, backflow prevention from the discharge line may be needed. The failure of automatic pump controls could cause a deviation in process pressure & temperature. Pumps operated with reduced or no flow can overheat and rupture. Pressure relief in the discharge piping should be provided where pumps can be over pressurized . Provisions may be made for pipeline expansion, movement, and temperature changes to avoid rupture. Valves & instruments that require servicing or other work should be accessible at ground level or from an operating platform. Operating vent & drain connections should be provided with double block valves, a block valve and plug, or blind flange for protection against releases.

Safe work practices and/or appropriate personal protective equipment is needed for exposure to hazards such as those related to liquids & vapors when opening or draining pumps, valves, and/or lines, and during product sampling, inspection, & maintenance activities.

30. Electric Power

Refineries may receive electricity from outside sources or produce their own power in captive power generation units with generators driven by steam turbines or gas engines. Electrical substations receive power from the utility or power plant for distribution throughout the facility. They are usually located in nonclassified areas, away from sources of vapor or cooling-tower water spray. Transformers, circuit breakers, and feed-circuit switches are usually located in substations. Substations feed power to distribution stations within the process unit areas. Distribution stations can be located in classified areas, provided that classification requirements are met. Distribution stations usually have a liquid-filled transformer and an oil-filled or air-break disconnect device ie, circuit breaker

OHS Considerations

Safe work practices and/or the use of appropriate personal protective equipment may be needed for avoiding exposures to noise, for exposure to hazards during inspection & maintenance activities, and when working around transformers & switches that may contain a dielectric fluid which requires special handling precautions.

Generators that are not properly classified and are located too close to process units may be a source of ignition if a spill occurs. Normal electrical safety precautions including dry footing, high voltage warning signs, and guarding must be taken to protect against electrocution. Lockout/tagout and other appropriate safe work practices must be established to prevent energization while work is being performed on electrical equipments.

Appendix 11.25: Glossary of Terms(Related to Pretoleum Refinery Engineering)

A

Abel tester: A closed-cup flash tester for kerosene and other oils

Absorption : The disappearance of one substance into another so that the absorbed substance loses its identifying characteristics, while the absorbing substance retains most of its original physical aspects. Used in refining to selectively remove specific components from process streams.

Acid Treatment: A process in which unfinished petroleum products such as gasoline, kerosene, and lubricating oil stocks are treated with sulfuric acid to improve color, odor, and other properties.

Acidity: The presence of acid-type constituents whose concentration is usually defined in terms of neutralization number. The constituents vary in nature and may or may not markedly influence the behavior of the oil.

Additive: Chemicals added to petroleum products in small amounts to improve quality or add special characteristics.

Adsorption : Adhesion of the molecules of gases or liquids to the surface of solid materials.

Air Fin Coolers : A radiator-like device used to cool or condense hot hydrocarbons; also called fin fans.

Air-Fuel Ratio: The ratio of air weight to fuel weight consumed in an internal combustion engine or furnace.

Alicyclic Hydrocarbons : Cyclic (ringed) hydrocarbons in which the rings are made up only of carbon atoms.

Aliphatic hydrocarbons : Hydrocarbons characterised by open-chain structures : ethane, butane, butene, acetylene, etc.

Alkylation: A Process using sulfuric or hydrofluoric acid as a catalyst to combine olefins (usually butylene) and isobutane to produce a high octane product known as alkylate.

Aniline Point: The aniline point of a petroleum product is the minimum equilibrium solution temperature with an equal volume of freshly distilled aniline.

API Gravity : An arbitrary scale expressing the density of petroleum products.

Aromatic : Organic compounds with one or more benzene rings.

Asphaltenes: The asphalt compounds soluble in carbon disulfide but insoluble in paraffin naphthas.

Ash: Inorganic residue remaining after ignition of combustible substances determined by definite prescribed methods.

ASTM: American Society for Testing Materials. Grade and quality specifications for petroleum products are determined by ASTM test methods.

Atmospheric Tower : A distillation unit operated at atmospheric pressure.

Atomization characteristics: The ability of an oil to be broken up into a fine spray by some mechanical means.

B

Barrel: A unit of volume measurement used for petroleum and its products. 1 barrel = 42 U.S. gallons or 35 British gallons or 160 litres. **Bbl:** Abbreviation for barrel.

Benzene: An unsaturated, six carbon, ring, basic aromatic compound.

Bleeder valve : A small flow valve connected to a fluid process vessel or line for the purpose of bleeding off small quantities of contained fluid. It is installed with a block valve to determine if the block valve is closed tightly.

Blending : The process of mixing two or more petroleum products with different properties to produce a finished product with desired characteristics.

Blowdown: The removal of hydrocarbons from a process unit, vessel, or line on a scheduled or emergency basis by the use of pressure through special piping and drums provided for this purpose.

Blower : Equipment for moving large volumes of gas against low pressure heads.

Boiling Range : The range of temperature (usually at atmospheric pressure) at which the boiling (or distillation) of a hydrocarbon liquid commences, proceeds, and finishes.

Bottoms : Tower bottoms are residue remaining in a distillation unit after the highest boiling-point material to be distilled has been removed. Tank bottoms are the heavy materials that accumulate in the bottom of storage tanks, usually comprised of oil, water, and foreign matter.

Bubble Tower : A fractionating (distillation) tower in which the rising vapors pass through layers of condensate, bubbling under caps on a series of plates.

C

Calorific Value: Amount of heat produced by the complete combustion of a unit weight of fuel. Usually expressed in calories per gram or BTU's per pound, the latter being numerically 1.8 times the former. **Catalyst**: A material that aids or promotes a chemical reaction between other substances but does not react itself. Catalysts increase reaction speeds and can provide control by increasing desirable reactions and decreasing undesirable reactions.

Catalytic Cracking : The process of breaking up heavier hydrocarbon molecules into lighter hydrocarbon fractions by use of heat and catalysts.

Cat Cracker: A large refinery vessel for processing reduced crudes or other feed-stocks in the presence of a catalyst, as opposed to the older method of thermal cracking, which employs heat and pressure only. Catalytic cracking is generally preferred since it produces less gas and other highly volatile byproducts. It produces a motor fuel of higher octane than the thermal process.

Catalytic Fines : Hard, abrasive crystalline particles of alumina, silica, and/or alumina silica that can be carried over from the fluidic catalytic cracking process of residual fuel stocks. Particle size can range from sub-micron to greater than sixty (60) microns in size. These particles become more common in the higher viscosity marine bunker fuels.

Caustic Wash: A process in which distillate is treated with sodium hydroxide to remove acidic contaminants that contribute to poor odor and stability.

Centrifuge: A machine using centrifugal force produced by high-speed rotation for separating materials of different densities. Applied to Diesel engine fuels and lubricating oils to remove moisture and other extraneous materials.

Cetane Index: An empirical measure of ignition quality. Defined as the percentage by volume of cetane in a mixture of cetane and methyl naphthalene which has the same ignition quality when used in an engine as a fuel under test.

CFR Diesel fuel testing unit: A standard engine employed in making cetane number tests of Diesel engine fuels.

C/H Ratio: Carbon/Hydrogen ratio

Clarifier: A machine used for a liquid-sludge separation in which the particles with a higher specific gravity are separated form the lower specific gravity of the liquid. A clarifier bowl has one outlet for the light phase oil; the heavier phase particles are retained on the bowl wall.

Cloud Point : Temperature at which wax begins to crystallize from a distillate fuel.

Coke : A high carbon-content residue remaining from the destructive distillation of petroleum residue.

Coking: A process for thermally converting and upgrading heavy residual into lighter products and by-product petroleum coke. Coking also is the removal of all lighter distillable hydrocarbons that leaves a residue of carbon in the bottom of units or as buildup or deposits on equipment and catalysts.

Condensate : The liquid hydrocarbon resulting from cooling vapors.

Condenser : A heat-transfer device that cools and condenses vapor by removing heat via a cooler medium such as water or lower temperature hydrocarbon streams.

Condenser Reflux: Condensate that is returned to the original unit to assist in giving increased conversion or recovery.

Cooler : A heat exchanger in which hot liquid hydrocarbon is passed through pipes immersed in cool water to lower its temperature.

Cutter stock: Flux Stock. A petroleum stock which is used to reduce the viscosity of a heavier residual stock by dilution.

Cracking : The breaking up of heavy molecular weight hydrocarbons into lighter hydrocarbon molecules by the application of heat and pressure, with or without the use of catalysts.

Crude Assay : A procedure for determining the general distillation and quality characteristics of crude oil.

Crude Oil: A naturally occurring mixture of hydrocarbons that usually includes small quantities of sulfur, nitrogen, and oxygen derivatives of hydrocarbons as well as trace metals.

Cycle Gas Oil: Cracked gas oil returned to a cracking unit.

D

Deasphalting: Process of removing asphaltic materials from reduced crude using liquid propane to dissolve nonasphaltic compounds.

Debutanizer : A fractionating column used to remove butane and lighter components from liquid streams.

DE-Ethanizer : A fractionating column designed to remove ethane and gases from heavier hydrocarbons.

Dehydrogenation: A reaction in which hydrogen atoms are eliminated from a molecule. Dehydrogenation is used to convert ethane, propane, and butane into olefins (ethylene, propylene, and butenes).

Demulsibility: The resistance of an oil to emulsification, or the ability of an oil to separate from any water with which it is mixed. The better the demulsibility rating, the more quickly the oil separates from water

Depentanizer : A fractionating column used to remove pentane and lighter fractions from hydrocarbon streams.

Depropanizer : A fractionating column for removing propane and lighter components from liquid streams.

Desalter: The desalter mixes the hydrocarbon stream with a small amount of fresh water (e.g. 10% by volume) forming a water-in-oil emulsion. The resulting emulsion is subjected to an electric field wherein the water is coalesced as an under flow from the upper flow of a relatively water-free, continuous hydrocarbon phase. The desalted hydrocarbon stream is produced at relatively low cost and has a very small residual salt content. The performance of this unit can be improved with a demulsifier .

Desalting : Removal of mineral salts (most chlorides, e.g., magnesium chloride and sodium chloride) from crude oil.

Desulfurization : A chemical treatment to remove sulfur or sulfur compounds from hydrocarbons.

Detonation: A violent explosion involving high-velocity pressure waves; in a gasoline engine, the spontaneous combustion of part of the compresses charge after spark occurs. Detonation usually produces a characteristic metallic sound, or knock.

Dewaxing: The removal of wax from petroleum products (usually lubricating oils and distillate fuels) by solvent absorption, chilling, and filtering.

Diesel index: Product of the API gravity and the aniline point (in degrees Fahrenheit) of a Diesel fuel, divided by 100; an indication of the ignition quality of the fuel.

Diethanolamine : A chemical ($C_4H_8\ 0_2N$) used to remove H_2S from gas streams.

Distillate : The products of distillation formed by condensing vapors.

Distillation: The process of heating a liquid to its boiling point and condensing and collecting the vapors

Doctor test: A qualitative method of detecting undesirable sulfur compounds in petroleum distillates, that is, of determining whether oil is "sour" or "sweet".

Downflow : Process in which the hydrocarbon stream flows from top to bottom.

Dry **Gas** : Natural gas with so little natural gas liquids that it is nearly all methane with some ethane.

E

Electrolytic process: A process that causes the decomposition of a chemical compound by the use of electricity.

Emulsion: A liquid mixture of two or more liquid substances not normally dissolved in one another, one liquid held in suspension in the other. Water-in-oil emulsions have water as the internal phase and oil as the external, while oil-in-water have oil as the internal phase and water as the external.

Engler viscosity: A viscosity obtained by dividing the out-flow time in seconds for 200 ml. of the material being tested, by the time in seconds for 200 ml. of water at 68F (20°C) to flow out of an Engler viscometer.

F

Feedstock : Stock from which meterial is taken to be fed (charged) into a processing unit. Final Boiling Point. The highest temperature indicated on the thermometer inserted in the flask during a standard laboratory distillation. This is generally the temperature at which no more vapor can be driven over into the condensing apparatus.

Fire Point: The lowest temperature at which an oil vaporizes rapidly enough to burn for at least 5 seconds after ignition, under standard conditions.

Flashing: The process in which a heated oil under pressure is suddenly vaporized in a tower by reducing pressure.

Flash Point: Lowest temperature at which a petroleum product will give off sufficient vapor so that the vapor-air mixture above the surface of the liquid will propagate a flame away from the source of ignition.

Flux : Lighter petroleum used to fluidize heavier residual so that it can be pumped.

Fouling : Accumulation of deposits in condensers, exchangers, etc.

Fraction: One of the portions of fractional distillation having a restricted boiling range.

Fractionating Column : Process unit that separates various fractions of petroleum by simple distillation, with the column tapped at various levels to separate and remove fractions according to their boiling ranges.

Fuel Gas : Refinery gas used for heating.

Fuel oil: The heavy distillates from the oil refining process; used as fuel for power stations, marine boilers.

Fungible: Interchangeable. Products which can be commingled for purposes of pipeline shipment.

G

Gas Oil: Middle distillate petroleum fraction with a boiling range of about 175-400°C ; usually includes diesel fuel, kerosene, heating oil, and light fuel oil.

Gasoline : A blend of napthas and other refinery products with sufficiently high octane and other desirable characteristics to be suitable for use as fuel in internal combustion engines.

H

Header : A manifold that distributes fluid from a series of smaller pipes or conduits.

Heat of Combustion (Gross): Total heat evolved during complete combustion of unit weight of a substance, usually expresses in Kcal/kg..

Heat of Combustion (Net): Gross heat of combustion minus the latent heat of condensation of any water produced.

Heavy crude: Crude oil with a high specific gravity and a low API gravity due to the presence of a high proportion of heavy hydrocarbon fractions and metallic content.

High-Line or High-Pressure GAS : High-pressure (7 kg/cm^2) gas from cracking unit distillate drums that is compressed and combined with low line gas as gas absorption feedstock.

Homogenizer: A mechanical device which is used to create a stable, uniform dispersion of an insoluble phase (asphaltenes) within a liquid phase (fuel oil).

HP: Horsepower

HHV: Higher heating value

Hydrocracking: A process used to convert heavier feedstock into lower boiling, higher value products. The process employs high pressure, high temperature, a catalyst, and hydrogen.

Hydrodesulfurization: A catalytic process in which the principal purpose in to remove sulfur from petroleum fractions in the presence of hydrogen.

Hydrofinishing : A catalytic treating process carried out in the presence of hydrogen to improve the properties of low viscosity index naphthenic and medium viscosity index napthenic oils. It is also applied to paraffin waxes and microcrystalline waxes for the removal of undesirable components. This process consumes hydrogen and is used in lieu of acid treating.

Hydroforming: Catalytic reforming of naphtha at elevated temperatures and moderate pressures in the presence of hydrogen to form high octane BTX aromatics for motor fuel or chemical manufacture. This process results in a net production of hydrogen and has rendered thermal reforming some what obsolete. It represents the total effect of numerous simultaneous reactions such as cracking, polymerization, dehydrogenation, and isomerization.

Hydrogenation : The chemical addition of hydrogen to a material in the presence of a catalyst.

Hydrometer: An instrument for determining the specific gravity of a liquid.

I

IBP: Initial Boiling Point. In a standard laboratory distillation, the temperature on the distillation thermometer at the moment the first drop of distillate falls from the condenser.

Inhibitor: Additive used to prevent or retard undesirable changes in the quality of the product,or in the condition of the equipment in which the product is used.

Isomerization : A reaction that catalytically converts straight-chain hydrocarbon molecules into branched-chain molecules of substantially higher octane number. The reaction rearranges the carbon skeleton of a molecule without adding or removing anything from the original material.

Iso-Octane: A hydrocarbon molecule (2,2,4-trimethylpentane) with excellent antiknock characteristics on which the octane number of 100 is based.

K

Kinematic Viscosity: The ratio of the absolute viscosity of a liquid to its specific gravity at the temperature at which the viscosity is measured. Expressed in Stokes or Centistokes.

Knockout Drum : A vessel where in suspended liquid is separated from gas or vapor.

L

Latent heat: Heat required to change the state of a unit weight of a substance from solid to liquid or from liquid to vapor without change of temperature.

Layering: This occurs in tanks when a high density fuel is mixed with a low density fuel.

Lean Oil: Absorbent oil fed to absorption towers in which gas is to be stripped. After absorbing the heavy ends from the gas, it becomes fat oil. when the heavy ends are subsequently stripped, the solvent again becomes lean oil.

LHV: Lower Heating Value

Lifting: Refers to tankers and barges taking on cargoes of oil or refined product at the terminal or transshipment point.

Light Crude: Crude oil with a low specific gravity and high API gravity due to the presence of a high proportion of light hydrocarbon fractions and low metallic compound.

Light Ends: The more volatile products of petroleum refining; eg. butane, propane, gasoline.

Long Ton: An weight measure equalling 2,240 pounds.

Low Line or Low Pressure Gas : Low-pressure (0.3 kg/cm^2) gas from atmospheric and vacuum distillation recovery systems that is collected in the gas plant for compression to higher pressures.

M

Marine Diesel Oil : Marine Diesel oil is a middle distillate fuel oil which can contain traces often percent (10%) or more residual fuel oil from transportation contamination and/or heavy fuel oil blending. The MDO does not require heated storage.

MCR: Maximum continuous rating

MDO: Marine Diesel Oil

Metric Ton: A weight measure equal to 1,000 kilograms, 2,204.62 pounds, and 0.9842 long tons.

Mg/L: Milligrams per liter = ppm (parts per million) - expresses a measure of the concentration by weight of a substance per unit volume.

Middle Distillate: Term applied to hydrocarbons in the so-called “middle range” of refinery distillation. Examples: heating oil, diesel fuels, and kerosene.

Micron: A unit of length. One millionth of a meter or one thousandth of a millimeter. One micron equals 0.00004 of an inch.

Motor Gasoline: A complex mixture of relatively volatile hydrocarbons with or without small quantities of additives, that have been blended to form a fuel suitable for use in spark-ignition engines.

MSDS: Material safety data sheet - a document that provides pertinent information and a profile of a particular hazardous substance or mixture. An MSDS is normally developed by the manufacturer or formulator of the hazardous substance or mixture. The MSDS is required to be made available to employees and operators whenever there is the likelihood of the hazardous substance or mixture being introduced into the workplace. Some manufacturers prepare MSDS for products that are NOT considered to be hazardous to show that the product or substance is NOT hazardous.

N

Naphtha : A general term used for low boiling hydrocarbon fractions that are a major component of gasoline. Aliphatic naphtha refers to those naphthas containing less than 0.1% benzene and with carbon numbers from C_3 through C_{16}. Aromatic naphthas have carbon numbers from C_6 through C_{16} and contain significant quantities of aromatic hydrocarbons such as benzene (> 0.1%), toluene, and xylene.

Naphthenes : Hydrocarbons (cycloalkanes) with the general formula C_nH_{2n}, in which the carbon atoms are arranged to form a ring.

Neutralization number: The number that expresses the weight in milligrams of an alkali needed to neutralize the acidic material in one gram of oil. The neutralization number of an oil is an indication of its acidity.

O

Octane Number: A number indicating the relative antiknock chacacteristics of gasoline.

Olefins : A family **of** unsaturated hydrocarbons with one carbon-carbon double bond and the general formula C_nH_{2n}.

Oxidation: Combining elemental compounds with oxygen to form a new compound. A part of the metabolic reaction.

Oxidizing agent: Any substance such as oxygen and chlorine , that can accept electrons. When oxygen or chlorine is added to wastewater, organic substances are oxidized. These oxidized organic substances are more stable and less likely to give off odors or to contain disease bacteria.

Outage: Space left in a product container to allow for expansion during the temperature changes it may undergo during shipment and application. Measurement of space that is NOT occupied in a drum.

Ozonation: The application of ozone to water, wastewater, or air, generally for the purposes of disinfection or odor control.

P

Paraffins : A family of saturated aliphatic hydrocarbons (alkanes) with the general formula $C_n H_{2n+2}$.

Particulate: Free suspended solids.

PAH: Polycyclic aromatic hydrocarbons.

PCB: Polychlorinated biphenyls; polychloro-biphenyls. Difficult to remediate chemical used in old-style transformers.

Pensky-Martens: A closed-cup test for flash points of oil.

Peristaltic pump: A type of positive displacement pump.

Petrochemical: An intermediate chemical derived from petroleum, hydrocarbon liquids or natural gas, such as: ethylene, propylene, benzene, toluene and xylene.

Petroleum: A generic name for hydrocarbons, including crude oil, natural gas liquids, natural gas and their products.

pH: pH is an expression of the intensity of the basic or acidic condition of a liquid. Mathematically, pH is the logarithm (base 10) of the reciprocal of the hydrogen ion concentration. The pH may range from 0 to 14, where 0 is most acidic, 14 most basic, and 7 is neutral. Natural waters usually have a pH between 6.5 and 8.5.

Phenol: An organic compound that is an alcohol derivative of benzene.

Pollution: The impairment (reduction) of water quality by agriculture, domestic or industrial wastes (including thermal and radioactive wastes) to such a degree as to hinder any beneficial use of the water or render it offensive to the senses of sight, taste, or smell or when sufficient amounts of waste creates or poses a potential threat to human health or the environment.

Polyforming: The thermal conversion of naphtha and gas oils into high quality gasoline at high temperatures and pressure in the presence of recirculated hydrocarbon gases.

Polymer: A chemical formed by the union of many monomers (a molecule of low molecular weight). Polymers are used with other chemical coagulants to aid in binding small suspended particles to form larger chemical flocs for easier removal from water. All polyelectrolytes are polymers, but not all polymers are polyelectrolytes.

Polymerization: Process of combining two or more simple molecules of the same type, called monomers, to form a single molecule having the same elements in the same proportion as in the original molecules, but having increased molecular weight. The product of the combination is a polymer.

Pour Point: Lowest temperature at which an oil will pour or flow under certain prescribed conditions.

ppm: Parts per million - the unit commonly used to designate the concentration of a substance in a wastewater in terms of weight ie. one pound per million pounds, etc. PPM is synonymous with the more commonly used term mg/L (milligrams per liter).

Preheater : Exchanger used to heat hydrocarbons before they are fed to a unit.

Pressure Regulating Valve : A valve that releases or holds process-system pressure (that is, opens or closes) either by preset spring tension or by actuation by a valve controller to assume any desired position between fully open and fully closed.

Purifier: A machine used for a liquid-liquid separation in which the two intermixed liquids which are insoluble in each other have different specific gravities. Solids with specific gravities higher than those of the liquids can be separated off at the same time. A purifier bowl has two outlets; one for the light phase liquid and one for the heavy phase liquid.

Pyrolysis Gasoline: A by-product from the manufacture of ethylene by steam cracking of hydrocarbon fractions such as nephtha or gas oil.

Q

Quench Oil : Oil injected into a product leaving a cracking or reforming heater to lower the temperature and stop the cracking process.

R

Raffinate : The product resulting from a solvent extraction process and consisting mainly of those components that are least soluble in the solvents. The product recovered from an extraction process is relatively free of aromatics, naphthenes, and other constituents that adversely affect physical parameters.

Ramsbottom coke: A carbon residue test originated by Dr. J.R. Ramsbottom in England.

Reactor: The vessel in which chemical reactions take place during a chemical conversion type of process.

Reagent: A pure chemical substance that is used to make new products or is used in chemical tests to measure, detect, or examine other substances.

Reboiler: An auxiliary unit of a fractionating tower designed to supply additional heat to the lower portion of the tower.

Recycle: The use of water or wastewater within (internally) a facility before it is discharged to a treatment system.

Recycle Gas : High hydrogen content gas returned to a unit for reprocessing.

Reduced Crude : A residual product remaining after the removal by distillation of an appreciable quantity of the more volatile components of crude oil.

Reducing agent: Any substance, such as the base metal (iron) or the sulfide ion that will readily donate (give up) electrons. The opposite of an oxidizing agent.

Redwood viscosity: The number of seconds required for 50 ml. of an oil to flow out of a standard Redwood viscosimeter at a definite temperature; British viscosity standard.

Refinery: A plant used to separate the various components present in crude oil and convert them into usable products or feedstock for other processes.

Reflux : The portion of the distillate returned to the fractionating column to assist in attaining better separtion into desired fractions.

Reformate : An upgraded naphtha resulting from catalytic or thermal reforming.

Reforming: The thermal or catalytic conversion of petroleum naphtha into more volatile products of higher octane number. It represents the total effect of numerous simultaneous reactions such as cracking, polymerization, dehydrogenation, and isomerization.

Regeneration : In a catalytic process the reactivation of the catalyst, sometimes done by burning off the coke deposits under carefully controlled conditions of temperature and oxygen content of the regeneration gas stream.

Fuel Oil: Heavy fuel oils produced from the non-volatile residue from the fractional distillation process. Heavy oils that are "leftovers" from various refining processes. Heavy black oils used in marine boilers and in heating plants.

S

Saybolt Furol viscosity: A viscosity test similar in nature to the Saybolt Universal viscosity test but one more appropriate for testing high viscosity oils. Certain transmission and gear oils, and heavy fuel oils are rated by this method. The results obtained are approximately 1/10th the viscosity which would be shown by the Saybolt Universal method.

Scrubbing : Purification of a gas of liquid by washing it in a tower.

SIT: Spontaneous Ignition Temperature. The temperature at which an oil ignites of its own accord in the presence of air or oxygen under standard conditions.

Short ton: An avoirdupois measure of weight equal to 2,000 lbs.

Slagging: Formation of hard deposits on boiler tubes and/or piston crowns, usually due to the presence of sodium, vanadium and sulfur.

Sludge: Deposits in fuel tanks and caused by the presence of wax, sand, scale, asphaltenes, tars, water, etc. The **"sludge" formed in a fuel oil** storage tank is mostly composed of heavy hydrocarbons. *Alken Even-Flo* eliminates this type of sludge by breaking the sludge into small particles and re-suspending them in the fuel for more efficient combustion. The **«sludge» formed in diesel** storage tanks is a combination of water with fungus and bacteria, which grow on the unevenly mixed water/fuel interface. Adding*Alken Even-Flo* to stored fuel promotes a clean separation of water and fuel, reducing the substrate upon which bacteria and fungus can grow. Since the bacteria and fungus bind to the separated water, they can be removed by draining the water from the storage tank. If draining the storage tank is impossible, *EF 905* will emulsify the water into tiny droplets and break the sludge into such small particles that they will no longer clog filters and will efficiently burn.

Soluble: Matter or compounds capable of dissolving into a solution.

Solvent: A substance, normally a liquid, which is capable of absorbing another liquid, gas, or solid to form a homogeneous mixture.

Solvent Extraction : The separation of materials of different chemical types and solubilities by selective solvent action.

Sour Gas : Natural gas that contains corrosive, sulfur bearing compounds such as hydrogen sulfide and mercaptans.

Specifications: Term referring to the properties of a given crude oil or petroleum product, which are "specified" since they often vary widely even within the same grade of product. In the normal process of negotiation, seller will guarantee buyer that product or crude to be sold will meet certain specified limits, and will agree to have such limits certified in writing. Generally the major qualities of oil for which a buyer would demand a guarantee are: API gravity (or specific gravity, in some cases), sulfur percentage measured by weight, pour point measured by degrees C maximum, viscosity min./max., BS&W percentage by weight, etc.

Specific gravity: Weight of a particle, substance or chemical solution in relation to an equal volume of water at 15 °C. Abbreviated as Sp.Gr.

Specific heat: The quantity of heat required to raise the temperature of a unit weight of a substance by 1 degree; usually expresses as calories/gram/ °C

SSF: Seconds Saybolt Furol

SSU: Seconds Saybolt Universal

Stabilize: To convert to a form that resists change. Organic material is stabilized by bacteria which convert the material to gases and other relatively inert substances. Stabilized organic material generally will not give off obnoxious odors.

Stabilization : A process for separating the gaseous and more volatile liquid hydrocarbons from crude petroleum or gasoline and leaving a stable (less volatile) liquid so that it can be handled or stored with less change in composition.

Static mixer: A motionless mixer which has a series of fixed, geometric elements enclosed within a tubular housing. The internal elements impart flow division and radial mixing to the media flowing through the housing to produce a uniform dilution of the production.

Stoke: The unit of kinematic viscosity

Straight-Run: Refers to a petroleum product produced by the primary distillation of crude oil, free of cracked components.

STP: Standard Temperature (25 ^{0}C) and Pressure (760 mm Mercury).

Straight Run Gasoline : Gasoline produced by the primary distillation of crude oil. It contains no cracked, polymerized, alkylated, reformed, or visbroken stock.

Stratification: Occurs in blended fuels that have a compatibility problem. It is usually experienced when paraffinic based oils are mixed with asphaltic based oils, causing asphaltenes to precipitate and settle to the bottom of the tank.

Stripping : The removal (by steam induced vaporization or flash evaporation) of the more volatile components from a cut or fraction.

Sulfuric Acid Treating : A refining process in which unfinished petroleum products such as gasoline, kerosene, and lubricating oil stocks are treated with sulfuric acid to improve their color, odor, and other characteristics.

Sulfurization : Combining sulfur compounds with petroleum lubricants.

Surfactant: Surface-active agent. The active agent in detergents that possesses a high cleaning ability. Used in a spray solution to improve its sticking and wetting properties when applied to plants, algae, or petroleum.

Sweetening : Processes that either remove obnoxious sulfur compounds (primarily hydrogen sulfide, mercaptans, and thiophens) from petroleum fractions or streams, or convert them, as in the case of mercaptans, to odorless disulfides to improve odor, color, and oxidation stability.

T

Tag-Robinson Colorimeter: An instrument used to determine the color of oils. Also a scale of color values.

Tail Gas : The lightest hydrocarbon gas released from a refining process.

TBN: Total Base Number. This is measured in mg. KOH needed to neutralize an acidic solution through a reverse titration. TBN is the ability of the product to neutralize acid. In a motor oil, this is a property which allows the oil to neutralize acids from combustion that would otherwise degrade the oil.

Thermal Cracking : The breaking up of heavy oil molecules into lighter fractions by the use of high temperature without the aid of catalysts.

Topped Crude Oil: Oil from which the light ends have been removed by a simple refining process. Also referred to as "reduced crude oil".

Total ExistentSediment: Combination of inorganic and hydrocarbon sediments existing in a fuel as delivered.

Toxic: A substance which is poisonous to a living organism.

Toxicity: The relative degree of being poisonous or toxic. A condition which may exist in wastes and will inhibit or destroy the growth or function of certain organisms.

Turnaround: A planned complete shutdown of an entire process or section of a refinery, or of an entire refinery to perform major maintenance, overhaul, and repair operations and to inspect, test, and replace process materials and equipment.

U

Ubbehohde viscosimeter: A suspended level apparatus for accurately determining the viscosity of a liquid.

Ullage: The amount which a tank or vessel lacks of being full.

V

Vanadium Inhibitor: An organic and/or inorganic metal bearing chemical intended to chemically and/or physically combine with the compounds formed during combustion of heavy fuel oil to improve the surface properties of the treated ash compounds.

Vacuum Distillation : The distillation of petroleum under vacuum which reduces the boiling temperature sufficiently to prevent cracking or decomposition of the feedstock.

Vapor : The gaseous phase of a substance that is a liquid at normal temperature and pressure.

Visbreaking: Viscosity breaking is a low-temperature cracking process used to reduce the viscosity or pour point of straight run residuum.

Viscometer: A device for determining the viscosity of oil. There are several methods or devices in general use. Basically, a fixed quantity of oil is allowed to pass through a fixed orifice at a specified temperature over a measured time span and then compared to a standard liquid such as a calibration oil or water.

Viscosity: Measure of the internal friction or resistance of an oil to flow. As the temperature of an oil is increased, its viscosity decreases and it is therefore able to flow more readily. Viscosity is measured on several different scales, including Redwood No. 1 at 100 ^{0}F, Engler Degrees, Saybolt Seconds, etc. The most common method for designation of viscosity is kinematic viscosity, measured in centistokes, cst @ 50 ^{0}C .

Volatile: A volatile substance is one that is capable of being evaporated or changed to a vapor at a relatively low temperature. Volatile substances also can be partially removed by air stripping.

W

Wet Gas : A gas containing a relatively high proportion of hydrocarbons that are recoverable as liquids.

12

Objective Type Q&A in Petroleum Refinery Engineering

1. Petroleum
   (a) is optically active.
   (b) constitutes mainly of olefins.
   (c) does not contain asphalt.
   (d) does not contain aromatics.
2. Main constituent of natural gas is
   (a) CH_4 (b) C_2H_2
   (c) C_2H_4 (d) C_2H_6
3. LPG stands for
   (a) liquid petroleum gas
   (b) liquified petrol gas
   (c) liquid petrol gas
   (d) liquefied petroleum gas
4. Which of the following is a naphthene?
   (a) Butene
   (b) Butadiene
   (c) Cyclohexane
   (d) Acetylene
5. Sweetening of petroleum product means the removal of
   (a) sulphur & its compounds
   (b) water
   (c) organic impurities
   (d) wax
6. Sour crude means the............... bearing crude.
   (a) asphalt
   (b) sulphur compounds
   (c) wax
   (d) nitrogen compounds
7. Octane number of gasoline is a measure of its
   (a) knocking tendency.
   (b) ignition delay.
   (c) ignition temperature.
   (d) smoke point.
8. Octane number of n-heptane is assumed to be
   (a) 100 (b) 0
   (c) 70 (d) ∞

ANSWERS

1. (a) 2. (a) 3. (d) 4. (c) 5. (a) 6. (b) 7. (a) 8. (b)

9. Tetra-ethyl lead is added in gasoline to
(a) increase its smoke point.
(b) reduce gum formation.
(c) reduce the pour point.
(d) increase its octane number.

10. Cetane number of a diesel fuel is the measure of its
(a) ignition delay
(b) smoke point
(c) viscosity
(d) oxidation stability

11. Good quality kerosene should have
(a) iow smoke point.
(b) high smoke point.
(c) high aromatics content.
(d) low paraffins content.

12. The octane number of aviation gasoline may be
(a) 79 *(b)* 87
(c) 97 *(d)* >100

13. Diesel index is defined as
(a) (°API) × (Aniline Point, °F)/100
(b) (°API) × (Aniline Point, °C)/100
(c) (°API) × (100)/Aniline Point, °F
(d) (°API) × (100)/Aniline Point, °C

14. Which of the following has the highest octane number?
(a) Aromatics *(b)* *i*-paraffins
(c) Naphthenes *(d)* Olefins
(e) *n*-paraffins

15. Which of the following has the lowest cetane number?
(a) Aromatics *(b)* *i*-paraffins
(c) Naphthene *(d)* Olefins
(e) n-paraffins

16. Viscosity index of a lubricating oil
(a) is the measure of its flash point.
(b) is the measure of variation of viscosity with temperature.
(c) should be low.
(d) none of these.

17. Pour point of a petrofuel is
(a) multiple of 3°F.
(b) multiple of 5°F.
(c) 5°C below the temperature at which oil ceases to flow.
(d) none of these.

18. Flash point of an oil is determined by the
(a) Pensky Martens apparatus.
(b) Ramsbottom apparatus.
(c) Saybolt viscometer.
(d) Conradson apparatus.

19. Reid vapour pressure of gasoline is the measure of its
(a) pour point
(b) cloud point
(c) vapour locking tendency
(d) carbon residue

20. Aniline point is the temperature at which
(a) equal weight of diesel & the aniline are completely miscible.
(b) equal weight of aniline & the test sample are completely miscible.
(c) equal volume of aniline & the test sample are completely miscible.
(d) aniline vaporises.

ANSWERS

9. *(d)*	10. *(a)*	11. *(b)*	12. *(d)*	13. *(a)*	14. *(a)*	15. *(a)*
16. *(b)*	17. *(c)*	18. *(a)*	19. *(c)*	20. *(c)*		

21. Petrolatum is
(a) same as petroleum ether.
(b) petroleum coke.
(c) a mixture of microcrystalline wax in viscous hydrocarbon liquids.
(d) none of these.

22. Presence of sulphur in gasoline
(a) leads to corrosion.
(b) increases lead susceptibility.
(c) decreases gum formation.
(d) helps during stabilisation.

23. Crude topping column operates at..........pressure.
(a) atmospheric (b) 10 atm
(c) vacuum (d) 3 atm.

24. The main aim of cracking is to produce
(a) gasoline (b) lube oil
(c) petrolatum (d) coke

25. Which of the following is the easiest to crack?
(a) Paraffins (b) Olefins
(c) Naphthenes (d) Aromatics

26. Cracking is
(a) an exothermic reaction.
(b) an endothermic reaction.
(c) favoured at very low temperature.
(d) none of these.

27. Visbreaking
(a) uses natural gas as feed.
(b) is carried out at atmospheric pressure.
(c) produces fuel oil of lower viscosity.
(d) produces gasoline only.

28. In catalytic cracking, the
(a) gasoline obtained has a very low octane number.
(b) pressure & temperature is very high.
(c) gasoline obtained has very high aromatic content.
(d) gasoline obtained has very high amount of gum forming compounds.

29. Catalyst used in the catalytic cracking is
(a) silica-alumina (b) silica gel
(c) vanadium pentoxide (d) nickel

30. Choose the correct statement.
(a) Coking tendency increases with increasing molecular weight.
(b) Coking tendency decreases with increasing molecular weight.
(c) Higher pressure enhances coke formation.
(d) Coking is an exothermic reaction.

31. Reforming converts
(a) olefins into paraffins
(b) naphthenes into aromatics
(c) naphthenes into olefins.
(d) naphthenes into paraffin.

32. The main reaction in reforming is the
(a) dehydrogenation of naphthenes.
(b) hydrogenation of naphthenes.
(c) hydrocracking of paraffins.
(d) saturation of olefins.

ANSWERS

21. (c) 22. (a) 23. (a) 24. (a) 25. (a) 26. (b) 27. (c)
28. (c) 29. (a) 30. (a) 31. (b) 32. (a)

33. Catalyst used in catalytic reforming is
(*a*) platinum on alumina
(*b*) nickel
(*c*) iron
(*d*) aluminium chloride

34. Higher pressure in the reforming reactor
(*a*) increases coke formation.
(*b*) increases the rate of reaction.
(*c*) produces high octane number gasoline.
(*d*) none of these.

35. Feed for reforming is generally
(*a*) naphtha or straight run gasoline.
(*b*) reduced crude.
(*c*) vacuum gas oil.
(*d*) atmospheric gas oil.

36. Polymerisation
(*a*) produces *i*-octane from cracked gases containing *i*-butane and butene.
(*b*) causes olefins to combine with each other.
(*c*) causes aromatics to combine with each other.
(*d*) is aimed at producing lubricating oil.

37. Catalyst used in the catalytic polymerisation is
(*a*) phosphoric acid on kieselghur
(*b*) aluminium chloride
(*c*) nickel
(*d*) vanadium pentaoxide

38. Alkylation
(*a*) causes olefins to combine with each other.
(*b*) causes olefins to combine with iso-paraffins.
(*c*) converts iso-paraffin into olefin.
(*d*) converts olefin into paraffin.

39. Catalyst used in alkylation process is
(*a*) sulphuric acid (*b*) nickel
(*c*) silica gel (*d*) alumina

40. Isomerisation converts the......... to i-paraffins.
(*a*) n- paraffins (*b*) olefins
(*c*) naphthenes (*d*) none of these

41. Catalyst used in the isomerisation is
(*a*) aluminium chloride
(*b*) alumina
(*c*) nickel
(*d*) phosphoric acid

42. Solvent used in the deasphalting process is
(*a*) furfurol (*b*) phenol
(*c*) propane (*d*) hexane

43. Which of the following fractions of petroleum contains maximum sulphur?
(*a*) Diesel (*b*) Gasoline
(*c*) Naphtha (*d*) Atmospheric residue

44. Most widely used solvent for dewaxing is
(*a*) methyl-ethyl-ketone (MEK)
(*b*) naphtha
(*c*) petroleum ether
(*d*) sodium plumbite

ANSWERS

33. (*a*)	34. (*d*)	35. (*a*)	36. (*b*)	37. (*a*)	38. (*b*)	39. (*a*)
40. (*a*)	41. (*a*)	42. (*a*)	43. (*d*)	44. (*a*)		

45. Solution used in Doctor's treatment for the removal of mercaptans is
(a) sodium hydroxide
(b) sodium plumbite
(c) cupric chloride
(d) potassium isobutyrate

46. In sweetening process, solutizer agent used with caustic alkali is
(a) potassium isobutyrate
(b) sodium plumbite
(c) methanol
(d) phenol

47. Performance number of a liquid fuel is related to its
(a) wax content
(b) spontaneous ignition temperature
(c) knocking tendency
(d) sulphur content

48. Cetane number of alpha methyl naphthalene is assumed to be
(a) 0 *(b)* 100
(c) 50 *(d)* ∞

49. Clay treatment is used to remove
(a) salt from the crude oil.
(b) colour & dissolved gases from cracked gasoline.
(c) wax from lube oil.
(d) none of these.

50. Butadiene is a/an
(a) di–olefin *(b)* naphthene
(c) aromatic *(d)* olefin

51. Higher viscosity of lubricating oil usually signifies
(a) lower Reid vapour pressure.
(b) higher acid number.
(c) higher flash point and fire point.
(d) lower flash point and fire point.

52. The most widely used crude topping column in refineries is the................column.
(a) bubble–cap
(b) packed bed
(c) fluidised bed
(d) perforated plate

53. Stabilisation of gasoline (petrol) means
(a) removal of dissolved gases from it.
(b) increasing its oxidation stability.
(c) improving its lead susceptibility.
(d) increasing its vapour pressure.

54. Most commonly used crude heater before the fractionation tower in a refinery is the heater.
(a) electric immersion *(b)* pipestill
(c) steam coil *(d)* none of these

55. Olefins are
(a) saturated hydrocarbons.
(b) unsaturated cyclic compounds (hydrocarbons).
(c) present in substantially good quantity in crude petroleum.
(d) none of these.

56. Which of the following has the highest gum forming tendency in gosoline?

ANSWERS

45. *(b)*	46. *(a)*	47. *(b)*	48. *(a)*	49. *(c)*	50. *(a)*	51. *(c)*
52. *(a)*	53. *(a)*	54. *(b)*	55. *(d)*	56. *(b)*		

(a) Paraffins (b) Diolefins
(c) Aromatics (d) Naphthenes

57. Gum formation in stored gasoline is mainly due to the
(a) alkylation of unsaturates.
(b) presence of sulphur.
(c) oxidation & polymerisation of unsaturates.
(d) higher aromatic content.

58. Which of the following is the most important property for a jet fuel?
(a) Cloud point
(b) Pour point
(c) Colour
(d) Freezing point

59. Boiling range of motor gasoline is an indication of the
(a) case of starting
(b) rate of acceleration
(c) vapour locking tendency
(d) all (a), (b) and (c)

60. Percentage of straight run gasoline in a typical crude oil may be around
(a) 6 (b) 18
(c) 38 (d) 52

61. Which of the following categories of gasoline has the highest lead susceptibility?
(a) Straight run gasoline
(b) Platinum reformed gasoline
(c) Catalytical cracked gasoline
(d) Polymer gasoline

62. Higher boiling fractions like atmospheric residue is distilled under vacuum at low temperature because at high temperature, there is a tendency of the predominance of
(a) thermal cracking
(b) gum formation
(c) coking
(d) discoloration

63. Smoke point of kerosene is the
(a) time after which smoking starts on burning.
(b) temperature at which smoking starts.
(c) maximum height of flame (in mm) without causing smoking, when burnt in a standard lamp.
(d) none of these.

64. Smoke point of a good burning kerosene may be around .... mm.
(a) 0–5 (b) 20–25
(c) 60–65 (d) 100–120

65. Flash point of motor gasoline may be around.................°C.
(a) 10 (b) 45
(c) 100 (d) 150

66. Which of the following fractions of a crude oil will have the maximum gravity API (*i.e.* °API)?
(a) Diesel
(b) Gasoline
(c) Atmospheric gas oil
(d) Vacuum gas oil

67. Petroleum coke is used mainly in the
(a) discoloration of yellow glycerine.
(b) sugar refining.

ANSWERS

57. (c) 58. (d) 59. (d) 60. (b) 61. (d) 62. (a) 63. (c)
64. (b) 65. (b) 66. (b) 67. (c)

(c) manufacture of carbon electrode.

(b) blast furnace for reduction of iron ore.

68. "Breathing loss" on storage of gasoline occurs due to the

(a) presence of unsaturated air.
(b) fluctuation of ambient temperature during day and night.
(c) both (a) and (b).
(d) neither (a) nor (b).

69. Pick out the additive property of a lube oil out of following.

(a) °API gravity
(b) Specific gravity
(c) Viscosity
(d) Flash point

70. A fuel oil consists of 4 fractions A, B, C and D. Their molar compositions and vapour pressures are given below :

Costituents	Mole %	Pure component vapour pressure in mm Hg
A	10	700
B	20	720
C	30	740
D	40	750
Total	100	

The vapour pressure of the fuel oil will be...............mm Hg.

(a) 736 (b) 727.5
(c) 512 (d) 767.8

71. The colour of gasoline is an indication of its

(a) octane number.
(b) lead susceptibility.
(c) gum forming tendency & thoroughness of refining.
(d) none of these.

72. Cetane number of diesel used in trucks may be about

(a) 5 (b) 14
(c) 35 (d) 85

73. Which one is preferred for aircraft engine?

(a) High viscosity index lube oil.
(b) Low viscosity index lube oil.
(c) High freezing point aviation fuel.
(d) None of these.

74. Aniline point of the diesel is a measure of its.................content.

(a) aromatic (b) paraffin
(c) olefin (d) naphthene

75. True boiling point apparatus is used for the

(a) determination of characterisation factor.
(b) evaluation of oil stocks.
(c) determination of true vapour pressure,
(d) none of these.

76. Older crude petroleum

(a) is light and better.
(b) gives more distillates.
(c) gives less tar.
(d) all (a), (b) and (c).

77. Which of the following is not a sulphur compound present in petroleum?

(a) Thiophenes (b) Mercaptans
(c) Sulphones (d) Pyroles

78. Pick out the correct statement.

(a) Paraffins have higher octane number than corresponding iso-paraffin.

ANSWERS

68. (c) 69. (b) 70. (a) 71. (c) 72. (c) 73. (a) 74. (a)
75. (b) 76. (d) 77. (b) 78. (c)

(b) Paraffins have lower smoke point than aromatics.
(c) Suitability of kerosene as a fuel & as an illuminant may be determined by char value test.
(d) Aviation fuel should have very high cloud point.

79. Name the hydrocarbon having the poorest oxidation stability.
(a) naphthene (b) olefin
(c) paraffin (d) aromatics

80. Presence of predominantly large quantity of aromatics (polynuclear) is not desirable in aviation fuel, because it has
(a) high pour point and low smoke point.
(b) low viscosity index.
(c) high self-ingnition temperature.
(d) all (a), (b) and (c).

81. Molecular weight of crude petroleum may be around
(a) 50 (b) 250
(c) 1500 (d) 5000

82. Higher vapour pressure of gasoline indicates
(a) low flash point
(b) high breathing loss
(c) both (a) and (b)
(d) neither (a) nor (b)

83. Flash point of an oil gives an idea of the
(a) nature of boiling point diagram of the system.
(b) amount of low boiling fraction present.
(c) explosion hazards.
(d) all (a), (b) and (c).

84. Paraffin base crude oil as compared to asphalt base crude gives
(a) higher yield of straight run gasoline.
(b) higher octane number gasoline.
(c) lower viscosity index lube oil.
(d) poorer yield of lube oil.

85. Aniline point is a property of the
(a) diesel (b) LPG
(c) naphtha (d) gasoline

86. Choose the correct statement regarding thermal cracking.
(a) Moderate changes in operating temperature does not change the depth of cracking.
(b) Increased residence time results in the decreased severity of cracking.
(c) At low pressure, the yield of lighter hydrocarbons are more.
(d) Greater depth of cracking gives lower octane number gasoline.

87. Pick out the wrong statement.
(a) Iso-paraffin crack faster than n-paraffin.
(b) Catalytic cracking is endothermic, but the regeneration of catalyst is exothermic.
(c) Rate of decomposition of olefins in catalytic cracking is

ANSWERS

79. (b) 80. (d) 81. (b) 82. (c) 83. (d) 84. (a) 85. (a)
86. (c) 87. (d)

slightly slower than the thermal cracking.

(d) None of these.

88. Which of the following reactions is undesirable in the production of catalytically reformed gasoline?

(a) Dehydrogenation of naphthene
(b) Dehydrogenation of lower paraffins
(c) Dehydrocyclisation of higher paraffins
(d) Isomerisation of paraffins

89. Which one of the following processes aims at producing higher yield of gaseous unsaturated hydrocarbons and aromatics like benzene & toluene?

(a) Reforming (b) Pyrolysis
(c) Alkylation (d) Hydrocracking

90. Dearomatisation of kerosene (by liquid sulphur dioxide extraction) is done to

(a) increase its smoke point.
(b) improve its oxidation stability.
(c) decrease the breathing loss.
(d) none of these.

91. Which of the following is the most suitable feed for platforming process (reforming)?

(a) Olefinic hydrocarbon
(b) Naphtha
(c) Fuel oil
(d) Atmospheric residue

92. Name the endothermic reaction out of the following:

(a) Catalytic cracking.
(b) Hydrocracking.
(c) Dehydrogeneration of naphthene to produce aromatic.
(d) Catalytic polymerisation.

93. Concentration of H2SO4 catalyst in alkylation is kept between 90-98%, because H2SO4 having concentration

(a) less than 90% promotes polymerisation.
(b) more than 98% promotes cracking.
(c) both (a) and (b).
(d) neither (a) nor (b).

94. In catalytic alkylation, higher iso-butane to olefin ratio gives

(a) low final boiling point product
(b) higher yield
(c) both (a) and (b)
(d) neither (a) nor (b)

95. Hydrofining is the most recent and effective method for the

(a) removal of sulphur.
(b) improvement of smoke point.
(c) reduction of breathing loss.
(d) improvement of viscosity index.

96. Pick out the undesirable property for a solvent meant for dewaxing of lube oil.

(a) Complete miscibility with oil.
(b) High solubility of wax in the solvent.
(c) Both (a) and (b).
(d) Neither (a) nor (b).

ANSWERS

88. (b) 89. (b) 90. (a) 91. (b) 92. (a) 93. (c) 94. (c)
95. (a) 96. (a)

97. Platforming is a............process.
(a) moving bed
(b) fluidised bed
(c) non-regenerative & fixed bed
(d) regenerative

98. Pick out the correct statement about catalytic polymerisation.
(a) H_2SO_4 polymerisation process gives gasoline rich in unsaturates.
(b) In H_2SO_4 polymerisation, H_3PO_4 is always used with 2% steam to prevent meta & ortho H3PO4 formation, which are inactive.
(c) both (a) and (b)
(d) neither (a) nor (b)

99. Mercapsol and Unisol processes are meant for the
(a) dearomatisation of kerosene
(b) removal of mercaptans
(c) dewaxing
(d) decoloration

100. The characterisation factor of crude petroleum oil is around
(a) 3 (b) 11
(c) 22 (d) 28

101. Research octane number refers to the
(a) low octane number motor fuels.
(b) high octane number motor fuels.
(c) high octane number aviation fuels.
(d) unleaded motor fuels.

102. Casing head gasoline is the liquid
(a) butane.
(b) propane.
(c) natural gas.
(d) gasoline separated from wet natural gas by compression.

103. The amount of tetraethyl lead added to improve the octane number of motor gasoline is around.................c.c per gallon of petrol.
(a) 3 (b) 300
(c) 3000 (d) 1000

104. Liquefied petroleum Gas (LPG) used for the household cooking comprises mainly of
(a) propane & butane
(b) butane & ethane
(c) methane & ethane
(d) methane & carbon monoxide

105. Gasoline extracted from natural gas (by compression and cooling) is called the................ gasoline.
(a) polymer
(b) unleaded
(c) casing head
(d) straight run

106. Paraffins are desirable in lubricating oil, as it has got high
(a) viscosity
(b) viscosity index
(c) smoke point
(d) pour point

107. Tetraethyl lead is added to the petrol to increase its octane number, because its octane number is

ANSWERS

97. (c) 98. (c) 99. (b) 100. (b) 101. (b) 102. (d) 103. (a)
104. (a) 105. (c) 106. (b) 107. (c)

(a) more than 100
(b) round about 100
(c) between 50 and 100
(d) less than 25

108. The characterisation factor of a crude oil is calculated as 12.5. It means that, it is
(a) paraffinic
(b) naphthenic
(c) intermediate
(d) none of these

109. The catalytic cracking of heavier petroleum fraction is done to produce mainly
(a) gasoline
(b) asphalt
(c) diesel oil
(d) tar

110. Extractor temperature is maintained at –20°C in Edeleanu process to reduce the............... of kerosene.
(a) smoke point
(b) paraffins
(c) aromatics
(d) naphthenes

111. Mercaptans is represented as
(a) R–COOH
(b) R–S–H
(c) R–S–R
(d) R–S–R'
where *R* and *R'* are alkyl groups

112. Which of the following additives improves the cetane number of diesel?
(a) Amyl nitrate
(b) Ethyl mercaptan
(c) Napthenic acid
(d) Tetra ethyl lead

113. With increase in the molecular weight of aromatic present in kerosene, its smoking tendency
(a) increases
(b) decreases
(c) remains same
(d) is unpredictable

114. Pentration test and softening point tests are important for the
(a) bitumen
(b) gasoline
(c) kerosene
(d) light gas oil

115. The general formula of naphthenes is
(a) C_nH_{2n+2}
(b) C_nH_{2n-6} (where, $n \geq 6$)
(c) C_nH_{n-4}
(d) same as that for olefins *i.e.* C_nH_{2n}

116. Which of the following constituents present in petroleum is responsible for ash formation?
(a) Nitrogen compounds
(b) Organometallic compounds
(c) Sulphur compounds
(d) Oxygen compounds

117. Choose the correct statement.
(a) Octane number of i-octane is zero.
(b) Octane number of paraffins increases with increasing number of carbon atoms.

ANSWERS

108. (a) 109. (a) 110. (c) 111. (b) 112. (a) 113. (b) 114. (a)
115. (d) 116. (b) 117. (c)

(c) Branched chain paraffins have higher octane number than straight chain paraffins with same number of carbon atoms.
(d) The aromaties have lower octane number than naphthenes with same number of carbon atoms.

118. For gasoline, the flash point (< 50°C) is determined by the
(a) Abel apparatus.
(b) Pensky-Marten's apparatus.
(c) Saybolt chromometer.
(d) none of these.

119. Carbon/hydrogen ratio (by weight) is maximum (out of following) for
(a) gasoline
(b) kerosene
(c) lightgas oil
(d) heavy fuel oil

120. Sulphur content in lighter and heavier petroleum products is generally determined respectively by
(a) lamp method and bomb method.
(b) bomb method and lamp method.
(c) bomb method and quartz tube method.
(d) quartz tube method and lamp method.

121. Flash point of diesel/kerosene (> 50°C) is determined by the
(a) Abel apparatus
(b) Pensky-Martens apparatus
(c) Saybolt chromometer
(d) none of these

122. Which is almost absent in crude petroleum?
(a) Olefins
(b) Mercaptans
(c) Naphthenes
(d) Cycloparaffins

123. Which of the following is desirable in petrol (gasoline) but undesirable in kerosene?
(a) Paraffins
(b) Aromatics
(c) Mercaptans
(d) Naphthenic acid

124. Which of the following is desirable in diesel and kerosene but is undersirable in gasoline?
(a) Aromatics
(b) Mercaptans
(c) Paraffins
(d) Naphthenic acid

125. Which of the following has maximum hydrogen/carbon ratio (by weight)?
(a) Naphtha
(b) Gasoline
(c) Diesel
(d) Fuel oil

126. With increase in the number of carbon and hydrogen atoms in hydrocarbon molecules, the density of petroleum products
(a) decreases
(b) increases

ANSWERS

118. (a) 119. (d) 120. (a) 121. (b) 122. (a) 123. (b) 124. (c)
125. (a) 126. (b)

(c) remains same
(d) unpredictable from the data

127. The yield of straight run LPG from crude oil is about ....weight percent.
(a) 20–25
(b) 1–1.5
(c) 10–15
(d) 0.1–0.2

128. Road grade bitumen is produced from vacuum residue by its
(a) aeration
(b) pyrolysis
(c) hydrogenation
(d) steam reforming

129. Which of the following processes consumes hydrogen?
(a) Fluid catalytic cracking
(b) Visbreaking
(c) Propane deasphalting
(d) None of these

130. The most suitable solvent for deasphalting vacuum residue is
(a) propane
(b) methyl ethyl ketone
(c) doctor's solution
(d) methanol amine

131. Which parameter is used for the grading of paraffin waxes?
(a) Specific gravity
(b) Melting point
(c) Viscosity
(d) Penetration number

132. The pressure (kg/cm^2) and temperature (°C) maintained in electrical desalters for crude oil are respectively
(a) 10 and 120
(b) 1 and 200
(c) 50 and 250
(d) 10 and 300

133. Quinoline is a/an................ compound.
(a) sulphur
(b) nitrogen
(c) oxygen
(d) none of these

134. In the atmospheric pressure crude distillation, the content of..................from lighter fraction to heavier ones.
(a) sulphur increases
(b) sulphur decreases
(c) nitrogen decreases
(d) none of these

135. Which of the following has the lowest flash point of all?
(a) Diesel
(b) Kerosene
(c) Petrol
(d) Furnace oil

136. Which of the following has the highest flash point of all?
(a) Diesel
(b) Kerosene
(c) Petrol
(d) Furnace oil

137. Which of the following has the minimum °API gravity of all?
(a) Diesel
(b) Kerosene
(c) Petrol
(d) Furnace oil

ANSWERS

127. (b) 128. (a) 129. (d) 130. (a) 131. (b) 132. (a) 133. (b)
134. (a) 135. (c) 136. (d) 137. (d)

138. Which of the following has the maximum °API gravity of all?
(*a*) Diesel
(*b*) Kerosene
(*c*) Petrol
(*d*) Furnace oil

139. With increase in density, the viscosity of petroleum products
(*a*) increases
(*b*) decreases
(*c*) remains same
(*d*) either (*a*) or (*b*)

140. Which of the following has the highest viscosity of all (at a given temperature)?
(*a*) Naphtha
(*b*) Fuel oil
(*c*) Light diesel oil
(*d*) Petrol

141. Which of the following has the lowest viscosity (at a given temperature) of all?
(*a*) Naphtha
(*b*) Kerosene
(*c*) Diesel
(*d*) Lube oil

142. 95% (by volume) of LPG at 760 mm Hg pressure will evaporate at..................°C.
(*a*) 2
(*b*) –40
(*c*) 30
(*d*) 55

143. LPG when brought to atmospheric pressure & temperature will be a
(*a*) liquid lighter than water.
(*b*) liquid heavier than water.
(*c*) gas lighter than air.
(*d*) gas heavier than air.

144. Antioxidants are added in petrol to
(*a*) impart colour to it, for easy identification.
(*b*) minimise the gum formation.
(*c*) prevent icing of the carburettor.'
(*d*) prevent the lead build up in engines.

145. Aromatics have the highest..............of all the hydrocarbons of same carbon atoms.
(*a*) smoke point
(*b*) octane number
(*c*) cetane number
(*d*) viscosity

146. Which of the following hydrocarbons of same carbon atoms has minimum smoking tendency?
(*a*) Paraffins
(*b*) Naphthenes
(*c*) Aromatics
(*d*) Iso-paraffins

147. High aniline point of diesel indicates that, it
(*a*) is highly aromatic.
(*b*) has a large ignition delay.
(*c*) is highly paraffinic.
(*d*) has a low diesel index.

148. The main use of heavy gas oil produced by the vacuum distillation unit is as a

ANSWERS

138. (*c*) 139. (*a*) 140. (*b*) 141. (*a*) 142. (*a*) 143. (*d*) 144. (*b*)
145. (*b*) 146. (*a*) 147. (*c*) 148. (*c*)

(a) blending component for kerosene.
(b) blending component for petrol.
(c) feedstock for fluid catalytic cracking unit.
(d) none of these.

149. The vacuum maintained in vacuum distillation unit for reduced crude is about............... mm Hg.
(a) 1.2 (b) 12
(c) 120 (d) 700

150. The doctor's solution comprises of sodium plumbite in
(a) alcohal
(b) water
(c) aqueous caustic soda
(d) soda ash

151. Hydrocracking employs
(a) high pressure & temperature.
(b) low pressure & temperature.
(c) high pressure and low temperature.
(d) high temperature and low pressure.

152. Petroleum coke is commercially produced by the................ process.
(a) hydrocracking
(b) visbreaking
(c) fluid catalytic cracking
(d) delayed coking

153. Phenols are added in gasoline to
(a) improve the octane number.
(b) act as an antioxidant.
(c) reduce its viscosity.
(d) increase its pour point.

154. Which is an anticing compound?
(a) Amyl nitrate
(b) Alcohols
(c) Mercaptans
(d) Pyridine

155. ....chloride present in crude petroleum as impurity is the most prolific producer of HC1 during distillation.
(a) Magnesium
(b) Potassium
(c) Calcium
(d) Sodium

156. Solvent used in Edeleanu process is
(a) furfural
(b) propane
(c) liquid SO_2
(d) phenol

157. C_nH_{2n} is the general formula for
(a) olefins
(b) naphthenes
(c) both (a) and (b)
(d) neither (a) nor (b)

158. Highest quality bitumen is produced from the................ crude oil.
(a) paraffinic
(b) naphthenic
(c) intermediate
(d) mixed

ANSWERS

149. (c) 150. (c) 151. (a) 152. (d) 153. (b) 154. (b) 155. (a)
156. (c) 157. (c) 158. (b)

159. Fuel oil is subjected to visbreaking to reduce its
(a) pour point
(b) viscosity
(c) pressure drop on pumping
(d) all (a), (b) and (c)

160. ....present in petrol are most succeptible to gum formation, thereby leading to choking of fuel lines.
(a) i–paraffins
(b) n–paraffins
(c) Aromatics
(d) Olefins

161. Which of the following contains maximum sulphur?
(a) Diesel
(b) Petrol
(c) Kerosene
(d) Fuel Oil

162. Presence of aromatics in
(a) diesel increases its cetane number.
(b) kerosene increases its smoke point.
(c) petrol increases its octane number.
(d) all (a), (b) and (c).

163. Which is the most undesirable component in kerosene?
(a) Aromatics
(b) i-paraffins
(c) n-paraffins
(d) Naphthenes

164. Which of the following theories of origin of petroleum does not explain the presence of nitrogen & sulphur compounds in crude oil?
(a) Modern theory
(b) Carbide theory
(c) Engler theory
(d) all (a), (b) and (c)

165. Carbon percentage (by weight) in crude petroleum may be about
(a) 65 (b) 75
(c) 85 (d) 95

166. Hydrogen percentage (by weight) in crude petroleum may be about
(a) 5 (b) 15
(c) 25 (d) 35

167. A petroleum well is called 'dry', if it contains
(a) very little oil.
(b) no natural gas.
(c) only natural gas.
(d) all (a), (b) and (c).

168. Mercaptans are
(a) low boiling sulphur compounds.
(b) added in LPG cylinders to detect gas leakage by its smell.
(c) undesirable in petrol, as they reduce its octane number.
(d) all (a), (b) and (c)

169. Petroleum deposits are detected by the
(a) oil seepage at the surface of the earth.
(b) measurement of density, elasticity and magnetic &

ANSWERS

159. (d) 160. (d) 161. (d) 162. (c) 163. (a) 164. (b) 165. (c)
166. (b) 167. (c) 168. (d) 169. (d)

electric properties of the rock in the crust of the earth.
(c) age & nature of rocks inside the crust of the earth.
(d) all (a), (b) and (c)

170. Main boring diameter for petroleum well is 20-30 cms in diameter, while the depth of the oil well may be about .... kms.
(a) 0.1 to 0.5
(b) 1.5 to 4.5
(c) 7.5 to 12.5
(d) 15–20

171. Water content in the crude oil as it comes out of oil well may be upto.........percent.
(a) 2 .
(b) 5
(c) 10
(d) 25

172. Operating condition in the electrical dehydrators for crude oil is about
(a) 6.5 kgf/cm² & 95°C.
(b) 1 atm. & 110°C.
(c) 20 kgf/cm² & 110°C.
(d) 50 atm. and 150°C.

173. Crude oil is transported inland from oil field to refineries, mainly by the
(a) road tankers
(b) rail tankers
(c) underground pipelines
(d) none of these

174. Lane and Garton classification of petroleum is based on its
(a) composition
(b) specific gravity
(c) optical properties
(d) viscosity

175. Liquefied Petroleum Gas (LPG) is mainly a mixture of
(a) propane & butane
(b) methane & ethane
(c) high boiling olefins
(d) high boiling naphthenes

176. A typical yield of diesel in straight run distillation of crude oil may be about..............percent,
(a) 8 (b) 18
(c) 28 (d) 35

177. Which is the most desirable component of a good quality kerosene?
(a) i-paraffins
(b) Aromatics
(c) n-paraffins
(d) Naphthenes

178. Which of the following petroleum products contain minimum sulphur?
(a) Naphtha
(b) Kerosene
(c) LSHS
(d) Furnace oil

179. A typical yield of kerosene in straight run distillation of crude oil may be about..............percent.
(a) 4 (b) 10
(c) 18 (d) 26

ANSWERS

170. (b) 171. (d) 172. (a) 173. (c) 174. (b) 175. (a) 176. (a)
177. (c) 178. (a) 179. (b)

180. Naphtha yield in straight run distillation of crude oil may be about................percent.
(a) 2 (b) 6
(c) 12 (d) 18

181. Crude oil produced by Indian oil fields are predominantly..................in nature.
(a) paraffinic
(b) naphthenic
(c) asphaltic
(d) mixed base

182. Crude oil is pumped by a....................pump.
(a) gear
(b) centrifugal
(c) screw
(d) reciprocating

183. Products drawn from the top to bottom of the crude oil distillation column has progressively increasing
(a) boiling points
(b) molecular weight
(c) C/H ratio
(d) all (a),(b) and (c)

184. Crude oil is subjected to vacuum distillation in the last stage, because
(a) high boiling point products like heavy fuel oil & lubricating oils are heat sensitive and may decompose.
(b) lighter/low boiling products are prone to thermal decomposition.
(c) high purity products can be obtained thereby.
(d) none of these.

185. Vacuum maintained in the vacuum distillation tower of the crude distillation plant is about...................mm Hg (absolute).
(a) 5–10
(b) 30–80
(c) 150–250
(d) 350–400

186. Which is the most ideal feed stock for 'coking' process used for the manufacture of petroleum coke?
(a) Naphtha
(b) Vacuum residu
(c) Light gas oil
(d) Diesel

187. Catalytic cracking compared to thermal cracking of residue of vacuum distillation of crude oil
(a) gives higher yield of petrol.
(b) lower octane number of petrol.
(c) higher sulphur content in the product.
(d) higher gum forming material in petrol.

188. Which of the following processes in oil refinery does not employ 'cracking'?
(a) Coking
(b) Visbreaking
(c) Pyrolysis
(d) None of these

ANSWERS

180. (b) 181. (c) 182. (b) 183. (d) 184. (a) 185. (b) 186. (b)
187. (a) 188. (d)

189. Pyrolysis of kerosene or natural gasoline is done to produce mainly the
(a) olefins and aromatics
(b) lighter paraffins
(c) stabilised gasoline
(a) diesel

190. Which of the following is used as a catalyst in fluidised bed catalytic cracking?
(a) Silica-magnesia
(b) Silica-alumina
(c) Bentonite clays
(d) All(a),(b)and (c)

191. In catalytic cracking process, olefins crack................times faster than in thermalcracking process.
(a) 100
(b) 200-300
(c) 1000-10000
(d) 10

192. Pressure & temperature maintained in catalytic cracking is about
(a) 2 atm & 500°C
(b) 10 atm & 500°C
(c) 30 atm & 200°C
(d) 50 atm. & 750°C

193. Pick out the wrong statement.
(a) Higher temperature is employed in visbreaking than in thermal cracking.
(b) Pyrolysis is a mild thermal cracking process.
(c) Lead suceptibility of petrol produced by catalytic process is more than that produced by thermal cracking.
(d) Operating pressure and temperature in thermal cracking process is more than that in catalytic cracking process.

194. Pick out the correct statement pertaining to catalytic cracking.
(a) With increase in the reactor pressure, octane number of gasoline decreases.
(b) With increase in the reactor temperature, gasoline yield decreases for a given conversion.
(c) Percentage conversion increases with increase in the catalyst to oil ratio.
(d) all (a), (b) and (c).

195. Thermofer catalytic cracking process is a................process.
(a) fixed bed
(b) moving bed
(c) fluidised bed
(d) non-catalytic

196. Which of the following is the most widely used cracking process in oil refineries?
(a) Dubbs process.
(b) T. C. C. moving bed process.
(c) Fluidised bed catalytic cracking process.
(d) Houdry's fixed bed process.

197. Reforming
(a) uses naphtha as feedstock.
(b) does not much affect the

ANSWERS

189. (a)	190. (d)	191. (c)	192. (a)	193. (a)	194. (d)	195. (d)
196. (c)	197. (d)					

molecular weight of the feed.
(c) improves the quality & yield of gasoline.
(d) all (a), (b) and (c).

198. Pick out the wrong statement.
(a) Lower boiling paraffins have higher octane number than higher paraffins.
(b) Dehydrogenation of naphthenes to aromatics is a desirable reaction in reforming process.
(c) The most ideal feedstock for thermal reforming is dearomatised kerosene.
(d) Catalyst used in catalytic reforming is platinum on silica-alumina base.

199. Operating temperature and pressure in catalytic reforming is about
(a) 1–5 Kgf/cm² & 200°C.
(b) 15–45 Kgf/cm² & 450 – 550°C.
(c) 50 – 75 kgf/cm² & 600 – 800°C.
(d) 5–10 kgf/cm² & 150 – 250°C.

200. Gasoline yield in catalytic reforming of naphtha may be about..............percent by weight.
(a) 85 (b) 65
(c) 50 (d) 98

201. Octane number (unleaded) of reformed gasoline may be upto
(a) 60 (b) 70
(c) 80 (d) 90

202. Which of the following is *a* non-regenerative fixed bed catalytic reforming process?
(a) Hydroforming
(b) Thermofer catalytic reforming
(c) Platforming
(d) Hyperforming

203. Feedstock for polymerisation is
(a) naphtha.
(b) cracked gases rich in C2 & C4 olefins.
(c) low boiling aromatics.
(d) none of these.

204. Catalyst used in catalytic polymerisation which produces polymer gasoline is
(a) H_2SO_4
(b) H_3PO_4
(c) both (a) and (b)
(d) $AlCl_3$

205. Pick out the wrong statement.
(a) Alkylation produces a larger iso-paraffin (having higher octane number) from the reaction of an olefin with smaller iso-paraffin.
(b) Catalytic alkylation processes use HF, AlC13 & H2SO4 as Catalysts.
(c) All the alkylation processes use very high temperature (> 1000°C).
(d) Gasoline having an octane number of 90 can be produced by alkylation process.

ANSWERS

198. (c) 199. (b) 200. (a) 201. (d) 202. (c) 203. (b) 204. (c)
205. (c)

206. ....converts *n*-paraffins to i-paraffins.
(*a*) Alkylation
(*b*) Polymerisation
(*c*) Isomerisation
(*d*) none of these

207. Catalyst used in isomerisation process is
(*a*) H_2SO_4 (*b*) H_3PO_4
(*c*) HF (*d*) $AlCl_3$

208. Octane number (unleaded) of gasoline produced by isomerisation of butane may be about
(*a*) 45 (*b*) 55
(*c*) 70 (*d*) 90

209. Doctor's solution used for sweetening of petroleum products is
(*a*) sodium plumbite
(*b*) sodium sulphite
(*c*) sodium thiosulphate
(*d*) lead sulphate

210. In solutizer sweetening process, solutizer solution used is
(*a*) methanol in Unisol process.
(*b*) naphthenic acid in Mercapsol process.
(*c*) both (*a*) and (*b*).
(*d*) neither (*a*) nor (*b*).

211. Catalytic desulphurisation process used for sweetening of straight run gasoline and kerosene uses...............as catalyst.
(*a*) bauxite
(*b*) fuller's earth
(*c*) activated clay
(*d*) all (*a*), (*b*) & (*c*)

212. In hydrofining catalytic desulphurisation process for sweetening of petroleum products, use of hydrogen
(*a*) enhances the desulphurisation process.
(*b*) minimises coke formation.
(*c*) both (*a*) and (*b*).
(*d*) neither (*a*) nor (*b*).

213. Waxes present in petroleum products
(*a*) can be separated out by distillation.
(*b*) are not soluble in them.
(*c*) crystallise out at low temperature.
(*d*) decrease their viscosity.

214. Solvent used for dewaxing of petroleum products are
(*a*) furfural
(*b*) methyl ethyl ketone (MEK)
(*c*) propane
(*d*) both (*b*) & (*c*)

215. Deoiling of wax is done by its
(*a*) heating
(*b*) cooling
(*c*) solvent extraction
(*d*) both (*b*) & (*c*)

216. Sulphuric acid treatment of petroleum products removes the.................materials.
(*a*) gum forming
(*b*) color forming
(*c*) asphaltic
(*d*) all (*a*), (*b*) & (*c*)

217. Clay treatment of petroleum products

ANSWERS

206. (*c*)	207. (*d*)	208. (*c*)	209. (*a*)	210. (*c*)	211. (*b*)	212. (*c*)
213. (*c*)	214. (*c*)	215. (*d*)	216. (*d*)	217. (*d*)		

(a) decolorises & stabilises cracked gasoline.
(b) desulphurise straight run gasoline & kerosene.
(c) adsorb arsenic from feedstock to catalytic reforming.
(d) all (a), (b) & (c).

218. Asphalts are
(a) low molecular weight & low boiling point compounds present in petroleum.
(b) desirable in catalytic cracking feedstock, because they produce coke.
(c) readily oxidisable and form carbonaceous sludge.
(d) all (a), (b) & (c).

219. Which of the following is used as a solvent in deasphalting of petroleum products?
(a) Furfural
(b) Propane
(c) Methyl ethyl ketone
(d) Liquid sulphur dioxide

220. Raw Kerosene has a smoke point of 15 mm. After it is subjected to dearomatisation by liquid SO_2 extraction (Edeleanu process), its smoke point may become .... mm.
(a) 5 (b) 10
(c) 25 (d) 100

221. Furfural solvent extraction is used for upgrading (by dissolving aromatics)
(a) naphtha
(b) lubricating oils
(c) wax
(d) cracking feedstock

222. Solvent used in duo-sol extraction for lube oil upgradation is a mixture of
(a) propane & phenol-cresol mixture.
(b) methyl ethyl ketone & glycol.
(c) phenol & furfural.
(d) propane & liquid sulphur dioxide.

223. Solvent used in the Udex (glycol) extraction process for removal of light aromatics from cracked naphtha is
(a) propane.
(b) diethylene glycol.
(c) aqueous solution (10% water) of diethylene glycol.
(d) methyl ethyl ketone (MEK).

224. Pick out the wrong statement.
(a) Higher specific gravity of petroleum products means higher C/H ratio.
(b) Aromatics have lower specific gravity than corresponding paraffins.
(c) Hydrocarbons of low specific gravity (e.g, paraffins) possess the maximum thermal energy per unit volume.
(d) Hydrocarbons of high specific gravity (e.g, aromat-

ANSWERS

218. (c) 219. (b) 220. (c) 221. (b) 222. (a) 223. (c) 224. (b)

ics) possess the maximum thermal energy per unit weight.

225. Increase in the specific gravity of petroleum products indicates
(a) decrease in paraffin content.
(b) increase in thermal energy per unit weight.
(c) increase in aromatic content.
(d) higher H/C ratio.

226. Molecular weight of petrol may be about
(a) 40–60
(b) 100–130
(c) 250–300
(d) 350–400

227. 1 centistoke is equal to.................Redwood I seconds.
(a) 1 (b) 4.08
(c) 0.408 (d) 40.8

228. Maximum viscosity of tar/PCM/fuel oil for easy and efficient atomisation in conventional burner is...............centistokes (or 100 Redwood I seconds).
(a) 5 (b) 25
(c) 50 (d) 100

229. Pick out the wrong statement.
(a) Multigrade lubricating oils have high viscosity index.
(b) Paraffinic oil has very high viscosity index.
(c) Naphthenic oil has very low viscosity index.
(d) High viscosity index means a large change in viscosity with change in temperature.

230. Petroleum liquid fuels having flash point greater than 66°C is considered as safe during storage and handling. Which of the following has flash point > 66°C?
(a) Naphtha
(b) Petrol
(c) Kerosene
(d) Heavy fuel oil

231. Flash point of atmospheric distillation residue is determined by...............apparates.
(a) Pensky-Martens (closed cup type).
(b) Abel.
(c) Cleveland (open cup type)
(d) none of these.

232. Diesel index (an alternative index for expressing the quality of diesel) is
(a) determined by using a test engine.
(b) not related to aniline point.
(c) equal to cetane number plus 3.
(d) all (a),(b)&(c).

233. Which of the following is an additive used for improving the cetane number of diesel?
(a) Tetraethyllead
(b) Tetramethyl lead

ANSWERS

225. (a) 226. (b) 227. (b) 228. (b) 229. (d) 230. (d) 231. (c)
232. (c) 233. (a)

(*c*) Ethyl nitrate or acetone
(*d*) None of these

234. Pick out the wrong statement.
(*a*) A pale color of petroleum product indicates lower viscosity.
(*b*) Color of petroleum products indicates the degree of refinement.
(*c*) Lighter petroleum distillates are lighter in color than the heavier residual oils.
(*d*) Flouroscene of oils helps to detect its adulteration.

235. Hydrogen content in petroleum products varies from 12 to 15% (by weight). As a result the difference between gross and net heating value of petroleum fuels varies in the range of..................kcal/kg.
(*a*) 600–750
(*b*) 250–350
(*c*) 1000–1500
(*d*) 2000–2500

236. Which of the following gasolines (unleaded) has the least octane number?
(*a*) Catalytically cracked gasoline
(*b*) Straight run gasoline
(*c*) Catalytically reformed gasoline
(*d*) Polymer gasoline

237. Diesel used in naval applications has a minimum cetane number of
(*a*) 25 (*b*) 35
(*c*) 45 (*d*) 65

238. Preheating temperature of medium viscosity furnace oil for better atomisation through burner is about................°C.
(*a*) 50 (*b*) 70
(*c*) 90 (*d*) 140

239. Which of the following does not require preheating during storage in the storage tank as well as during atomisation through burners?
(*a*) PCM
(*b*) Tar
(*c*) Light diesel oil
(*d*) Low viscosity furnace oil

240. Maximum use of petroleum coke is in
(*a*) adsorption refining operation.
(*b*) fuel gas manufacture.
(*c*) carbon electrode manufacture.
(*d*) iron ore reduction.

241. ....treatment is done for appreciable improvement in viscosity index of lubricating oil.
(*a*) Acid
(*b*) Solvent extraction
(*c*) Alkali
(*d*) Clay

242. ....test is done to find out the softening point of bitumen.
(*a*) Impact

ANSWERS

234. (*a*) 235. (*a*) 236. (*b*) 237. (*c*) 238. (*c*) 239. (*c*) 240. (*c*)
241. (*b*) 242. (*b*)

(b) Ball and ring
(c) Flame
(d) Viscosity

243. Smoke point of a test sample of kerosene is found to be 15 mm. On removal of...........from it, the smoke point rises to 25 mm.
(a) n–paraffins
(b) olefins
(c) aromatics
(d) none of these

244. Complete removal of .... from gasoline is done by Unisol process using caustic soda and methyl alcohol.
(a) waxes
(b) mercaptans
(c) asphalt
(d) iolefins

245. The most important property for a jet fuel is its
(a) viscosity
(b) freezing point
(c) calorific value
(d) flash point

246. The best method of determining sulphur in crude oil is by the...............method.
(a) Kjeldah
(b) Duma
(c) Bomb calorimeter
(d) Junkar's calorimeter

247. ....base crude oil is also called asphaltic crude.
(a) Paraffinic
(b) Naphthenic
(c) Mixed
(d) Aromatic

248. Liquefied Petroleum Gas (LPG) in domestic use cylinders is in the liquid form. The density of LPG in liquid form is about..............of that of water (i.e. 1 gm/c.c).
(a) one fourth
(b) one third
(c) half
(d) one eighth

249. Straight run naphtha is converted into high octane number petrol (gasoline) by catalytic
(a) cracking
(b) polymerisation
(c) reforming
(d) isomerisation

250. An upper limit of oil content is limited to about............... percent for achieving efficient and satisfactory level of wax sweating.
(a) 5 (b) 15
(c) 40 (d) 60

251. Pick out the wrong statement.
(a) Pensky-Marten apparatus is used for determining flash points above 50°C.
(b) Characterisation factor of paraffinic crude oil is more than 12.
(c) Abel apparatus is used for determining flash points below 50°C.
(d) An oil having high susceptibility to change in

ANSWERS

243. (c) 244. (b) 245. (b) 246. (c) 247. (b) 248. (c) 249. (c)
250. (c) 251. (d)

viscosity with temperature changes, has a high viscosity index.

252. Crude oils containing more than kg of total salts (expressed in terms of NaCl) per thousand barrel is called a 'salty crude'.
(a) 1 *(b)* 5
(c) 15 *(d)* 25

253. Smoke volatility index is equal to smoke point plus
(a) 0.42 × (% distilled at 204°C)
(b) 5 mm
(c) 0.84 × *(%* distilled at 204°C)
(d) 10 mm. approximately

254. ....determination is not a very significant and important test for gasoline.
(a) Gum & sulphur content
(b) Viscosity
(c) Octane number
(d) Reid vapor pressure

255. ....is not an important refinery process for upgrading the quality of lubricating oil.
(a) Deoiling
(b) Solvent refinning
(c) Clay treatment
(d) Hydrotreatment

256. Though increased pressure has a retarding effect on cracking reaction, yet in actual process, a positive pressure of 10-15 kgf/cm^2 is maintained during cracking mainly to
(a) increase the yield of light distillates.
(b) suppress coke formation.
(c) enhance the octane number of gasoline.
(d) reduce gum content in gasoline.

257. The solvent used in Barisol dewaxing process is
(a) hexane
(b) furfural
(c) benzol and ethylene dichloride
(d) methyl ethyl ketone (MEK)

258. 'Solvent naphtha' used mostly as a solvent in paints and perfumery is produced by the.................of virgin naphtha into small boiling range cuts.
(a) steam reforming
(b) distillation
(c) desulphurisation
(d) none of these

259. Deoiling is the process of removal of oil from wax. It is done by the...............process.
(a) solvent extraction
(b) sweating
(c) resettling
(d) all *(a)*, *(b)* & *(c)*

260. Aniline point test of an oil qualitatively indicates the .... content of an oil.
(a) paraffin *(b)* olefin
(c) aromatic *(d)* naphthene

261. Glycol added to petrol acts as a/an.................agent.
(a) anti-knocking
(b) anti-icing

ANSWERS

252. *(b)* 253. *(a)* 254. *(b)* 255. *(a)* 256. *(b)* 257. *(c)* 258. *(b)*
259. *(d)* 260. *(c)* 261. *(b)*

(c) anti-gum forming
(d) dewaxing

262. Bottom product of atmospheric pressure crude oil distillation column is termed as
(a) reduced crude
(b) heavy ends
(c) asphalt
(d) residuum

263. The condesate obtained on compression of wet natural gas is termed as
(a) liquefied natural gasoline
(b) natural gasoline
(c) liquid natural gas
(d) none of these

264. Waxy crudes are treated with chemical additives mainly to
(a) depress its pour point
(b) dissolve wax
(c) precipitate wax
(d) remove wax

265. The terminology used for the bottom most product from the vacuum crude distillation unit is
(a) residual crude
(b) residuum
(c) reduced crude
(d) petrolatum

266. The coking process normally mostly used in Indian oil refineries is the.................coking process.
(a) delayed
(b) flexi
(c) fluid
(d) contact

267. The main purpose of recycling the byproduct hydrogen gas in the reformer reactor is to
(a) obviate catalyst poisoning.
(b) maintain the reaction temperature.
(c) sustain the reactor pressure.
(d) hydrogenate the feed stock.

268. Removal of light fractions from crude oil is called its
(a) sweetening
(b) dehydration
(c) stabilisation
(d) visbreaking

269. Testing of the knocking characteristics of petrofuels is done in a..................engine.
(a) Carnot
(b) CFR (Co-operative fuel research)
(c) Stirling
(d) diesel

270. Which is the most effective catalyst used in catalytic cracking of petroleum products?
(a) Iron oxide
(b) Nickel
(c) Vanadium pentoxide
(d) Zeolite

271. Illuminating characteristics of kerosene is expressed by its
(a) smoke point
(b) aniline point
(c) luminosity number
(d) aromatic content

ANSWERS

262. (a) 263. (b) 264. (a) 265. (b) 266. (a) 267. (c) 268. (c)
269. (b) 270. (d) 271. (c)

272. The most commonly used feed stock for the reforming reactor is
(a) heavy fuel oil
(b) residuum
(c) straight run gasoline
(d) casing head gasoline

273. Petroleum is believed to have originated from................. sources.
(a) vegetable
(b) animal
(c) both *(a)* and *(b)*
(d) neither *(a)* nor *(b)*

274. Which of the following processes is used for the production of petroleum coke?
(a) Stabilisation
(b) Visbreaking
(c) Cracking
(d) Reforming

275. Both asphalt and wax are produced by.................base crude oils.
(a) naphthenic
(b) asphalt
(c) paraffin
(d) mixed

276. Octane numbers of motor gasoline used in India and America are respectively
(a) 87&94
(b) 94&87
(c) 94 & 100
(d) 83 & 100

277. The first crude oil refinery of India is located at
(a) Naharkatiya
(b) Digboi
(c) Kochin
(d) Madras

278. Visbreaking process is used mainly for making
(a) high cetane diesel
(b) high octane gasoline
(c) fuel oil
(d) smoke free kerosene

279. Smoke point of kerosene expresses its
(a) burning characteristics.
(b) luminosity characteristics.
(c) aromatic content directly.
(d) lamp wick wetting characteristics.

280. Natural gasoline is produced
(a) from oil wells
(b) in oil refineries
(c) by natural gas stripping
(d) none of these

281. Absolute vapor pressure of petrofuels is found by Reid bomb which is heated in water bath to 100°F. In Reid apparatus, the ratio of the volume of air chamber to that of the liquid fuel chamber is
(a) 1 *(b)* 2
(c) 3 *(d)* 4

282. Aniline point of high speed diesel may be about..................°C.
(a) 35 *(b)* 70
(c) 105 *(d)* 150

283. Natural gas recovered along with crude oil from oil wells

ANSWERS

272. *(c)* 273. *(c)* 274. *(c)* 275. *(d)* 276. *(a)* 277. *(b)* 278. *(c)*
279. *(a)* 280. *(c)* 281. *(d)* 282. *(b)* 283. *(b)*

is called wet natural gas which has a higher .... compared to the dry natural gas.
(a) unsaturated hydrocarbon content
(b) calorific value
(c) quantity of propane
(d) quantity of butane

284. LSHS is a type of furnace oil, which
(a) is more viscous than high viscosity furnace oil (HVFO).
(b) stands for low sulphur heavy stock.
(c) is an ideal fuel for metallurgical furnaces due to its lower sulphur content (< 1%).
(d) all (a), (b) and (c).

285. Cetane number of high speed diesel must be≥
(a) 30 (b) 45
(c) 75 (d) 95

286. A good lubricant should have high
(a) viscosity index
(b) volatility
(c) pour point
(d) none of these

287. A *multigrade* lubricating oil means an oil having high
(a) viscosity index
(b) viscosity
(c) aniline point
(d) flash point

288. Maximum sulphur percentage in low sulphur heavy stock (LSHS) furnace oil is about
(a) 0.1 (b) 1
(c) 2.5 (d) 3.5

289. Pressure maintained in the high pressure primary tower of a three stage crude oil distillation system is about .... kg/cm^2.
(a) 1.5 (b) 3
(c) 6 (d) 12

290. Octane number of gasoline produced by two stage fluidised catalytic cracking process is
(a) 80 (b) 87
(c) 92 (d) 97

291. Pick out the wrong statement.
(a) Aromatics have higher specific gravity than paraffins.
(b) Gross calorific value (GCV) of petrofuels is equal to (12400 - 2100 ρ^2) where, ρ is the specific gravity of the fuel at 15.5°C.
(c) Heavier petrofuels have higher GCV on weight basis (i.e., Kcal/kg) but lower GCV on volume basis (*i.e.,* Kcal/litre).
(d) Higher specific gravity of petrofuels means higher C/H ratio.

292. During electrical desalting of crude oil, the electrical conductivity of a mixture of crude oil and water (which ranges between 3 to *8%* water)

ANSWERS

284. (d) 285. (b) 286. (a) 287. (a) 288. (b) 289. (b) 290. (d)
291. (c) 292. (b)

.... with increase in the amount of water.

(a) decreases
(b) increases
(c) remains unchanged
(d) decreases linearly

293. Which of the following is not an important property of fuel oil/furnace oil?

(a) Sulphur content
(b) Viscosity
(c) Aniline point
(d) Flash point

294. Pick out the wrong statement about the smoking tendency of various hydrocarbon constituents of kerosene.

(a) Smoking tendency of hydrocarbons increases in the order : paraffins → iso-paraffins → naphthenes → aromatics.
(b) Smoking tendency of paraffins increases with decrease in its molecular weight.
(c) Smoking tendency of naphthenes decreases with its increasing molecular weight & also on addition of double bond.
(d) Smoking tendency of aromatics decreases with increase in its molecular weight.

295. Ethyl mercaptan is added to the Doctor negative LPG for facilitating the detection of its leakage (by bad odour) to the extent of about................ppm.

(a) 1 (b) 50
(c) 5000 (d) 10000

296. Water separometer index (modified) (WSIM) of a petrofuel is the measure of its

(a) emulsification tendency.
(b) water separation characteristics.
(c) water content.
(d) water absorbing capacity from atmosphere.

297. Refractive index of a petrofuel which is the ratio of velocity of light in air to its velocity in the petrofuel gives an indication if its

(a) molecular weight
(b) aromatics content
(c) both *a* & *b*
(d) neither *a* nor *b*

298. Specific gravity of a petroleum product gives an indication of its

(a) degree of refinement.
(b) hydrocarbon content type (aromatic or paraffinic).
(c) ease of atomisation.
(d) sulphur content.

299. High aniline point of a petrofuel (say diesel) indicates that

(a) it is highly aromatic in nature.
(b) it is highly paraffinic in nature.
(c) it has a very low diesel index.

ANSWERS

293. (c) 294. (b) 295. (b) 296. (b) 297. (c) 298. (b) 299. (b)

(*d*) its ignition quality is very poor.

300. Char value of Kerosene is the amount of charred oil deposition on the wick obtained after burning it in a standard wick lamp at a standard rate for 24 hours. Char value of a good quality kerosene should be less than...............mg/kg of kerosene.

(*a*) 1 (*b*) 20
(*c*) 100 (*d*) 500

301. Straight run petrol as compared to methyl/ethyl alcohol has
(*a*) lower calorific value.
(*b*) lower octane number.
(*c*) higher specific gravity.
(*d*) higher ignition temperature.

302. Which of the following tests is not done for transformer oil?
(*a*) Flash point and acid value.
(*b*) Aniline point.
(*c*) Dielectric strength.
(*d*) Copper strip corrosion test.

303. In case of liquid petrofuels, momentary combustion is observed at its
(*a*) flash point.
(*b*) preheating temperature corresponding to viscosity of 25 centistokes.
(*c*) flame temperature.
(*d*) fire point.

304. Iso-octane is used as a reference substance in the definition of octane number and it is assigned an octane number value of 100. Iso-octane is chemically known as
(*a*) alpha-methyl naphthalene.
(*b*) 2–2–4 tri methyl pentane.
(*c*) 1, 3 butadiene.
(*d*) tetra methyl ethylene.

305. Penetration test determines the.............of the grease.
(*a*) stiffness
(*b*) lubricating properties (e.g. oilness)
(*c*) service temperature
(*d*) variation in viscosity with temperature

306. Tanks used for the storage of petroleum products (which are inflammable) should be painted with a.................paint.
(*a*) black (*b*) white
(*c*) red (*d*) yellow

307. Pour point and freezing point is equal for
(*a*) petrol
(*b*) diesel
(*c*) water
(*d*) crude petroleum

308. Mercaptans are added to liquefied petroleum gas (LPG) to
(*a*) reduce its cost.
(*b*) narrow down its explosion limit.
(*c*) assist in checking its leakage from cylinder.
(*d*) increase its calorific value.

309. Detergent is added as an additive in engine lubricating oil to

ANSWERS

300. (*b*) 301. (*b*) 302. (*b*) 303. (*a*) 304. (*b*) 305. (*a*) 306. (*b*)
307. (*c*) 308. (*c*) 309. (*a*)

(a) reduce deposit formation.
(b) keep contaminants in suspension.
(c) increase oxidation stability.
(d) prevent rusting.

310. Aromatics are desired constituents of
(a) lubricating oil
(b) diesel
(c) kerosene
(d) petrol

311. Ash percentage in the petrofuels is in the range of
(a) 0.03–0.07 (b) 0.5–1.0
(c) 2–3 (d) 4–5

312. LPG vapor is heavier than air. Liquid LPG evaporates to produce about................times volume of the gas.
(a) 50 (b) 100
(c) 250 (d) 500

313. Petrochemicals are chemicals
(a) obtained from petroleum and natural gas and made on large scale.
(b) used for tertiary oil recovery.
(c) added to crude petroleum for processing before it is fed to the distillation column for refining.
(d) added to gasoline and other fuels in order to improve their properties.

314. Dry natural gas contains mainly
(a) methane (plus small amount of other components)
(b) ethane
(c) propane
(d) butane

315. Light gasoline has an approximate boiling point range of................°C.
(a) 20 to 75
(b) 75 to 200
(c) 200 to 250
(d) 250 to 350

316. In petrochemical industry, any petroleum fraction having an approximate boiling point range between 20 and 200°C which is used as a feedstock is called
(a) Kerosene
(b) gas oil
(c) naphtha
(d) none of these

317. Naptha having an approximate boiling point range of 20 to 200°C contains compounds having.................carbon atoms.
(a) 1 to 6 (b) 4 to 12
(c) 8 to 16 (d) ≥ 16

318. A good quality gasoline should have correct volatility characteristics and correct burning characteristics. Correct burning characteristics are expressed by
(a) aniline point
(b) cetane number

ANSWERS

310. (d) 311. (a) 312. (c) 313. (a) 314. (a) 315. (a) 316. (c)
317. (b) 318. (c)

(c) octane number
(d) none of these

319. In petrochemical industry, thermal cracking or steam cracking process is used to produce mainly
(a) ethane *(b)* ethylene
(c) benzene *(d)* styrene

320. Which of the following processes is employed in petrochemical industry to produce benzene, toluene and xylene (BTX)?
(a) Thermal cracking
(b) Catalytic reforming
(c) Steam reforming
(d) Catalytic cracking

321. Main feed stocks used for the manufacture of ethylene by thermal cracking process are
(a) ethane and propane
(b) naphtha and gas oil
(c) butane to a minor extent,
(d) all *a,b* & c

322. In petrochemical industry thermal cracking of hydrocarbons to produce ethylene is accomplished in cracking furnace and the feed to the cracking furnace is a mixture of hydrocarbon and.................which acts as an inert diluent.
(a) nitrogen
(b) carbon dioxide
(c) steam
(d) air

323. Under otherwise uniform conditions yield of ethylene from a thermal cracker will be........ that obtained from a catalytic cracker.
(a) lower than
(b) equal to
(c) higher than
(d) none of these (thermal cracking and catalytic cracking are two distinctly different processes involving different fundamental cracking reactions. Performance wise the two processes can never be compared).

324. Under otherwise uniform conditions, branched chain alkanes and cycloalkanes, on thermal cracking, will yield
(a) higher amount of ethylene compared to the straight chain alkanes.
(b) almost the same amount of ethylene compared to the straight chain alkanes.
(c) lower amount of ethylene compared to the straight chain alkanes.
(d) ethylene and other products; the amount of ethylene produced cannot be compared with the amount of ethylene produced from straight-chain alkane feed simply because such a comparison does not exist and there is no method by which; such an information can be obtained theoretically.

ANSWERS

319. *(b)* 320. *(b)* 321. *(d)* 322. *(c)* 323. *(c)* 324. *(c)*

325. Why cracking furnace are not operated at reduced absolute pressure?

(a) Because creating and maintaining reduced pressure in the furnace is a problem.

(b) If leakage occurs, air could be sucked into the cracking tubes and an explosive mixture could be formed.

(c) The rate of cracking is too large to be managed.

(d) Severe carbon formation occurs and this carbon gets deposited on the furnace tubes causing fouling.

326. Aromatic hydrocarbons are separated from alkanes and cycloalkanes by

(a) fractional distillation

(b) solvent extraction

(c) adsorption onto activated carbon

(d) azeotropic distillation

327. Important reactions occuring in catalytic reforming are

(a) dehydrogenation, dehydro isomerization and dehydrocyclization

(b) oxidation, isomerization and cyclization

(c) catalytic cracking, hydrogenation

(d) thermal cracking, cyclisation

328. Feed to a catalytic reformer is

(a) hydrocarbon vapor.

(b) liquid hydrocarbon and hydrogen gas.

(c) hydrocarbon vapor and nitrogen in a molar ratio of 1:4.

(d) hydrocarbon vapor and nitrogen in a molar ratio of around 1:6.

329. Reforming catalysts get deactivated because of coke deposition on the surface. Catalyst regeneration is accomplished by

(a) washing the catalyst with an organic solvent and subsequently drying it by hot air.

(b) grinding the deactivated catalyst, separating carbon from the catalyst by floatation and then remarking the catalyst.

(c) simply burning off the coke deposited on the catalyst surface.

(d) none of these

330. Of the isomeric xylenes, which one is of least importance (from user point of view)?

(a) ortho – xylene

(b) meta – xylene

(c) para – xylene

(d) all are of equal importance

331. Reformate from a catalytic reformer will have an octane number..............that of the feed hydrocarbon.

(a) higher than

(b) lower than

ANSWERS

325. *(b)* 326. *(b)* 327. *(a)* 328. *(c)* 329. *(c)* 330. *(b)* 331. *(a)*

(*c*) approximately equal to
(*d*) unpredictable

332. Cyclohexane is made by catalytic hydrogenation of
(*a*) acetylene (*b*) ethylene
(*c*) benzene (*d*) phenol

333. Cyclohexanol is produced commercially by
(*a*) hydrogenation of cyclohexanone
(*b*) oxidation of cyclohexane
(*c*) oxidation of benzene
(*d*) decomposition of benzoic acid

334. Cyclohexanol & cyclohexanone, produced commercially by oxidation of cyclohexane are used in the manufacture of
(*a*) nylon intermediates
(*b*) terylene intermediates
(*c*) polycarbonates
(*d*) alkyd resins

335. Cyclohexanol/cyclohexanone mixture, on oxidation with nitric acid, produces
(*a*) benzoic acid
(*b*) terephthalic acid
(*c*) adipic acid
(*d*) none of these

336. Adipic acid, on reaction with ammonia in presence of a dehydrating catalyst, produces
(*a*) ethylene diamine
(*b*) caprolactum
(*c*) hexamethylene diamine
(*d*) adiponitrile

337. Which of the following statement is not true in case of isomerisation of butane?
(*a*) Increase in LHSV will tend to increase in isobutene yield
(*b*) Increase in temperature will increase isobutane yield
(*c*) Increase in temperature will decrease isobutane yield
(*d*) Reducing hydrogen to hydrocarbon ratio will be decreased

338. For highly paraffinic crude oil, the characterisation factor will be in range of
(*a*) 11.5-12.5 (*b*) 12.5-13.0
(*c*) 13.5-14.0 (*d*) 14.5 -15.0

339. During the desalting of crude oil which of the statement is not true?
(*a*) Desalter pressure is maintained above vaporisation pressure of crude oil
(*b*) Desalter pressure is maintained below vaporisation pressure of crude oil
(*c*) Low voltage is preferable for better desalting
(*d*) Lower temperature is preferable

340. Which of the following statement is not true in case of catalytic reforming?
(*a*) High temperature results in loss of reformate yield
(*b*) Highly naphthenic stocks require high space velocity
(*c*) High paraffinic stock requires low space velocity
(*d*) Presence of water decrease the hydrocracking activity

ANSWERS

332. (*c*) 333. (*b*) 334. (*a*) 335. (*c*) 336. (*d*) 337. (*d*) 338. (*b*)
339. (*a*) 340. (*d*)

341. If the crude is paraffinic then its characterization factor will be
(a) 5 (b) 8
(c) 11 (d) above 12

342. Ring number is used to express
(a) TBP
(b) Thermoviscosity
(c) Aromatic content
(d) Boiling point

343. Smoke Volatility index is expressed as
(a) Smoke point =0.42×(percentage distilled at 204°C)
(b) Smoke point = 1.42×(percentage distilled at 204°C)
(c) Smoke point = 2.42×(percentage distilled at 204°C)
(d) Smoke point = 3.42×(percentage distilled at 204°C)

344. If diesel has cetane number of 50 then the diesel index will be
(a) 36 (b) 46
(c) 56 (d) 66

345. Bromine number is measure of
(a) Paraffins
(b) Unsaturates
(c) Saturates
(d) Aromatics

346. Which of the following expression is not true in case of fluid catalytic cracking?
(a) Combustion factor = (regenerator dense phase temperature-riser out let temperature)/coke yield
(b) Combustion factor = (regenerator dense phase temperature-riser out let temperature)/delta coke
(c) Combustion factor = (riser out let temperature-regenerator dense phase temperature)/delta coke
(d) Combustion factor = (riser out let temperature-regenerator dense phase temperature)/coke yield

347. Aviation Fuel Contains
(a) Light Naphtha
(b) Medium Naphtha
(c) Kerosene
(d) Diesel

348. Which of the following is for the manufacture of linear alkyl benzene
(a) Kerosene
(b) Naphtha
(c) Gas Oil
(d) Diesel

349. Pyrolysis gasoline is obtained from
(a) Catalytic cracking
(b) Gasification
(c) Steam cracking
(d) Thermal cracking

350. Which of the crude has higher sulphur content?
(a) Assam crude
(b) Gujarat crude
(c) Rajasthan crude
(d) Bombay high crude

ANSWERS

341. (d) 342. (b) 343. (a) 344. (c) 345. (b) 346. (a) 347. (c)
348. (a) 349. (c) 350. (c)

351. Which of the following will have higher API gravity?
(a) Gas Oil, (b) Gasoline
(c) Crude Oil (d) Kerosene

352. Aniline point is related to
(a) Octane number
(b) Diesel number
(c) Smoke
(d) Point pour point

353. VGC of crude oil is around
(a) 0.1- 0.2 (b) 0.3 - 0.4
(c) 0.5- 0.7 (d) 0.8 - 0.9

354. Characterization factor for paraffin base is around
(a) 2 (b) 9
(c) 11 (d) 12

355. Which of the following is not thermal cracking?
(a) Fluid coking
(b) Flexi coking
(c) Uniflex
(d) MSCC

356. If a fuel has high aniline point then which of the statement is true?
(a) Fuel has low paraffin
(b) Fuel has high aromatics
(c) Fuel has low diesel index
(d) Fuel has high diesel index

357. Propane in petroleum refining is used for the removal of
(a) Asphalt
(b) Wax
(c) Sulphur compound
(d) Aromatics

358. As per the Bharat II norms for gasoline, the benzene content should be
(a) 1 percent (b) 2 percent
(c) 3 percent (d) 4 percent

359. Which of the following type of crude oil contains lowest wax content?
(a) Asphalt base
(b) Paraffin based
(c) Mixed based crude
(d) Naphthenic based

360. Which of the following hydrocarbons having API gravity mentioned below will have highest sulphur content?
(a) 10 (b) 20
(c) 30 (d) 40

361. Which of the following is in Edeleanu process?
(a) Propane
(b) Dimethyl formamide
(c) SO_2
(d) Furfural

362. Which of the following will have highest API gravity?
(a) Ethane (b) Propane
(c) Butane (d) Heptane

363. Which one is not carbon rejection technique for residue up-gradation?
(a) Solvent deasphalting
(b) Gasification
(c) FCC
(d) Delayed coking

364. Which of the following has higher characterization factor?
(a) Paraffinic base crude oil
(b) Naphthenic base
(c) Intermediate base
(d) Cracked gasoline

ANSWERS

351. (b)	352. (b)	353. (d)	354. (d)	355. (d)	356. (d)	357. (a)
358. (a)	359. (b)	360. (a)	361. (c)	362. (a)	363. (b)	364. (d)

365. If a crude has specific gravity of 0.8576 and characterisation factor of 11.66 then its average boiling point in °C will be
(a) 238.8 *(b)* 298.8
(c) 398.8 *(d)* 498.8

366. As per the Euro III norms for gasoline the benzene content should be
(a) 1 percent *(b)* 2 percent
(c) 3 percent *(d)* 4 percent

367. In case of catalytic reforming dehydrogenation of naphthenes to aromatics is
(a) Favoured by increase in temperature and decrease in pressure
(b) Favoured by increase in temperature an increase in pressure
(c) Favoured by decrease in temperature and decrease in pressure
(d) None of these

368. Udex process is used for extraction of
(a) Aromatics
(b) Butenes
(c) Wax
(d) Paraffins from Kerosene

369. Which of the following statement is not true in case of catalytic reforming?
(a) High temperature results in loss of reformate yield
(b) Highly naphthenic stock require high space velocity
(c) High paraffinic stock requires low space velocity
(d) Presence of water decreases the hydrocracking activity hydrocracking

370. In Alkylation process, which of the following is not used as catalyst?
(a) HF
(b) H_2SO_4
(c) Solid acid catalyst
(d) Mixed acid (H_2SO_4 and HNO_3)

371. Isochronal test is used for measuring
(a) Smoke point
(b) Asphalt content
(c) Wax content
(d) Back pressure of testing well

372. For extraction of which of the following UDEX process is used?
(a) Aromatics
(b) Butenes
(c) Wax
(d) Paraffins from Kerosene

373. In case of which of the following process Snow balling term is used?
(a) Fluidised bed cracking
(b) Crude oil distillation
(c) Hydro desulphurisation
(d) Aromatic extraction

374. Knocking is the term used for
(a) Diesel
(b) Gasoline
(c) Jet fuel
(d) Kerosene

ANSWERS

365. *(b)* 366. *(a)* 367. *(a)* 368. *(c)* 369. *(d)* 370. *(d)* 371. *(d)*
372. *(a)* 373. *(a)* 374. *(b)*

375. Crude oil is made up mainly of –
(*a*) alkanes, cycloalkanes and aromatics
(*b*) alkanes, aldehydes and ketones
(*c*) alkanes, alkenes and alkynes
(*d*) None of these

376. Naphthenes are the same as
(*a*) branched alkanes
(*b*) cycloalkanes
(*c*) alkenes
(*d*) alkynes

377. Cycloaikanes present in crude oil contain:
(*a*) only four membered ring compounds,
(*b*) only five membered ring compounds,
(*c*) only five and six membered ring compounds,
(*d*) none of the foregoing,

378. Dry natural gas contains mainly
(*a*) methane (plus small amount other components)
(*b*) ethane
(*c*) propane
(*d*) butane

379. Wet natural gas contains substantial amount of
(*a*) ethane
(*b*) propane
(*c*) butane
(*d*) all of the above

380. Light gasoline has an approximate boiling point range of –
(*a*) 20 to 75 °C
(*b*) 75 to 200 °C
(*c*) 200 to 250 °C
(*d*) 250 to 350 °C

381. Naphtha having an approximate boiling point range of 20 to 200 °C contain compounds having
(*a*) 1 to 6 carbon atoms,
(*b*) 4 to 12 carbon atoms,
(*c*) 8 to 16 carbon atoms,
(*d*) 16 carbon atoms and higher,

382. A good quality gasoline should have correct volatility characteristics and correct burning characteristics. Correct burning characteristics are expressed by
(*a*) aniline point,
(*b*) cetane number,
(*c*) octane number,
(*d*) none of the foregoing,

383. As the molecular weight of the hydrocarbon feed to the cracking furnace increases (from ethane to gas oil), the proportion of steam in the feed
(*a*) decreases
(*b*) increases
(*c*) remains unchanged (at a constant value)

384. Under otherwise uniform conditions yield of ethylene from a thermal cracker will be
(*a*) lower than that obtained from a catalytic cracker,
(*b*) equal than that obtained from a catalytic cracker,
(*c*) higher than that obtained from a catalytic cracker,

ANSWERS

375. (*a*)	376. (*b*)	377. (*c*)	378. (*a*)	379. (*d*)	380. (*a*)	381. (*b*)
382. (*c*)	383. (*b*)	384. (*c*)				

(*d*) none of the above; thermal cracking and catalytic cracking are two distinctly different processes involving different fundamental cracking reactions. Performance wise the two processes can never be compared.

385. Catalytic cracking involves –

(*a*) free radical chain reactions as in thermal cracking,

(*b*) reactions involving carbocations,

(*c*) reactions involving carbanions,

(*d*) reactions chemistry of which still remains unknown.

386. Under otherwise uniform conditions, branched chain alkanes and cycloalkanes, on thermal cracking, will yield –

(*a*) higher amount of ethylene compared to the straight chain alkanes,

(*b*) almost the same amount of ethylene compared to the straight chain alkanes,

(*c*) lower amount of ethylene compared to the straight chain alkanes,

(*d*) ethylene and other products; the amount of ethylene produced cannot be compared with the amount of ethylene produced from straight-chain alkane feed simply because such a comparison does not exist in literature and there is no method by which ;such an information; can be obtained theoretically.

387. You know thermal cracking involves free radical chain reactions. Now in order to achieve highest yield of ethylene from a thermal cracker which of the following reactions will you try to promote?

(*a*) B-scission reaction

(*b*) chain transfer reactions

(*c*) chain termination reactions,

(*d*) all of the above

388. In reactions involved in thermal cracking process chain transfer and chain termination reactions are bimolecular whereas B-scission reactions are unimolecular. In order to increase the yield of ethylene, the rates of chain transfer and chain termination reactions should be reduced relative to the rates of B-scission reactions and this may be achieved by carrying out thermal cracking at reduced pressure. Thermal crackers or cracking furnaces, however, are not operated at reduced absolute pressure because of safety considerations. Instead the hydrocarbon partial pressure in the cracking

ANSWERS

385. (*b*) 386. (*c*) 387. (*b*) 388. (*a*)

furnace is maintained at a low value by using –

(a) steam as a diluent,
(b) nitrogen as a diluent,
(c) CO_2 as a diluent,
(d) helium as a diluent,

389. Why cracking furnaces are not operated at reduced absolute pressure ?

(a) Because creating and maintaining reduced pressure in the furnace is a problem
(b) If leakage occurs, air could be sucked into the cracking tubes and an explosive mixture could be formed
(c) The rate of cracking is too large to be managed
(d) Severe carbon formation occurs and this carbon gets deposited on the furnace tubes causing fouling.

ANSWERS

389. (b)

References/Bibliography/ Further reading

1. R.A.meyors, Handbook of Petroleum Refining Processes ,Mcgraw Hill Book Co. Inc. New York .
2. R. J .Hengstebeck , Petroleum Processing, Mcgraw Hill Book Co. New York.
3. W.L.Nelson, Petroleum Refinery Engineering, Mcgraw Hill Book Co. New York .
4. G.D.Hobson and W.Pohl, Modern Petroleum Technology ,Institute of Petroleum,London.
5. Standard Methods For Testing Petroleum and Its Products, Institute of Petroleum,London.
6. G.N.sarkar,Advanced Petroleum Refinery,Khanna Publishers,Delhi.
7. Ram Prasad, Petroleum Refining Technology,Khanna Publishers,Delhi.
8. Williams and Jones,Liquid Fuels,Pergamon Press,London
9. Gary, James H., Handwerk, Glenn E., and Kaiser, Mark J.; Petroleum Refining *Technology and Economics;* Fifth Edition; CRC Press; 2007
10. Leffler, William L.; *Petroleum Refining in Nontechnical Language*; Third Edition; PennWell Corp.; 2000
11. Maples, Robert E.; *Petroleum Refinery Process Economics*; 2nd Edition; PennWell Corp.; 2000
12. Parkash, Surinder; *Refining Processes Handbook*; Gulf Professional Publishing; 2003
13. Rao, B.K.B.(1990). Modern Petroleum Refining Processes (2nd Edition Ed.) Oxford & IBH Publishers. ISBN 81-204-0481-5.
14. James. G. Speight (2006). The Chemistry and Technology of Petroleum (4th ed.). CRC Press. ISBN 0-8493-9067-2.
15. Jones D.S.J., Elements of Petroleum Processing, John Wiley & Sons, 1995
16. Ullmann F., Encyclopedia of Industrial Chemistry, Wiley – VCH, 1999-2012
17. Kirk R. E., Othmer D. F., Encyclopedia of Chemical Technology, John Wiley and Sons, 1999-2012

Subject Index